Electronics
Practical

6.   Adjust the amplitude of both the input signals such that amplitude of carrier signal is more than that of message signal.

7.   Trace the PWM output.

8.   Connect channel 1 of CRO across the carrier signal and take the trace.

9.   Reconnect channel 1 of CRO across the message signal. Measure its frequency and take the trace.

10. Connect the circuit shown in Fig. 8.4.6 on breadboard.

11. Connect the PWM output to the input of demodulator circuit.

12. Reconnect channel 2 of CRO at the output of demodulator circuit and take the trace of demodulated signal and measure the output frequency.

13. Compare the frequency of message signal and demodulated signal.

14. Change the frequency of message and carrier signals and repeat the above steps.

OBSERVATION TABLE

Table 8.4.1-Observation table for pulse width demodulation

S.No.	f_m (Hz)	f_c (kHz)	Demodulated signal frequency
1			
2			
3			

OBSERVATIONS

Attach the traces for pulse width modulation and demodulation.

RESULT

Pulse width modulation and demodulation circuit have been designed successfully. The demodulated and the modulating signal frequency comes out to be almost same.

DISCUSSION

Advantages

1.   Noise is less as compared to PAM since PWM contains information in its width and not in the amplitude.

Electronics Practical

Swati Gupta
Shweta Gupta

Alpha Science International Ltd.
Oxford, U.K.

Swati Gupta
Department of Physics and Electronics
Hansraj College
University of Delhi
New Delhi

Shweta Gupta
Department of Electronics
Bhaskaracharya College of Applied Sciences
University of Delhi
New Delhi

Copyright © 2016

ALPHA SCIENCE INTERNATIONAL LTD.
7200 The Quorum, Oxford Business Park North
Garsington Road, Oxford OX4 2JZ, U.K.

www.alphasci.com

ISBN 978-1-78322-224-4

Preface

This book has been written to cater the requirements of various disciplines at undergraduate level viz., B.Sc/B.Tech – Electronics, Computer Science, Physics, Instrumentation, etc. It consists of eight chapters based on Network Analysis, Semiconductor Devices, Analog Electronics–I, Digital Electronics, Analog Electronics–II, Signal and Systems, Microprocessor 8086 and Communication.

The first chapter contains experiments on Network Analysis that includes verification of network theorems, designing of RC circuits such as differentiator, integrator, low pass and high pass filter and series and parallel LCR circuits.

The second chapter comprises of experiments based on study of characteristics of various semiconductor devices such as diode, transistors (BJT, UJT, FET) and power device like SCR.

In the third chapter, experiments based on Analog Electronics have been discussed. The detailed experimental procedure along with elaborated theory has been discussed to study rectifiers, clipping and clamping circuits, various feedback configuration for transistors, oscillators, BJT and FET amplifiers.

The fourth chapter covers experiments on Digital Electronics such as designing of adders, subtractors, multiplexers, flip flops, registers and counters.

The fifth chapter contains experiments based on Operational Amplifiers. Designing of inverting and non-inverting configuration of operational amplifier has been discussed. Besides this, designing of various circuits such as integrator, differentiator, oscillators, mono-stable and astable multivibrator has been discussed. The designing of various filters, namely, first and second order low and high pass filters, band pass, and band reject filters has also been discussed in this chapter.

The sixth chapter covers experiments based on signals and systems. The well known open source SCILAB software has been used to write the various programs such as generation of continuous and discrete time signals, convolution of signals, Fourier Series representation and Fourier Transform of continuous time signals, Discrete Time Fourier Analysis, etc.

The seventh chapter covers assembly language programming using Microprocessor 8086. The well descriptive introduction of 8086 microprocessor has been given so that the readers can develop comprehensive understanding. The assembly language programs for addition, subtraction, multiplication, transfer of a block, generation of Fibonacci series, sorting of hexadecimal numbers, finding square root and factorial of a number, generation of digital clock have been written.

The eighth chapter includes practical based on Communication Electronics. Experiments on various analog and digital modulation techniques such as AM, FM, PAM, PWM, PPM, PCM, DM have been discussed. Practical on TDM, ASK, PSK, FSK have also been mentioned.

In order to elucidate the readers of this book about the experiment, each chapter is consisting of elaborated theory, well descriptive illustrations, and systematically written procedure with relevant graphs.

We hereby extend our profound gratitude to the Almighty for his persistent blessings and exemplary guidance during the documentation of this book. We also thank all our teachers and colleagues for their consistent cooperation. We also wholeheartedly express our sincere thanks to our publisher M/s Narosa Publishing House Pvt. Ltd. for their acceptance to bring out this book. Lastly, we convey our heartfelt thanks to our family for their incessant support and motivation in writing this book.

Swati Gupta
Shweta Gupta

Contents

1

Experiments on Network Analysis

LIST OF EXPERIMENTS

1. Verify the Thevenin, Norton and Superposition Theorem.

2. Verify the Maximum Power Transfer Theorem.

3. RC Circuits: Time Constant, Differentiator, Integrator.

4. Design a Low Pass RC Filter and Study its Frequency Response.

5. Design a High Pass RC Filter and Study its Frequency Response.

6. To Study the Generation of Lissajous Figures.

7. To Measure the Z-Parameters of a Two-Port Network.

8. To Study the Frequency Response of a Series LCR Circuit and Determine its (a) Resonant Frequency (b) Impedance at Resonance (c) Quality Factor (d) Band Width.

9. To Study the Frequency Response of a Parallel LCR Circuit and Determine its (a) Resonant Frequency (b) Impedance at Resonance (c) Quality Factor (d) Band Width.

Experiment 1

THEVENIN, NORTON and SUPERPOSITION THEOREMS

AIM

Verify the Thevenin, Norton and Superposition Theorem.

APPARATUS REQUIRED

Breadboard, resistors, 5V and 10V dc power supply, multimeter, ammeter, connecting wires.

THEORY

Thevenin Theorem

Thevenin theorem states that "Any two terminal linear network containing energy sources (current sources or voltage sources or more precisely generators) and impedances can be replaced with an equivalent circuit consisting of a single voltage source E_{TH} in series with the impedance Z_{TH}" as shown in Fig. 1.1.1.

The value of E_{TH} is the open circuit voltage measured between the terminals of the network and Z_{TH} is the impedance measured between the terminals by eliminating all the energy sources and retaining their internal impedances in the circuit.

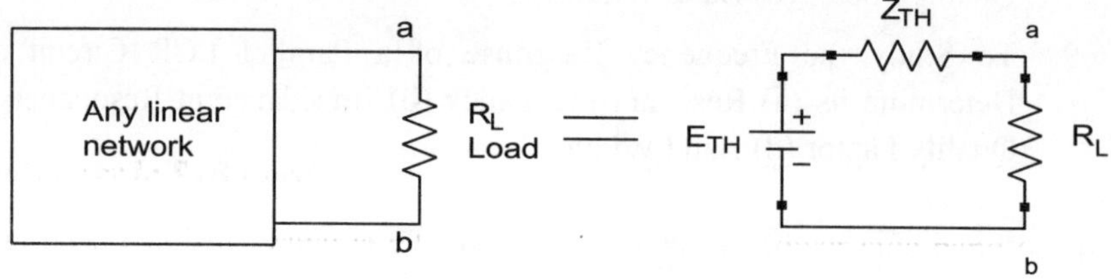

Fig. 1.1.1 A two terminal linear network replaced by its thevenin equivalent

Hence, by using thevenin theorem, any complex linear circuit can be replaced by an equivalent circuit with a single voltage source E_{TH} in series with single impedance Z_{TH}.

Steps to find the thevenin equivalent of a given circuit

Consider the circuit shown in Fig. 1.1.2.

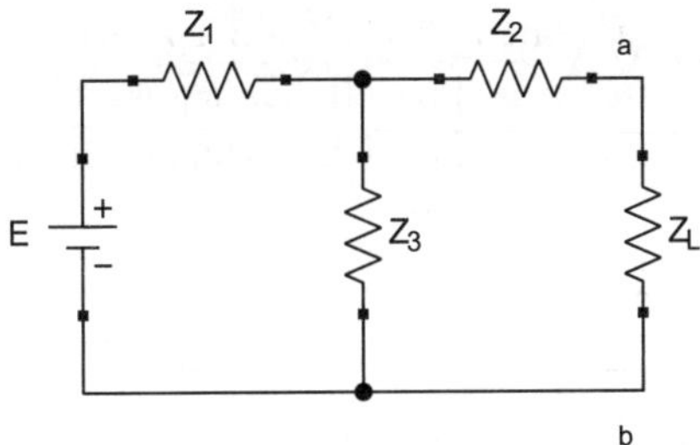

Fig. 1.1.2 Given circuit for thevenin theorem

1. Find Thevenin voltage

- Remove the load impedance Z_L.

- Find the thevenin voltage which is also known as open circuit voltage, E_{TH} that appears across the terminals a and b as shown in Fig. 1.1.3.

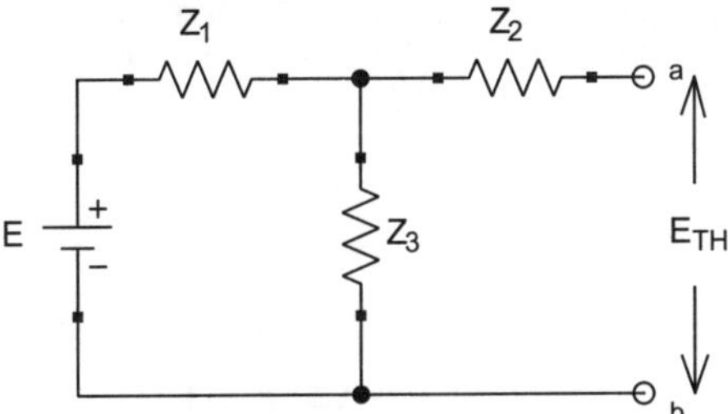

Fig. 1.1.3 Thevenin voltage measurement

- Since terminals a and b are open, no current will flow through Z_2, hence thevenin voltage is equal to voltage across Z_3.

$$E_{TH} = \frac{EZ_3}{(Z_1 + Z_3)}$$

2. Find Thevenin impedance

- Remove the load impedance Z_L.

- Replace all energy sources with their internal impedances. In case, the internal impedances of the energy sources are not given, take their ideal value *i.e.* zero in case of voltage source and infinite in case of current source. For *e.g.* short the voltage source and open the current source.

- Find the thevenin impedance Z_{TH} of the whole network as looked into from these two open terminals a and b as shown in Fig. 1.1.4.

$$Z_{TH} = Z_2 + (Z_1 || Z_3)$$

$$\Rightarrow \qquad Z_{TH} = Z_2 + \frac{Z_1 Z_3}{(Z_1 + Z_3)}$$

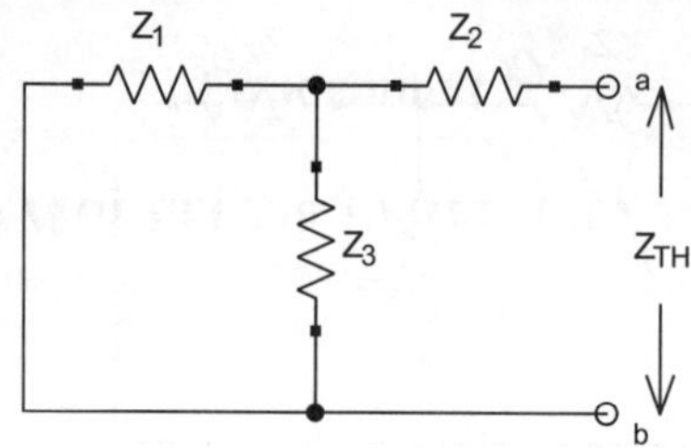

Fig. 1.1.4 Thevenin impedance measurement

3. Equivalent circuit

- Replace the entire network by a single thevenin source E_{TH} in series with thevenin impedance Z_{TH} as shown in Fig. 1.1.5.

- Connect the load impedance Z_L back to its terminal from where it was previously removed.

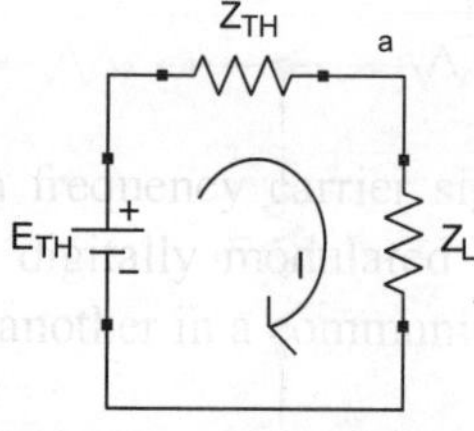

Fig. 1.1.5 Equivalent circuit

- Calculate the current flowing through the load impedance Z_L by using the equation

$$I = \frac{E_{TH}}{Z_{TH} + Z_L}$$

Example

Consider the circuit as shown in Fig. 1.1.6.

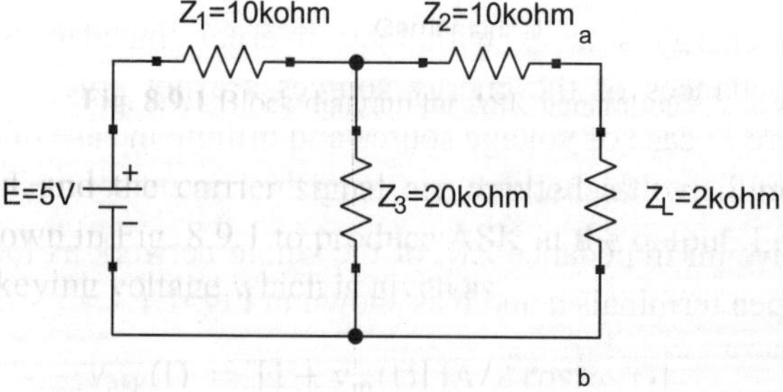

Fig. 1.1.6 Example

Using Thevenin theorem in the given circuit,

$$E_{TH} = \frac{EZ_3}{(Z_1 + Z_3)} = \frac{5 \times 20}{20 + 10} = 3.33 \text{ V}$$

$$Z_{TH} = Z_2 + \frac{Z_1 Z_3}{(Z_1 + Z_3)} = 10k + \frac{10k \times 20k}{10k + 20k} = 16.67 \text{ k}\Omega$$

$$I = \frac{E_{TH}}{Z_{TH} + Z_L} = \frac{3.33}{16.67k + 2k} = 0.18 \text{ mA}$$

Equivalent circuit is shown in Fig. 1.1.7.

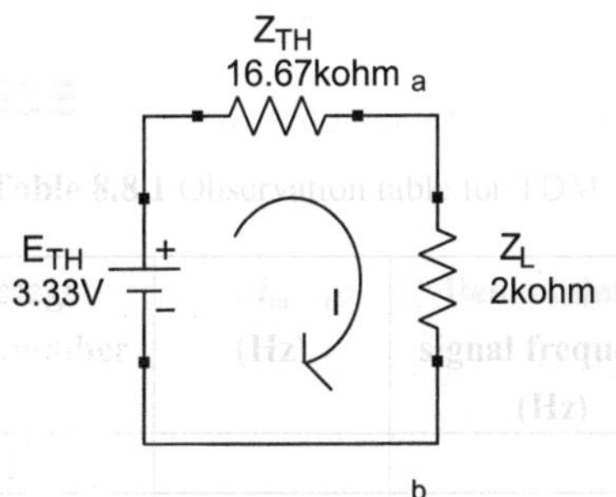

Fig. 1.1.7 Equivalent circuit

Norton Theorem

Norton theorem states that "Any two terminal linear network containing energy sources (voltage source or current source or more precisely generators) and impedances can be replaced with an equivalent circuit consisting of a single current source I_N in parallel with equivalent impedance Z_N" as shown in Fig. 1.1.8. The value of I_N is the short circuit current flowing through the terminals of the network and Z_N is the impedance measured between the terminals by eliminating all the energy sources and retaining their internal impedances in the circuit.

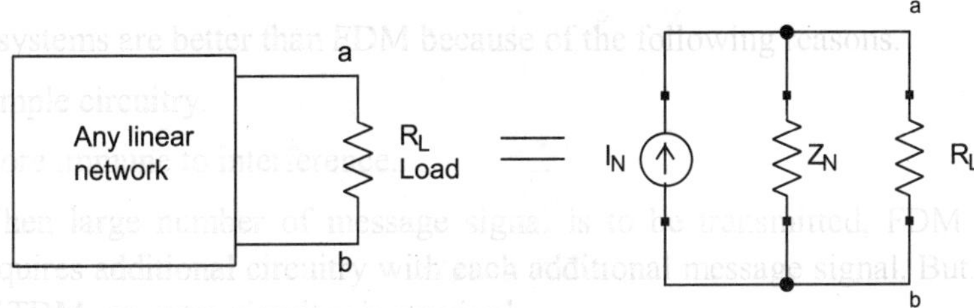

Fig. 1.1.8 A two terminal linear network replaced by its equivalent Norton equivalent

Hence, by using Norton theorem, any complex linear circuit can be replaced by an equivalent circuit with a single current source I_N in parallel with impedance Z_N.

Steps to find the norton equivalent of a given circuit

Consider the circuit shown in Fig. 1.1.9.

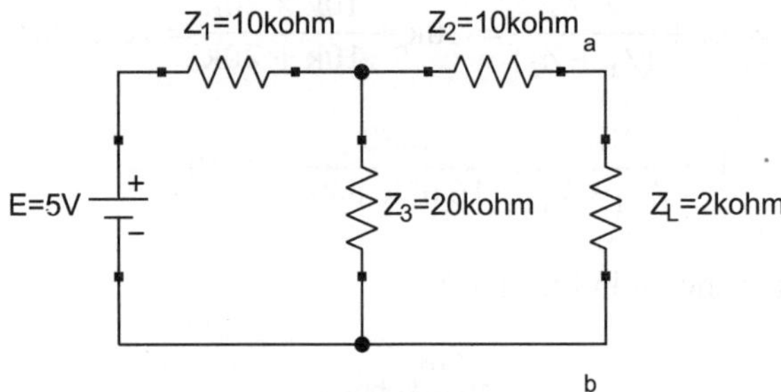

Fig. 1.1.9 Circuit for Norton theorem

1. Find Norton current

- Replace 2kΩ load impedance Z_L with a short.

- Find the Norton current which is also known as short circuit current, I_N that is flowing through the terminals a and b as shown in Fig. 1.1.10.

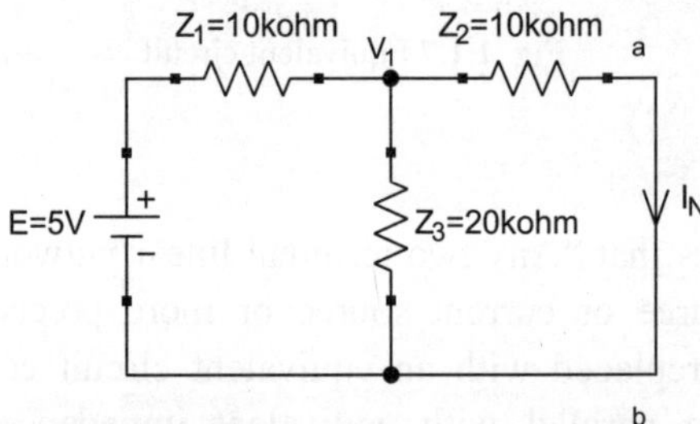

Fig. 1.1.10 Norton current measurement

To find the Norton current, let us find the voltage V_1 using nodal analysis. Applying the Kirchhoff's current law, we get

$$\frac{V_1 - 5}{10k} + \frac{V_1}{20k} + \frac{V_1}{10k} = 0$$

$$\Rightarrow \quad \frac{2V_1 - 10 + V_1 + 2V_1}{20k} = 0$$

$$\Rightarrow \quad 5V_1 = 10$$

$$\Rightarrow \quad V_1 = 2V$$

Hence, Norton current will be

$$I_N = \frac{V_1}{10k} = \frac{2}{10k} = 0.2mA$$

2. Find Norton impedance

- Remove the 2kΩ load impedance Z_L.

- Replace all energy sources with their internal impedances. For *e.g.* short the voltage source and open the current source. In case, the internal impedances of the energy sources are not given, take their ideal value *i.e.* zero in case of voltage source and infinite in case of current source. In the given circuit, replace 5V voltage source with a short.

- Find the Norton impedance Z_N of whole network as looked into from these two open terminals a and b as shown in Fig. 1.1.11.

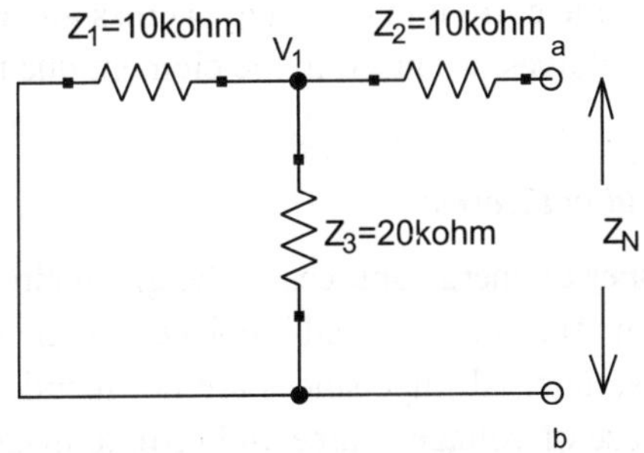

Fig. 1.1.11 Norton impedance measurement

$$Z_N = Z_2 + (Z_1 || Z_3)$$

$$\Rightarrow \qquad Z_N = Z_2 + \frac{Z_1 Z_3}{(Z_1 + Z_3)}$$

$$\Rightarrow \qquad Z_N = 10k + \frac{10k \times 20k}{10k + 20k} = 16.67k\Omega$$

3. Equivalent circuit

- Replace the entire network by a single Norton current source I_N parallel with the equivalent impedance Z_N.

- Connect the load impedance Z_L back to its terminal from where it was previously removed as shown in Fig. 1.1.12.

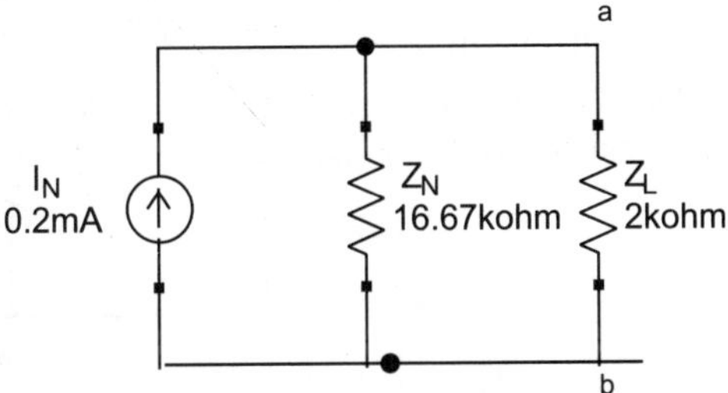

Fig. 1.1.12 Norton equivalent circuit

- Calculate the current flowing through the load impedance Z_L by using the equation

$$I_L = \frac{I_N \cdot Z_N}{Z_N + Z_L}$$

$$\Rightarrow \qquad I_L = \frac{0.2 * 16.67}{16.67k + 2k} = 0.18 \text{mA}$$

Superposition Theorem

Superposition theorem states that "In any linear bilateral network containing two or more energy sources, the current in or voltage across any element is the vector sum of the currents or voltages produced in the element due to each source in the network".

Steps to apply superposition theorem

1. If there are N number of energy sources in the given circuit, then, disconnect N-1 sources from the circuit and replace them with their internal impedances. In case internal impedances are not mentioned, take their ideal value *i.e.* zero in case of voltage source and infinite in case of current source or we can say short circuit the voltage sources and open circuit the current sources.

2. Find the current flowing through the load impedance due to the energy source which was remained connected in the circuit as in Step 1. Let this current be represented as I_1.

3. Remove the energy source which was connected in step 2 and now, connect the other sources one by one and measure the corresponding current flowing through the load impedance. Let these currents be denoted as I_2, I_3, I_4,....I_N.

4. The total current flowing through the load impedance will be the algebraic sum of the currents obtained in step 2 and 3.

5. The voltage across the load impedance can be calculated by multiplying the load impedance with the net current flowing through it.

Example

Consider the circuit shown in Fig. 1.1.13 and find the current I flowing through 20Ω resistor (a-b terminal).

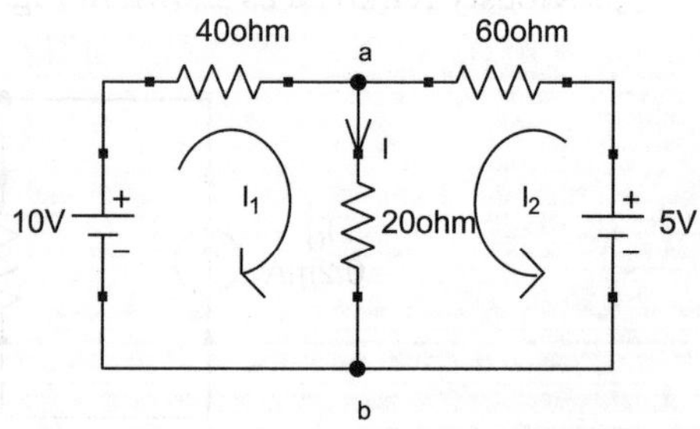

Fig. 1.1.13 Circuit for superposition theorem

1. Calculate current I without using theorem *i.e.* with both the energy sources in the circuit.

Applying KVL in mesh 1,

$$(40 + 20)I_1 + 20I_2 \;=\; 10$$

$\Rightarrow$
$$3I_1 + I_2 = 0.5 \qquad\qquad -\text{ eqn 1}$$

Applying KVL in mesh 2,

$$20I_1 + 80I_2 = 5$$

$\Rightarrow$
$$2I_1 + 8I_2 = 0.5 \qquad\qquad -\text{ eqn 2}$$

Solving equations 1 and 2, we get

$$I_1 = 0.16A \text{ and } I_2 = 0.02A$$

Hence current through 20Ω resistor is

$$I = I_1 + I_2 = 0.18A$$

2. Now, calculate current I using the theorem *i.e.* with one energy source in the circuit at a time.

- Replace 5V voltage source with its internal impedance. Since, internal impedance is not mentioned, replace it by a short as shown in Fig. 1.1.14.

- Calculate current $I_1{}'$ and $I_2{}'$ by using Kirchhoff's voltage law.

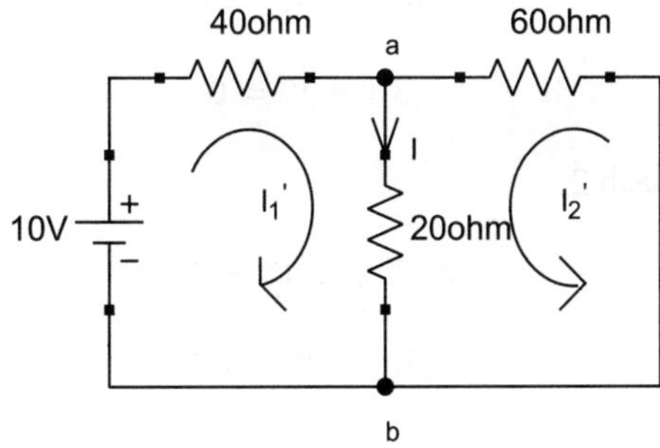

Fig. 1.1.14 Circuit with 5V source eliminated

Applying KVL in mesh 1,

$$60I_1' + 20I_2' = 10$$

$\Rightarrow$
$$3I_1' + I_2' = 0.5 \qquad\qquad -\text{ eqn 3}$$

Applying KVL in mesh 2,

$$20I_1' + 80I_2' = 0$$

$\Rightarrow$
$$I_1' + 4I_2' = 0 \qquad\qquad - \text{eqn 4}$$

Solving equations 3 and 4, we get

$$I_1' = 0.182A \text{ and } I_2' = -0.045A$$

- Now, reconnect the voltage source 5V in the circuit.
- Replace 10V voltage source with its internal impedance. Since, internal impedance is not mentioned, replace it by a short as shown in Fig. 1.1.15.
- Calculate current I_1'' and I_2'' by using Kirchhoff's voltage law.

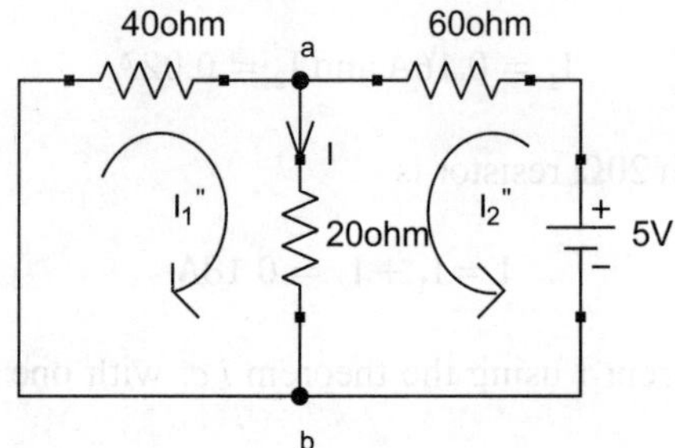

Fig. 1.1.15 Circuit with 10V source eliminated

Applying KVL in mesh 1,

$$60I_1'' + 20I_2'' = 0$$

$\Rightarrow$
$$3I_1'' + I_2'' = 0 \qquad\qquad - \text{eqn 5}$$

Applying KVL in mesh 2,

$$20I_1'' + 80I_2'' = 5$$

$\Rightarrow$
$$2I_1'' + 8I_2'' = 0.5 \qquad\qquad - \text{eqn 6}$$

Solving equations 5 and 6, we get

$$I_1'' = -0.023A \text{ and } I_2'' = 0.068A$$

- Now, superposition theorem states that the total current I_1 in mesh 1 is the algebraic sum of I_1' and I_1'' *i.e.*

$$I_1 = I_1' + I_1'' = 0.182 - 0.023 = 0.16A$$

- And total current I_2 in mesh 2 is the algebraic sum of I_2' and I_2'' *i.e.*

$$I_2 = I_2' + I_2'' = -0.045 + 0.068 = 0.023A$$

- Hence, total current through 20Ω resistor becomes

$$I = I_1 + I_2 = 0.18A$$

This value is same as the value obtained without using the theorem.

PROCEDURE

Thevenin Theorem

1. Connect the circuit as shown in Fig. 1.1.6 on breadboard.
2. Remove the load resistance and measure the thevenin voltage E_{TH} by connecting the multimeter between terminals a and b.
3. Replace voltage source by a short circuit and measure the thevenin impedance Z_{TH} across terminals a and b with the multimeter.
4. Now, reconnect the voltage source and load resistance in the circuit.
5. Connect ammeter in series with load resistance and note the value of current I flowing through it.
6. Compare the theoretical and practical values of thevenin voltage E_{TH}, thevenin impedance Z_{TH} and load current I.

Norton Theorem

1. Connect the circuit as shown in Fig. 1.1.9 on breadboard.
2. Short the load resistance and measure the Norton current I_N by connecting the ammeter at terminals a and b.
3. Replace the voltage source with a short circuit and measure the Norton impedance Z_N by connecting multimeter across terminals a and b.
4. Now, reconnect the load resistance and voltage source in the circuit.
5. Connect ammeter in series with the load resistance and note down the value of current I_L flowing through it.
6. Compare the theoretical and practical values of Norton current I_N, Norton impedance Z_N and load current I_L.

Superposition Theorem

1. Connect the circuit as shown in Fig. 1.1.13 on breadboard.
2. Using ammeter measure current I_1, I_2 and I flowing through 40Ω, 60Ω and 20Ω resistors respectively.

3. Keep 10V source and replace the 5V source with a short circuit.

4. Using ammeter measure current I_1', I_2' and I' flowing through 40Ω, 60Ω and 20Ω resistors respectively.

5. Keep 5V source and replace the 10V source with a short circuit.

6. Using ammeter measure current I_1'', I_2'' and I'' flowing through 40Ω, 60Ω and 20Ω resistors respectively.

7. Calculate $I' = I_1' + I_2'$ and $I'' = I_1'' + I_2''$

8. Net current I should be equal to the sum of I' and I''.

9. Compare the theoretical and practical values of current I.

OBSERVATION TABLE

Thevenin Theorem

Table 1.1.1 Observation table for thevenin theorem

S.No.	Parameter	Theoretical value	Practical value
1.	E_{TH}	3.33V	
2.	Z_{TH}	16.67kΩ	
3.	I	0.18mA	

Norton Theorem

Table 1.1.2 Observation table for norton theorem

S.No.	Parameter	Theoretical value	Practical value
1.	I_N	0.2mA	
2.	Z_N	16.67kΩ	
3.	I_L	0.18mA	

Superposition Theorem

Table 1.1.3 Observation table for superposition theorem

S.No.		Parameter	Theoretical value	Practical value
1.	Without using theorem	I_1	0.16A	
		I_2	0.02A	
		I	0.18A	
2.	With 10V Source	I_1'	0.182A	
		I_2'	-0.045A	
		I'	0.137A	
3.	With 5V Source	I_1''	-0.023A	
		I_2''	0.068A	
		I''	0.045A	

Table 1.1.4 Comparison of theoretical and practical values for superposition theorem

Total Current	Theoretical value	Practical value	
		Without theorem	With theorem
I	0.18A		

RESULT

Thevenin, Norton and Superposition theorems have been verified successfully.

DISCUSSION

Theorems are used to convert a complex circuit into a simple circuit. For *e.g.* if a complex circuit is given and only one particular resistor known as load resistor changes, then each time with the change in load resistance, re-calculations of currents and voltages in the complex circuit are required. Hence, a better way is to reduce that complex circuit into an equivalent simple circuit connected with the load resistance. This can be accomplished by using Thevenin and Norton theorem.

Superposition theorems are widely used in the circuits where ac is superimposed with dc source. To see the combined effect of both ac and dc sources, analysis of circuit is done with dc input alone and then with ac input. Both the results are then combined to see the complete effect of ac and dc sources acting together.

The main limitation of superposition theorem is that it can be applied to linear circuits only. It can be used only to measure voltage and current in a circuit but not power. Power is a non-linear function of current *i.e.* $P \propto I^2$ and if we use superposition theorem to find power across resistor, we will get $(I' + I'')^2 = I'^2 + I''^2$. This equation is not true. Hence, superposition theorem cannot be applied to those circuits which have non-ohmic components.

Experiment 2

MAXIMUM POWER TRANSFER THEOREM

AIM

Verify Maximum Power Transfer Theorem.

APPARATUS REQUIRED

Breadboard, resistors - 2.2kΩ, 1kΩ, 3.3kΩ, 5V dc power supply, multimeter, connecting wires, resistance box.

THEORY

Maximum Power Transfer Theorem

The theorem states that "In any active network, maximum power will be delivered from a source to a load when the load impedance Z_L is the complex conjugate of an equivalent impedance of the network Z_{TH} which is measured from the terminals of the load".

If circuit consists of pure resistances, then maximum power transfer takes place when load resistance R_L is equal to the equivalent resistance of the network R_{TH}.

Proof

To prove maximum power transfer theorem, consider the resistive circuit with voltage source V_S having source resistance R_S connected to a variable load resistance R_L as shown in Fig. 1.2.1.

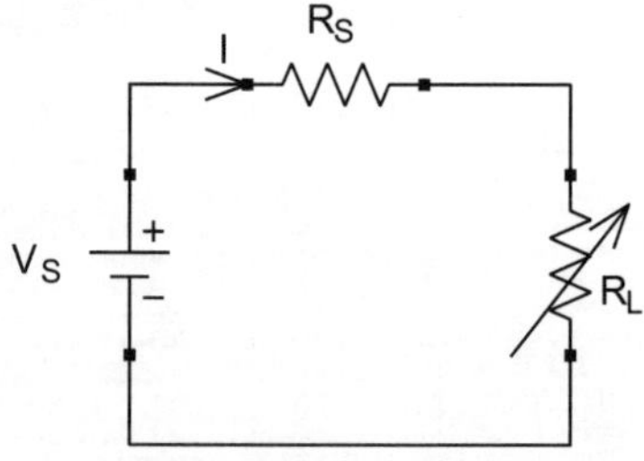

Fig 1.2.1 Resistive circuit

Current I flowing through the circuit is given as

$$I = \frac{V_S}{R_S + R_L}$$

Power P delivered from source to the load will be

$$P = I^2 . R_L$$

$$\Rightarrow \qquad P = \frac{V_S^2 . R_L}{(R_S + R_L)^2} \qquad \qquad - \text{eqn } 1$$

To find the value of R_L for which maximum power will be transferred from source to the load, we differentiate the above equation with respect to R_L and equate it to zero.

i.e.

$$\frac{dP}{dR_L} = 0$$

$$\Rightarrow \qquad \frac{d}{dR_L}\left[\frac{V_S^2 . R_L}{(R_S + R_L)^2}\right] = 0$$

$$\Rightarrow \qquad \frac{V_S^2\{(R_S + R_L)^2 - 2R_L(R_S + R_L)\}}{(R_S + R_L)^2} = 0$$

$$\Rightarrow \qquad (R_S + R_L)^2 - 2R_L(R_S + R_L) = 0$$

$$\Rightarrow \qquad (R_S + R_L)^2 = 2R_L(R_S + R_L)$$

$$\Rightarrow \qquad R_S + R_L = 2R_L$$

$$\Rightarrow \qquad R_S = R_L$$

Hence, maximum power will be transferred when load resistance is equal to the source resistance. Putting $R_L = R_S$ in equation 1, we get maximum power as,

$$P = \frac{V_S^2 . R_S}{(R_S + R_S)^2}$$

$$\Rightarrow \qquad P = \frac{V_S^2}{4R_S} \qquad \qquad - \text{eqn } 2$$

Steps to apply maximum power transfer theorem

Consider the circuit as shown in Fig. 1.2.2.

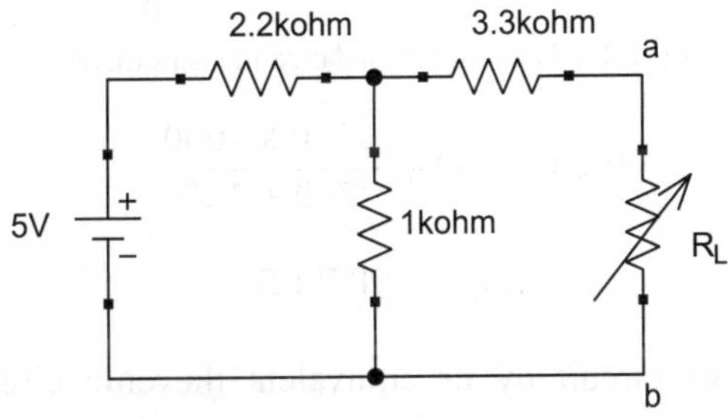

Fig. 1.2.2 Circuit for maximum power transfer theorem

1. Temporarily remove the load resistance R_L from the network through which maximum power is to be calculated.

2. Find the open circuit voltage or thevenin voltage V_{TH} that appears across the two terminals a and b from where the load resistance was initially removed as shown in Fig. 1.2.3.

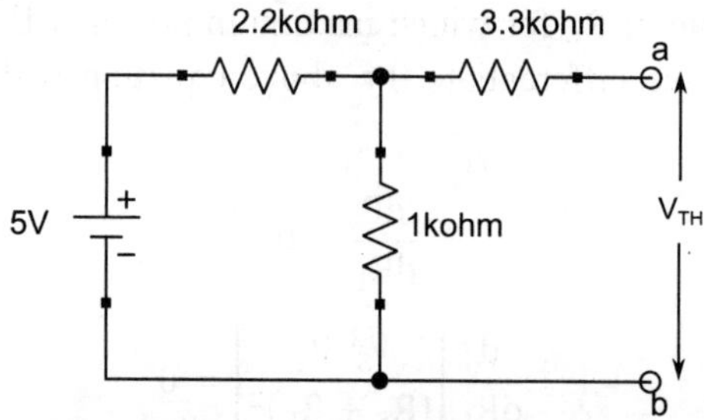

Fig. 1.2.3 Thevenin voltage measurement

$$V_{th} = 5 \, X \, \frac{1}{1 + 2.2}$$

$\Rightarrow$ $$V_{th} = 1.5625V$$

3. Replace all the energy sources by their internal impedances. In case, the internal impedances are not mentioned, take their ideal value *i.e.* zero for voltage source and infinite for current source. In the given circuit, replace 5V voltage source by a short. Now, compute the impedance or resistance of the whole network R_{TH} as looked into from these two open terminals a and b as shown in Fig. 1.2.4.

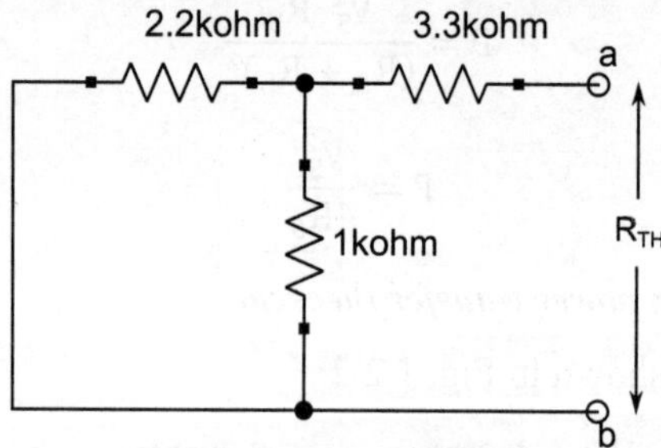

Fig. 1.2.4 Thevenin impedance measurement

$$R_{th} = 3300 + \frac{2200 \, X \, 1000}{2200 + 1000}$$

$\Rightarrow$ $$R_{th} = 3.9875K\Omega$$

4. Replace the given circuit by its equivalent thevenin circuit as shown in Fig. 1.2.5.

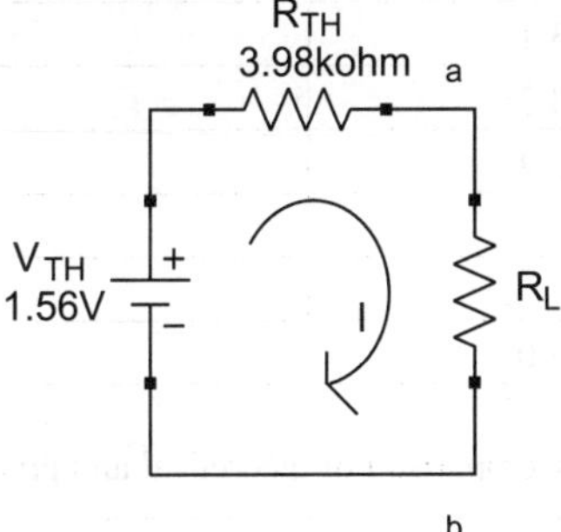

Fig. 1.2.5 Equivalent circuit

5. Maximum power will be delivered to the load impedance only when the load impedance R_L is equal to R_{TH}. By using equation 2, we get maximum power as

$$P = \frac{V_{th}^2}{4R_{th}} = \frac{(1.56)^2}{4 \times 3.98k} = 0.153mW$$

PROCEDURE

1. Connect the circuit as shown in Fig. 1.2.2 on breadboard.
2. Remove the load resistance from the circuit and measure thevenin voltage V_{TH} by connecting the multimeter between terminals a and b.
3. Replace 5V voltage source by a short and measure the equivalent resistance R_{TH} of the circuit by connecting multimeter between terminals a and b.
4. Now, reconnect 5V voltage source in the circuit and connect resistance box in place of load resistance.
5. For different value of load resistance (starting from values below R_{TH} to above R_{TH} *i.e.* 2.5kΩ to 5kΩ), measure the voltage V_L across R_L by connecting multimeter across it. Note the readings in a tabular form.
6. Calculate the power delivered to the load in each case.
7. Draw the graph between load resistance and power delivered to the load.
8. Find the value of maximum power delivered to load resistance graphically.
9. Compare the theoretical and practical values of V_{TH}, R_{TH} and P_{max}.

OBSERVATION TABLE

Table 1.2.1 Observation table for maximum power transfer theorem

S.No.	Load Resistance R_L (kΩ)	Voltage V_L (V)	Power = V_L^2/R_L (mW)
1.	2.5		
2.	2.7		
3.	2.9		

4.	3.1		
5.	3.3		
6.	3.4		
...	..		
...	..		
...	5.0		

Table 1.2.2 Comparison of theoretical and practical values

S.No.	Parameter	Theoretical value	Practical value
1.	Open circuit Voltage V_{TH}	1.56V	
2.	Equivalent Resistance R_{TH}	3.98kΩ	
3.	Maximum Power Pmax	0.153mW	

GRAPH

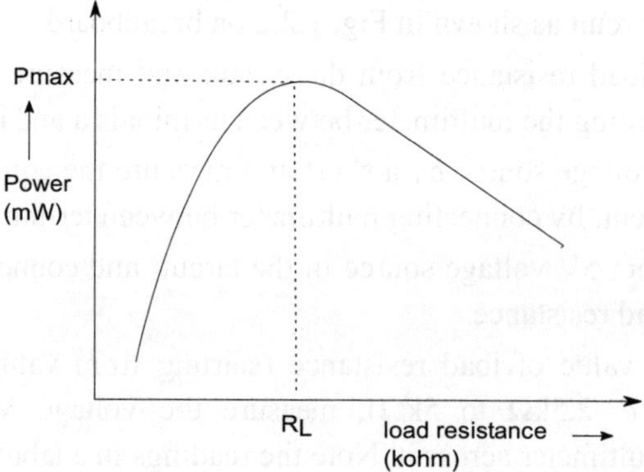

Fig. 1.2.6 Graph between power *vs.* load resistance

RESULT

Maximum power theorem has been verified successfully. The variation between load resistance and power delivered has been studied. It is found that power delivered to the load is the maximum when load resistance is equal to the equivalent circuit resistance.

DISCUSSION

The Maximum power transfer theorem is used to maximize the power delivered to the load. For *e.g.* in audio sound system, output amplifier is used with the loud speaker to amplify the signals and hence loud speaker appears as a load for the output amplifier. The output impedance of amplifier is very high and input impedance of loud speaker is low. But according to the theorem, maximum power will be transferred only if the two impedances are equal. Therefore, in order to do so, an output transformer is used to match the two impedances.

The theorem finds applications in radio, TV transmitters and receivers circuits. Maximum power transfer occurs at 50% efficiency; hence it is not always desirable.

Experiment 3

TIME CONSTANT, DIFFERENTIATOR and INTEGRATOR

AIM (TIME CONSTANT)

RC Circuit: Time Constant.

APPARATUS REQUIRED

Resistor - 100kΩ, capacitor - 100μF, 5V dc power supply, voltmeter, Morse key, stop watch, connecting wires.

THEORY

Capacitor

A capacitor is a passive device that can store energy in form of electric field. It consists of two conducting plates which are separated by an insulator or dielectric. As the capacitor is connected to dc source, current starts flowing through the circuit and as a result both conducting plates of the capacitor get equal and opposite charges with some potential difference across it. This process is called as charging of the capacitor.

As the potential difference reaches the maximum value *i.e.* value of dc voltage source, the capacitor is fully charged. Once it is fully charged, no more current can flow through it. As the capacitor is disconnected from the dc source, it starts discharging itself through the resistor connected to it and the potential difference across it gradually becomes zero. This process is called as discharging of the capacitor.

Charging of a capacitor

Consider an RC circuit with a dc voltage source as shown in Fig. 1.3.1.

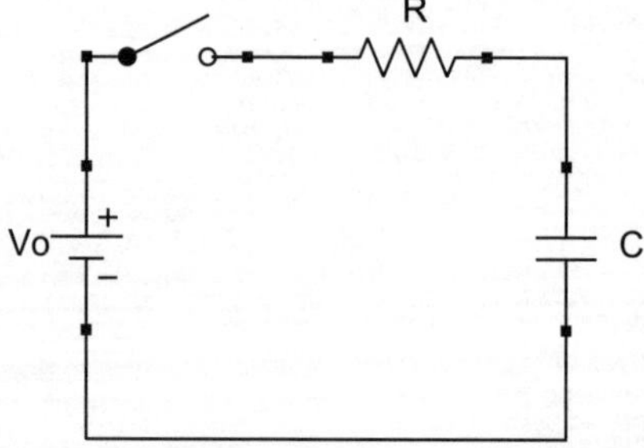

Fig. 1.3.1 Circuit for charging of capacitor

As the key is closed, let I be the current that starts flowing through the circuit and thus capacitor starts charging itself to the voltage V_O. Applying KVL in the loop,

$$V_O = V_C + I.R$$

$$\Rightarrow \quad V_O = \frac{Q}{C} + R.\frac{dQ}{dt}$$

Where,

V_C: voltage across the capacitor, Q: charge on the capacitor.

Rewriting the above equation, we get

$$V_O - \frac{Q}{C} = R.\frac{dQ}{dt}$$

$$\Rightarrow \quad \frac{dQ}{V_O - \frac{Q}{C}} = \frac{dt}{R}$$

Taking integral on both sides,

$$\int \frac{dQ}{V_O - \frac{Q}{C}} = \int \frac{dt}{R}$$

$$\Rightarrow \quad -C \ln\left[V_O - \frac{Q}{C}\right] = \frac{t}{R} + A$$

Where, A: constant of integration.

Since, at time $t = 0$, the initial charge on capacitor $Q = 0$.

$$\therefore \quad A = -C \ln V_O$$

Putting the value of constant A in the above equation, we get

$$-C \ln\left[V_O - \frac{Q}{C}\right] = \frac{t}{R} - C \ln V_O$$

$$\Rightarrow \quad C\left[\ln V_O - \ln\left[V_O - \frac{Q}{C}\right]\right] = \frac{t}{R}$$

$$\Rightarrow \quad \ln \frac{V_O}{V_O - \frac{Q}{C}} = \frac{t}{R.C}$$

$$\Rightarrow \qquad \ln\frac{V_0 - \frac{Q}{C}}{V_0} = -\frac{t}{R.C}$$

$$\Rightarrow \qquad 1 - \frac{Q}{C.V_0} = e^{-t/R.C}$$

$$\Rightarrow \qquad \frac{Q}{C.V_0} = 1 - e^{-t/R.C}$$

$$\Rightarrow \qquad \frac{V_C}{V_0} = 1 - e^{-t/R.C}$$

$$\Rightarrow \qquad V_C = V_0\left(1 - e^{-t/R.C}\right)$$

The above equation shows the charging equation across the capacitor with time t. When time t = RC, the voltage across capacitor becomes

$$V_C = V_0\left(1 - e^{-RC/RC}\right)$$

$$\Rightarrow \qquad V_C = V_0\left(1 - \frac{1}{e}\right)$$

$$\Rightarrow \qquad V_C = 0.632\, V_0 \qquad\qquad - \text{eqn } 1$$

Discharging of a capacitor

Consider an RC circuit without any voltage source as shown in Fig. 1.3.2. Assuming that the capacitor is initially charged, capacitor will now discharge through resistor R.

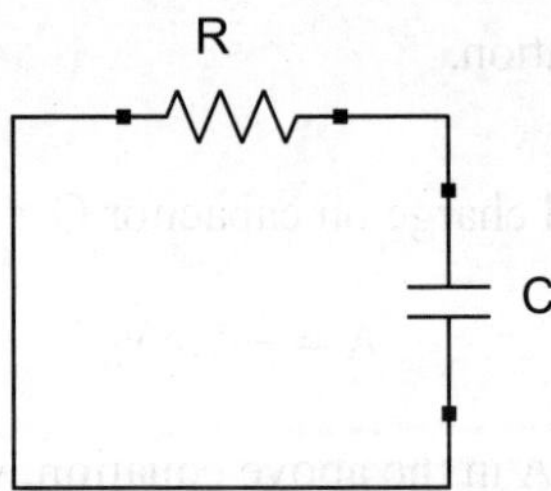

Fig. 1.3.2 Circuit for discharging of capacitor

Applying KVL in the loop, we get

$$V_C + I.R = 0$$

Where,

I: current flowing through the circuit, V_C: voltage across the capacitor.

We can write the above equation as

$$\frac{Q}{C} + R.\frac{dQ}{dt} = 0$$

$$\Rightarrow \qquad R.\frac{dQ}{dt} = -\frac{Q}{C}$$

$$\Rightarrow \qquad \frac{dQ}{Q} = -\frac{dt}{R.C}$$

Where,

Q: charge on the capacitor.

Taking integral on both sides,

$$\int \frac{dQ}{Q} = -\int \frac{dt}{R.C}$$

$$\Rightarrow \qquad \ln Q = -\frac{t}{R.C} + B$$

Where, B: constant of integration.

Since, at time t = 0, the capacitor is assumed to be fully charged. Hence $Q = Q_0$ and voltage across capacitor is V_O. Therefore, putting this initial value, we get constant as

$$B = \ln Q_0$$

Putting the value of constant in the above equation,

$$\ln \frac{Q}{Q_0} = -\frac{t}{R.C}$$

$$\Rightarrow \qquad \frac{Q}{Q_0} = e^{-t/RC}$$

$$\Rightarrow \qquad Q = Q_0.e^{-t/RC}$$

Since,

$$V_C = \frac{Q}{C} \text{ and } V_O = \frac{Q_0}{C}$$

Putting these values on the above equation, we get

$$V_C = V_O.e^{-t/RC}$$

The above equation represents the discharging of a capacitor through a resistance R. When time t = RC, the voltage across capacitor becomes

$$V_C = V_0 \cdot e^{-RC/RC}$$

$\Rightarrow$
$$V_C = V_0\left(\tfrac{1}{e}\right)$$

$\Rightarrow$
$$V_C = 0.368\, V_0 \qquad\qquad - \text{eqn 2}$$

Time constant

The time constant of circuit is equal to the product of circuit resistance and capacitance *i.e.* RC. It is defined as the time required for capacitor to charge to 0.632 times the maximum charge (refer eqn1) or discharge to 0.37 times the initial value (refer eqn2). Smaller is the value of resistance or capacitance, smaller will be the value of time constant. Hence, charging and discharging rate of the capacitor will be faster.

CIRCUIT DIAGRAM

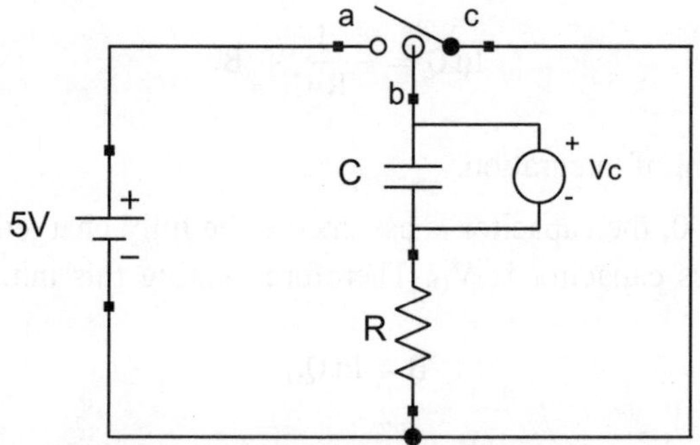

Fig. 1.3.3 Circuit for time constant of capacitor

PROCEDURE

Charging of Capacitor

1. Connect the circuit as shown in Fig. 1.3.3 on breadboard.
2. Press the Morse key to make connection between a and b terminals such that capacitor is connected to 5V source and it starts charging.
3. Start the stop watch and note the time till the voltage across the capacitor reaches to 0.1 V.
4. Release the key such that terminals b and c are connected and discharge the capacitor through resistor R such that voltage across capacitor becomes 0V and reset the stop watch.
5. Again charge the capacitor to 0.2V by pressing the Morse key and measure the corresponding time by stop watch.
6. Again reset the stop watch and release the key to discharge the capacitor to 0V.

7. Repeat the above steps to take at least 10 observations for different values of V_C *i.e.* $V_C = 0.3V$, $0.4V$ and so on.

8. Plot the graph between time taken by the capacitor to charge and the corresponding voltage across it.

9. From the curve, determine the time constant as the time corresponding to the capacitor voltage $V_C = 0.632\ V_O$.

Discharging of Capacitor

1. Connect the circuit as shown in Fig. 1.3.3 on breadboard.

2. Press the Morse key to make connection between a and b terminals such that capacitor charges to its maximum voltage V_O say 5V.

3. Release the key such that terminals b and c are connected and discharge the capacitor through resistor R.

4. Start the stop watch and note the time till capacitor discharges to $V_C = 4.5V$.

5. Again charge the capacitor to its maximum value by pressing the key and reset the stop watch.

6. Release the key and note the time till capacitor discharges to 4V.

7. Repeat the above steps to take at least 10 observations for different values of V_C *i.e.* $V_C = 3.5V$, $3V$ and so on.

8. Plot the graph between discharging time of the capacitor and the corresponding voltage across it.

9. From the curve, determine the time constant as the time corresponding to the capacitor voltage $V_C = 0.37\ V_O$.

CALCULATION

Theoretical Time Constant $= R.C = 100k\Omega \times 100\mu F = 10sec$

OBSERVATION TABLE

Table 1.3.1 Observation table for charging and discharging time of capacitor

S.No.	Charging		Discharging	
	Time (sec)	Voltage Vc (volts)	Time (sec)	Voltage Vc (volts)
1.				
2.				
3.				
...				

Table 1.3.2 RC time constant

Parameter	Charging	Discharging
Time Constant RC		

GRAPH

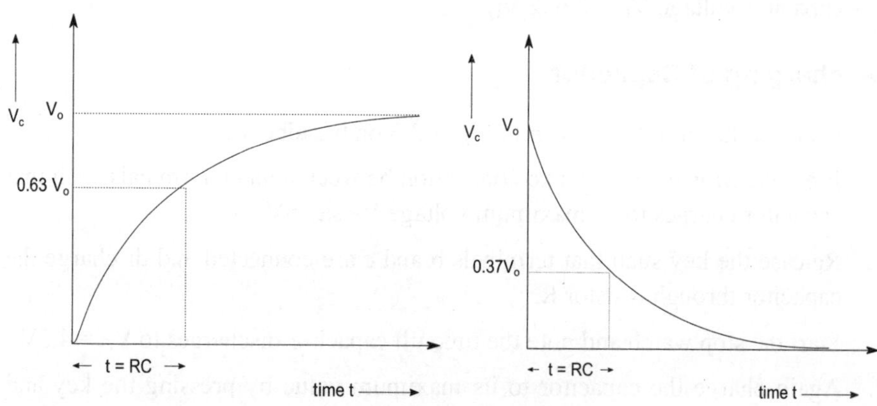

Fig. 1.3.4 Charging and Discharging curve

RESULT

The charging and discharging curves of a capacitor in given RC circuit have
been drawn successfully. The theoretical and graphical value of time constant
RC is found to be …… and …… respectively.

DISCUSSION

Capacitor is one of the most important passive components used in almost all the
electronic devices. Capacitors in combination with resistors are used in
frequency selective filters. They are also used as coupling, dc blocking and
decoupling capacitors. Decoupling capacitors are used to decouple one part of
the circuit to the other which helps in shunting the noise caused by circuit
elements reducing its effect on other parts of the circuit. Capacitors in
combination with inductors and resistors act as a tuned circuit which helps in
selecting or rejecting a particular frequency.

AIM (DIFFERENTIATOR)

RC Circuit: Differentiator.

APPARATUS REQUIRED

Breadboard, connecting wires, resistor - 10kΩ, capacitor - 0.01μF, function
generator, cathode ray oscilloscope, probes.

THEORY

Differentiator

A RC differentiator is a passive wave shaping circuit that gives an output voltage proportional to the time derivative of the input *i.e.*

$$\text{output } \alpha \; \frac{d(\text{input})}{dt}$$

A basic differentiating or differentiator circuit is shown in Fig. 1.3.5.

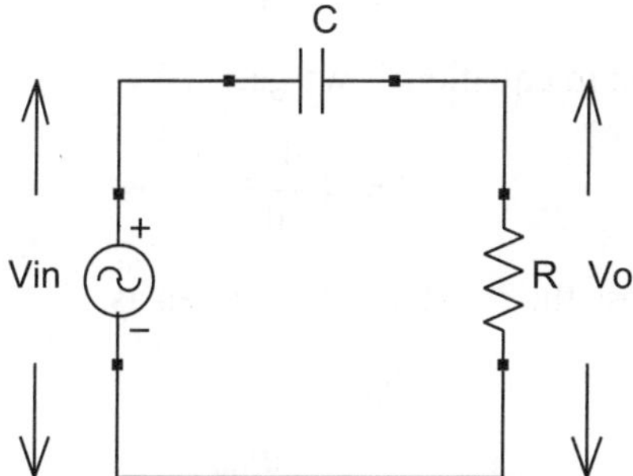

Fig. 1.3.5 Circuit for differentiator

The output is taken across the resistor and the circuit is designed such that the voltage across resistor is the derivative of the input voltage.

The conditions necessary to be fulfilled for the circuit to act as a differentiator:

1. The time constant RC of the circuit should be much smaller than the time period T of the input signal *i.e.* RC << T.
2. The value of capacitive reactance Xc should be greater than or equal to 10 times the resistance R *i.e.* Xc ≥ 10R.

Equations involved

The instantaneous charge q on the capacitor is given as

$$q = CV_C$$

Where,

Vc: voltage drop across capacitor, C: capacitance.

Let i be the current flowing through the circuit which is given as

$$i = \frac{dq}{dt}$$

$$\Rightarrow \qquad i = \frac{d(CV_C)}{dt}$$

$$\Rightarrow \qquad\qquad i = C\frac{dV_C}{dt} \qquad\qquad - \text{eqn 3}$$

Input voltage can be written as

$$V_{in} = V_C + V_R$$

Where, V_R: voltage across resistor R.

Since, $X_C \gg R$ therefore, the voltage across the capacitor may be considered to be equal to the input voltage *i.e.*

$$V_C = V_{in}$$

Using the above relation in equation 3, we get

$$\therefore \qquad\qquad i = C\frac{dV_{in}}{dt}$$

The output voltage across the resistance R is given as

$$V_o = i.\,R$$

$$\Rightarrow \qquad\qquad V_o = RC\frac{dV_{in}}{dt}$$

Hence,

$$\text{output voltage} \ \alpha \ \frac{d(\text{input voltage})}{dt}$$

CALCULATIONS

Choose the value of R and C.

Let R = 10kΩ, C = 0.01μF

Time constant RC = 10kΩ × 0.01μF = 0.1ms

PROCEDURE

1. Design the differentiator by choosing the appropriate values of R and C.

2. Connect the circuit as shown in Fig. 1.3.5 on breadboard.

3. Connect the function generator to give the input signal using a probe.

4. Select the input waveform as square wave with frequency of 500Hz such that the time period of input signal is much greater than the time constant RC *i.e.* T >> RC and Xc ≥ 10R.

5. Connect channel 1 of CRO across the input and channel 2 across the resistor to view the input and output waveforms respectively.

6. Trace the input and output waveforms for all three cases T >> RC, T == RC, T << RC by changing the input frequency to 500 Hz, 10 kHz and 15 kHz.

7. Change the input signal to sine and triangular wave and trace the output waveforms for different conditions.

INPUT AND OUTPUT WAVEFORMS

1. Square wave

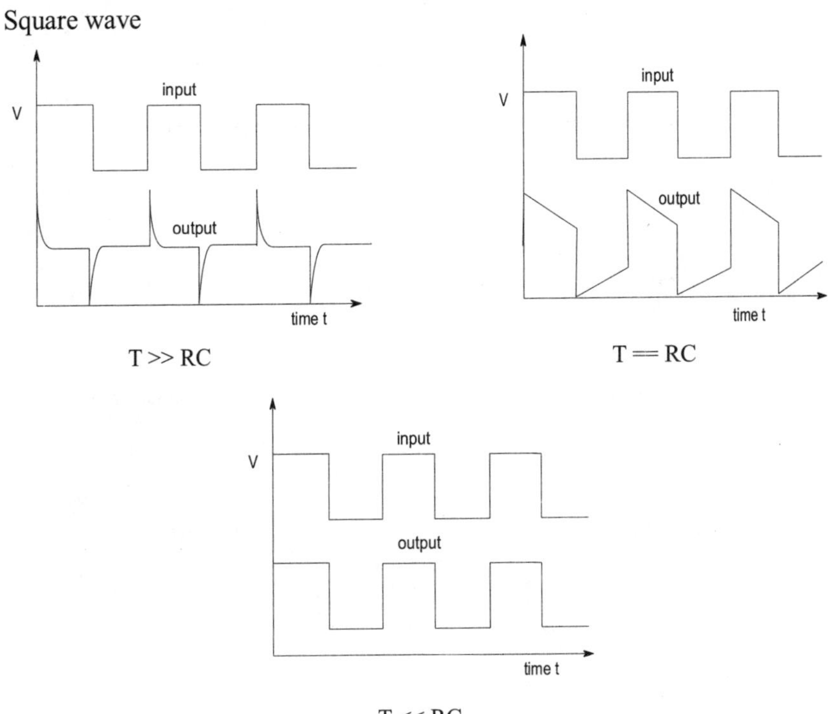

Fig. 1.3.6 Output waveforms for square wave input

2. Triangular wave

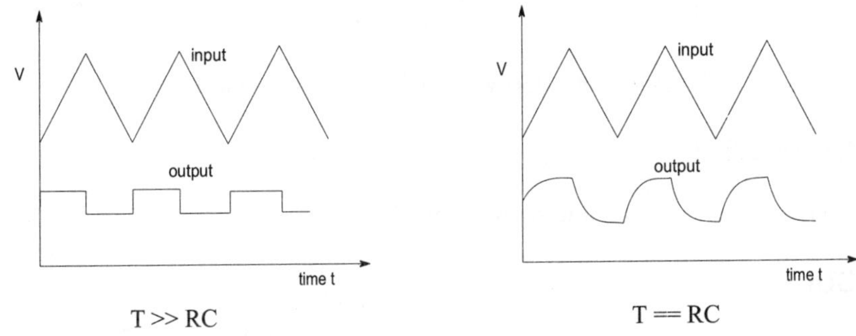

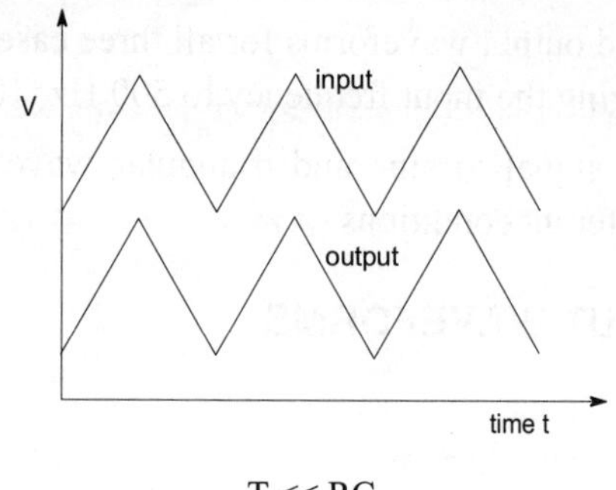

$$T \ll RC$$

Fig. 1.3.7 Output waveforms for triangular wave input

3. Sine wave

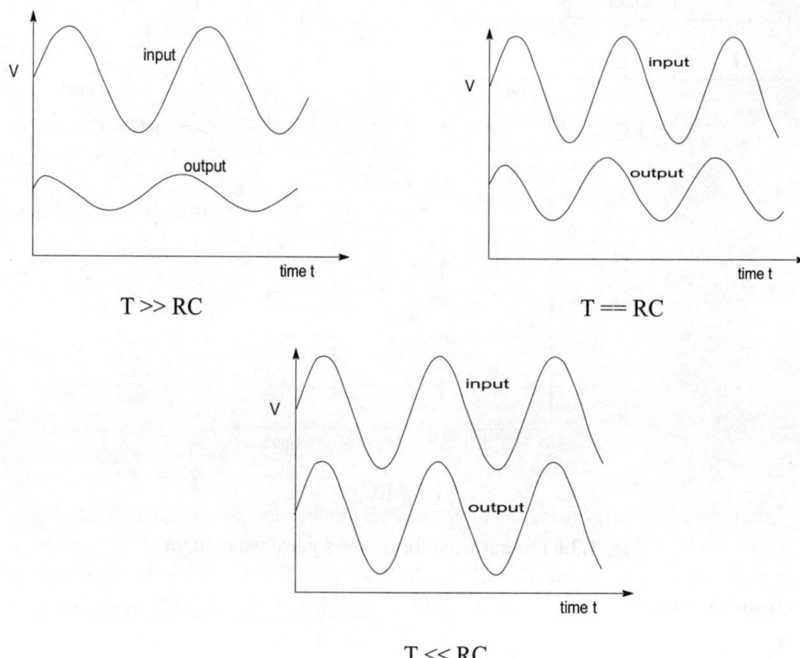

Fig. 1.3.8 Output waveforms for sine wave input

OBSERVATIONS

Attach the traces for input and output waveforms.

RESULT

The differentiator circuit has been designed successfully and traces for different input signals have been taken.

DISCUSSION

A differentiator circuit produces an output directly proportional to the derivative of the input. When a dc input is applied to the differentiator circuit, it produces a zero output (since derivative of constant is zero). For a square wave input, it produces sharp narrow pulses when the amplitude of input changes from low to high or high to low and zero output when the amplitude is constant. For a triangular wave input, it produces a constant voltage when the amplitude of input changes at a constant rate and thus produces a square wave output.

Differentiator also acts as a high pass filter. They are widely used to generate a square wave output from triangular wave input, step from a ramp input and spiked from square input. Either the positive or negative spike can be used as trigger pulses or synchronization pulses in television and cathode ray oscilloscopes.

They are also used to convert PWM into PPM by converting PWM into spikes and then the negative spikes trigger the mono-stable multivibrator to produce PPM pulses.

AIM (INTEGRATOR)

RC Circuit: Integrator.

APPARATUS REQUIRED

Breadboard, connecting wires, resistor - 20kΩ, capacitor - 0.01μF, function generator, cathode ray oscilloscope, probes.

THEORY

Integrator

An integrator is a wave shaping circuit designed to give an output voltage proportional to the integral of the input *i.e.*

$$\text{output } \alpha \int \text{input}$$

A basic integrating or integrator circuit is shown in Fig. 1.3.9.

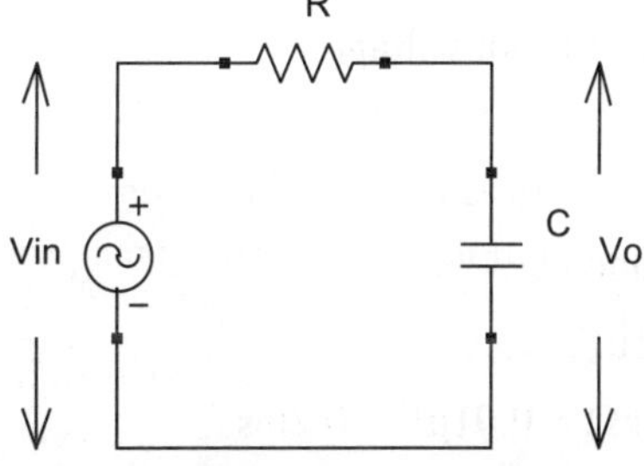

Fig. 1.3.9 Circuit for integrator

The output is taken across the capacitor and the circuit is designed such that the voltage across capacitor is the derivative of the input voltage.

The conditions necessary to be fulfilled for the circuit to act as an integrator:

1. The time constant RC of the circuit should be much larger than the time period T of the input signal *i.e.* $RC \gg T$.

2. The value of resistance should be chosen such that $R \geq 10X_C$, where X_C is the capacitive reactance.

Equations involved

Input voltage can be written as

$$V_{in} = V_C + V_R$$

Where,

V_R: voltage across resistor R, V_C: voltage across capacitor C.

Since, $R \gg X_C$ therefore, the voltage across the resistor may be considered to be equal to the input voltage.

$$\therefore \qquad V_{in} = V_R$$

Let i be the current flowing in the circuit, then the instantaneous charge q on the capacitor is given as

$$q = \int i \, dt$$

$$\Rightarrow \qquad q = \frac{1}{R} \int V_{in} \, dt$$

The output voltage across the capacitor is given as

$$V_o = \frac{q}{C}$$

$$\Rightarrow \qquad V_o = \frac{1}{RC} \int V_{in} \, dt$$

Hence, output voltage $\alpha \int$ input voltage.

CALCULATIONS

Choose the value of R and C.

Let $R = 20k\Omega$, $C = 0.01\mu F$

Time constant $RC = 20k\Omega \times 0.01\mu F = 0.2ms$

PROCEDURE

1. Design the integrator by choosing the appropriate values of R and C.

2. Connect the circuit as shown in Fig. 1.3.9 on breadboard.

3. Connect the function generator to give the input signal using a probe.

4. Select the input waveform as square wave with frequency of 15kHz such that the time period of input signal is much lower than the time constant RC *i.e.* $T \ll RC$ and $R \geq 10Xc$.

5. Connect channel 1 of CRO across the input and channel 2 across the capacitor to view the input and output waveforms respectively.

6. Trace the input and output waveforms for all three cases $T \gg RC$, $T == RC$, $T \ll RC$ by changing the input frequency to 500 Hz, 5 kHz and 15 kHz.

7. Change the input signal to sine and triangular wave and trace the output waveforms for different conditions.

INPUT AND OUTPUT WAVEFORMS

1. Square wave

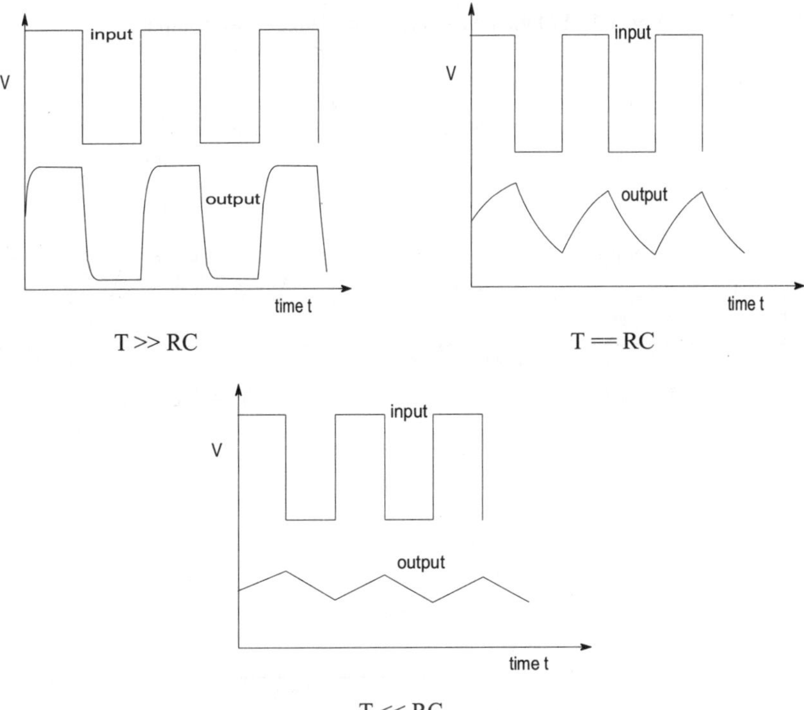

Fig. 1.3.10 Output waveforms for square wave input

2. Triangular wave

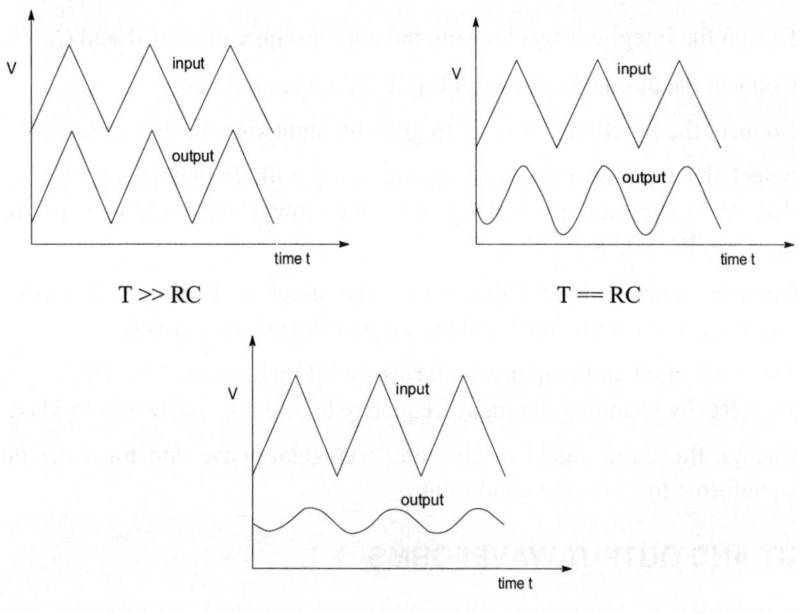

Fig. 1.3.11 Output waveforms for triangular wave input

3. Sine wave

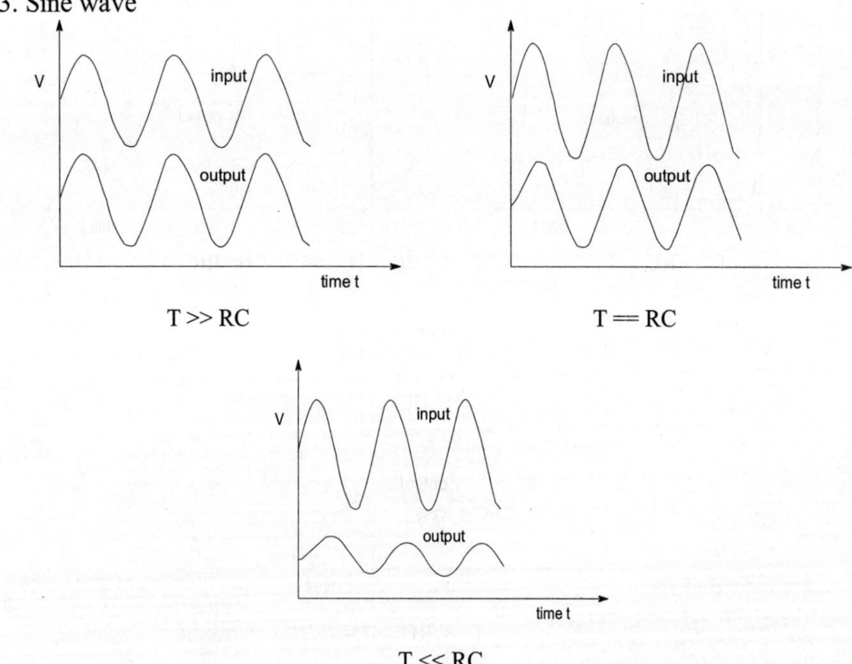

Fig. 1.3.12 Output waveforms for sine wave input

OBSERVATIONS

Attach the traces for input and output waveforms.

RESULT

The integrator circuit has been designed successfully and traces for different input signals have been taken.

DISCUSSION

Integrators act as a low pass filter. They are widely used to generate a triangular wave from square wave, saw tooth wave from a rectangular wave. They are also used in wave shaping circuits; analog computers to perform mathematical integration.

Experiment 4

LOW PASS RC FILTER

AIM

Design a Low Pass RC filter and Study its Frequency Response.

APPARATUS REQUIRED

Breadboard, connecting wires, resistor - 7.9kΩ, capacitor - 0.01μF, function generator, cathode ray oscilloscope, probes.

THEORY

Filters

These are the electronic circuits which are used to pass the wanted frequency components and reject the unwanted one. They are of different types:

1. Passive and Active filters

    *   Passive filters use passive elements like resistors, capacitors and inductors. They do not depend on any external power supply for their operation.

    *   Active filters use amplifying components like operational amplifiers. They need an external power source for their operation.

2. High pass, Low pass, Band pass, Band reject, All pass filter

    *   High pass filter passes the frequencies which are above the cut-off frequency while rejecting the lower ones.

    *   Low pass filter passes the frequencies which are below the cut-off frequency while rejecting the higher ones.

    *   Band pass filter passes a band of frequencies while rejecting the others that lie outside this band.

    *   Band reject filter rejects a band of frequencies while passing the others that lie outside this band.

    *   An All pass filter passes all the frequencies without any attenuation but introduces the phase change in the output.

Low Pass Filter

A low pass RC filter is a passive circuit that passes low frequency signals while attenuating signals with frequencies higher than the cut-off frequency f_C. Since, it is designed by using passive components it does not amplify the signal and thus its output is always less than or equal to the input.

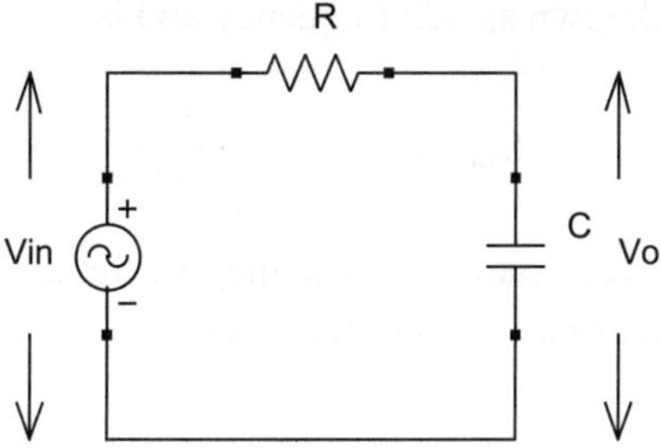

Fig. 1.4.1 Circuit for low pass filter

The output is taken across the capacitor. As the input frequency changes, the capacitive reactance changes, hence gain varies with the change in frequency of the applied signal.

Working

Capacitive reactance is given by the formulae:

$$X_c = \frac{1}{\omega C}$$

a. DC signal

For dc, the reactance of capacitor is infinite; hence it acts as an open switch. The complete voltage drop appears across the capacitor and hence output voltage is same as the input voltage.

b. Low frequencies *i.e.* $f < f_C$

For low frequencies, the capacitive reactance is very high which results in a high voltage drop across the capacitor. Almost negligible voltage appears across the resistance R. Thus, output voltage is almost same as input voltage.

c. At cut off frequency *i.e.* $f = f_C$

As the frequency increases, the reactance of capacitor keeps on decreasing which results in decrease of voltage drop across it. Hence, there is a fall in the output voltage. Therefore, gain (ratio of output voltage to input voltage) also decreases. At cut-off frequency, the gain becomes 0.707 times the maximum gain or 3dB down the maximum gain.

d. For high frequencies *i.e.* $f > f_C$

At very high frequencies, reactance offered by capacitor becomes very less such that output voltage across capacitor becomes very small as compared to the input voltage. Hence, output keeps on falling as frequency increases.

Hence, low pass filter blocks the high frequency signals while allowing the low frequency signals to pass through it.

Cut-off frequency

It is the frequency at which the output voltage is 0.707 times the maximum voltage V_{max}. It is also known as 3db frequency and is given by

$$f_{cut-off} \text{ or } f_C = \frac{1}{2\pi RC}$$

Power at cut-off frequency is 50% the maximum value, hence cut off frequency is also called as half power frequency.

Equations involved

The current I flowing through the circuit is given by

$$I = \frac{V_{in}}{R - jX_C}$$

Where,

R: resistance, C: capacitance, X_C: capacitive reactance, V_{in}: input voltage.

The output voltage V_O across the capacitor is given by

$$V_o = \frac{V_{in} \cdot (-jX_C)}{R - jX_C}$$

$$\Rightarrow \qquad V_o = \frac{V_{in}}{1 - \dfrac{R}{jX_C}}$$

$$\Rightarrow \qquad V_o = \frac{V_{in}}{1 + j2\pi fRC}$$

Where, f: frequency of the applied signal.

The voltage gain A of the circuit is given by

$$|A| = \left| \frac{V_o}{V_{in}} \right|$$

$$\Rightarrow \qquad |A| = \left| \frac{1}{1 + j2\pi fRC} \right|$$

$$\Rightarrow \qquad |A| = \frac{1}{\sqrt{1 + \left(\dfrac{f}{f_C}\right)^2}}$$

Where,

f_C: cut-off frequency

As f→0, |A| →1

At f = f$_C$, |A| = 0.707

To express voltage gain in dB scale,

$$20\log(|A|) = 20\log\left(\frac{1}{\sqrt{1+\left(\frac{f}{f_c}\right)^2}}\right)$$

$$\Rightarrow \qquad 20\log(|A|) = 20\log(1) - 10\log\left(1+\frac{f^2}{f_c^2}\right)$$

$$\Rightarrow \qquad 20\log(|A|) = -10\log\left(1+\frac{f^2}{f_c^2}\right)$$

At cut-off frequency *i.e.* when f = fc

$$20\log(|A|) = -3\text{dB}$$

Hence, at cut-off frequency, gain is 3dB down the maximum value.

Frequency response

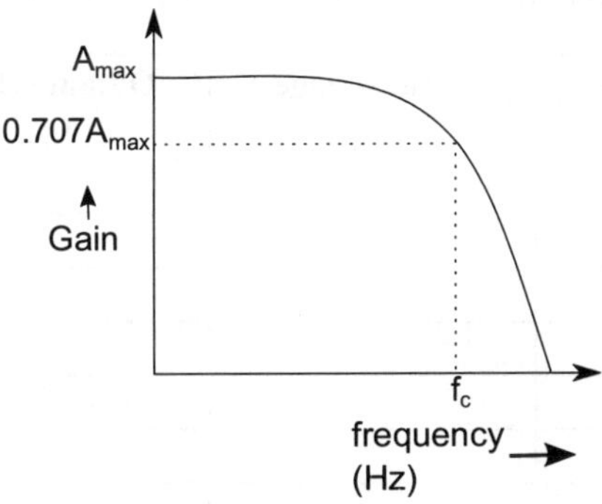

Fig. 1.4.2 Graph between gain vs frequency

CALCULATIONS

Given cut-off frequency f$_C$ = 2 KHz (say)

Fix C = 0.01 μF

Find R by using the formula for cut-off frequency which is given as

$$f_C = \frac{1}{2\pi RC}$$

$$\Rightarrow \qquad R = \frac{1}{2\pi \times 2000 \times 0.01 \times 10^{-6}}$$

$$\Rightarrow \qquad R = 7.9k\Omega$$

PROCEDURE

1. Connect the circuit as shown in Fig. 1.4.1 on breadboard.
2. Connect function generator to the input of circuit using a probe. Set it to sine wave with fixed amplitude.
3. Connect channel 1 of CRO at the input and channel 2 across the capacitor to view the input and output waveforms respectively.
4. Vary the frequency of input signal from 100Hz to 100 kHz.
5. Measure the input and output voltage through CRO and note the readings in a tabular form.
6. Plot the graph on semi-log graph paper between gain and input frequency.
7. Find the cut-off frequency graphically where gain becomes 0.707 times the max value.
8. Compare the theoretical and practical value of cut-off frequency.

OBSERVATION TABLE

Table 1.4.1 Observation table for low pass filter

S.No.	Frequency (Hz)	Input voltage V_{in}	Output voltage V_{out}	Gain V_{out} / V_{in}
1.	100			
2.	200			
3.	300			
4.	400			
5.	500			
...	...			

OBSERVATIONS

f_C (theoretical) = ...

f_C (practical) = ...

RESULT

The low pass filter for cut-off frequency of 2 kHz has been designed and the frequency response has been drawn successfully. The cut-off frequency comes out to be ......

DISCUSSION

Electronic filters are widely used to reject the unwanted frequencies while allowing the desired frequencies to pass through it. They are used in electronic circuits as anti-aliasing filters, digital filters *etc.* Low pass filters are also act as integrator when time period of the input signal T is much less than the time constant RC of the circuit *i.e.* T << RC. Combinations of low pass and high pass filters are used in tune controls of radios and record players.

Experiment 5

HIGH PASS RC FILTER

AIM

Design a High Pass RC filter and Study its Frequency Response.

APPARATUS REQUIRED

Breadboard, connecting wires, resistor - $3.2k\Omega$, capacitor - $0.01\mu F$, function generator, cathode ray oscilloscope, probes.

THEORY

Filters

These are the electronic circuits which are used to pass the wanted frequency components and reject the unwanted one. They are of different types:

1. Passive and Active filters

   - Passive filters use passive elements like resistors, capacitors and inductors. They do not depend on any external power supply for their operation.

   - Active filters use amplifying components like operational amplifiers. They need an external power source for their operation.

2. High pass, Low pass, Band pass, Band reject, All pass filter

   - High pass filter passes the frequencies which are above the cut-off frequency while rejecting the lower ones.

   - Low pass filter passes the frequencies which are below the cut-off frequency while rejecting the higher ones.

   - Band pass filter passes a band of frequencies while rejecting the others that lie outside this band.

   - Band reject filter rejects a band of frequencies while passing the others that lie outside this band.

   - An All pass filter passes all the frequencies without any attenuation but introduces the phase change in the output.

High Pass Filter

A high pass RC filter is a passive circuit that passes high frequency signals while attenuating signals with frequencies lower than the cut-off frequency. Since, it is designed by using passive components it does not amplify the signal and thus its output is always less than or equal to the input.

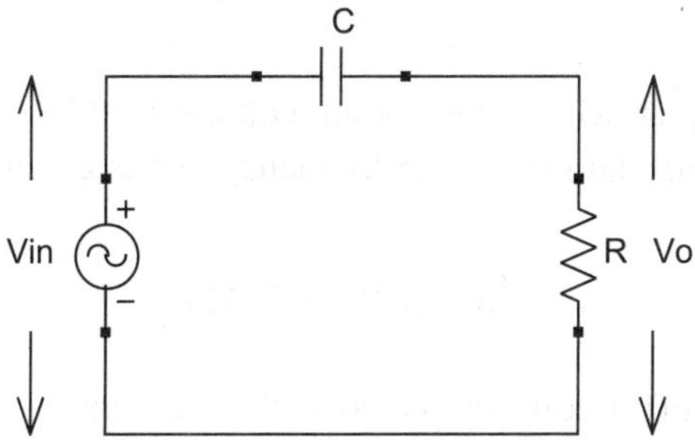

Fig. 1.5.1 Circuit for high pass filter

The output is taken across the resistor. As the input frequency changes, the capacitive reactance changes, hence gain varies with the change in frequency of the applied signal.

Working

Capacitive reactance is given by the formulae:

$$X_c = \frac{1}{\omega C}$$

a. DC signal

For dc, the reactance of capacitor is infinite; hence it acts as an open switch. The whole voltage drop appears across the capacitor and no current flows through the circuit. Hence, output voltage across resistor is zero.

b. Low frequencies *i.e.* $f < f_C$

For low frequencies, the capacitor offers very high reactance which results in a high voltage drop across it. Therefore, very less current flows through the circuit and hence voltage drop across the resistor is also less. Thus, output voltage is very small as compared to the input voltage.

c. At cut off frequency *i.e.* $f = f_C$

As the frequency increases, the reactance of capacitor decreases which results in decrease of voltage drop across it. This results in increase of the circuit current and hence output voltage across resistor also increases. Therefore, the gain (ratio of output voltage to input voltage) also increases. At cut-off frequency, the gain becomes 0.707 times the maximum gain or 3dB down the maximum gain.

d. For high frequencies *i.e.* $f > f_C$

For frequencies greater than the cut-off frequency, the capacitive reactance becomes very small. As a result, the voltage drop across the capacitor is almost negligible and the complete input voltage appears across the resistor. Hence at very high frequencies, the output voltage is almost same as the input voltage.

Hence, a high pass filter blocks the low frequency signals while allowing the high frequency signals to pass through it.

Cut-off frequency

It is the frequency at which the output voltage is 0.707 times the maximum voltage V_{max}. It is also known as 3db frequency and is given by

$$f_{cut-off} \text{ or } f_C = \frac{1}{2\pi RC}$$

Power at cut-off frequency is 50% the maximum value, hence cut off frequency is also called as half power frequency.

Equations involved

The current I flowing through the circuit is given by

$$I = \frac{V_{in}}{R - jX_C}$$

Where, R: resistance, C: capacitance, X_C: capacitive reactance, V_{in}: input voltage.

The output voltage V_O across the resistor is given by

$$V_o = \frac{V_{in} \cdot R}{R - jX_C}$$

$$\Rightarrow \qquad V_o = \frac{V_{in}}{1 - \dfrac{jX_C}{R}}$$

$$\Rightarrow \qquad V_o = \frac{V_{in}}{1 - j/2\pi fRC}$$

Where, f: frequency of the applied signal

The voltage gain A of the circuit is given by

$$|A| = \left| \frac{V_o}{V_{in}} \right|$$

$$\Rightarrow \qquad |A| = \left| \frac{1}{1 - j/2\pi fRC} \right|$$

$$\Rightarrow \qquad |A| = \frac{1}{\sqrt{1 + \left(\dfrac{f_c}{f} \right)^2}}$$

Where, f_C: cut-off frequency

As $f \to \infty$, $|A| \to 1$

At $f = f_C$, $|A| = 0.707$

To express voltage gain in dB scale,

$$20 \log(|A|) = 20 \log \left(\frac{1}{\sqrt{1 + \left(\frac{f_C}{f}\right)^2}} \right)$$

$\Rightarrow \qquad 20 \log(|A|) = 20 \log(1) - 10 \log \left(1 + \frac{f_c^2}{f^2} \right)$

$\Rightarrow \qquad 20 \log(|A|) = -10 \log \left(1 + \frac{f_c^2}{f^2} \right)$

At cut-off frequency *i.e.* when $f = fc$

$$20 \log(|A|) = -3 \text{dB}$$

Hence, at cut-off frequency, gain is 3dB down the maximum value.

Frequency response

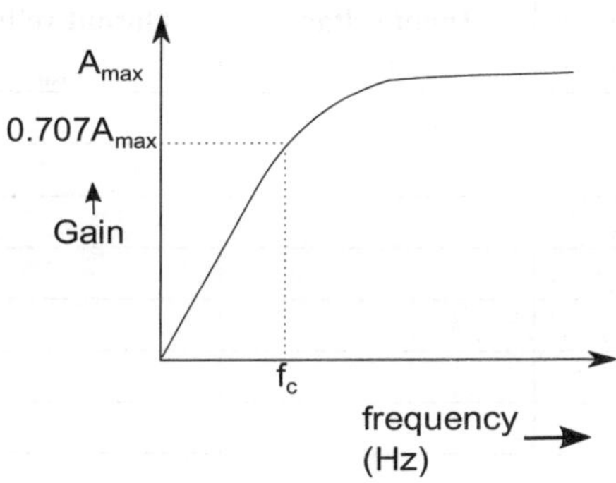

Fig. 1.5.2 Graph between gain vs frequency

CALCULATIONS

Given cut-off frequency $f_C = 5$ KHz (say)

Fix $C = 0.01\ \mu F$

Find R by using the formula for cut-off frequency which is given as

$$f_C = \frac{1}{2\pi RC}$$

$$\Rightarrow \qquad R = \frac{1}{2\pi \times 5000 \times 0.01 \times 10^{-6}}$$

$$\Rightarrow \qquad R = 3.2\text{k}\Omega$$

PROCEDURE

1. Connect the circuit as shown in Fig. 1.5.1 on breadboard.
2. Connect function generator to the input of circuit using a probe. Set it to sine wave with fixed amplitude.
3. Connect channel 1 of CRO at the input and channel 2 across the resistor to view the input and output waveforms respectively.
4. Vary the frequency of input signal from 100Hz to 100 kHz.
5. Measure the input and output voltage through CRO and note the readings in a tabular form.
6. Plot the graph on semi-log graph paper between gain and input frequency.
7. Find the cut-off frequency graphically where gain becomes 0.707 times the max value.
8. Compare the theoretical and practical value of cut-off frequency.

OBSERVATION TABLE

Table 1.5.1 Observation table for high pass filter

S.No.	Frequency (Hz)	Input voltage V_{in}	Output voltage V_{out}	Gain V_{out} / V_{in}
1.	100			
2.	200			
3.	300			
4.	400			
5.	500			
...	...			

OBSERVATIONS

f_C (theoretical) = ...

f_C (practical) = ...

RESULT

The high pass filter for cut-off frequency of 5 kHz has been designed and the frequency response has been drawn successfully. The cut-off frequency comes to be ......

DISCUSSION

Electronic filters are widely used to reject the unwanted frequencies while allowing the desired frequencies to pass through it.

High pass filters are widely used in audio power amplifiers to prevent the amplification of dc signal which can harm the amplifier. High pass filter also acts as a differentiator when time period of the input signal T is much greater than the time constant RC of the circuit *i.e.* $T \gg RC$. Combinations of low pass and high pass filters are used in tune controls of radios and record players.

Experiment 6

LISSAJOUS FIGURE

AIM

To Measure the Unknown Frequency of an AC Signal and Find the Phase Difference between the Two AC Signals using Lissajous Figure.

APPARATUS REQUIRED

Breadboard, two function generators, cathode ray oscilloscope, capacitor - $0.1\mu F$, resistance box, probes, connecting wires.

THEORY

Lissajous Figure

Lissajous figures are the patterns produced by the intersection of two curves of simple harmonic motion which are at right angles to each other. The Lissajous figures can be viewed on a cathode ray oscilloscope when two sinusoidal voltages are simultaneously applied on the horizontal and vertical inputs of an oscilloscope. Depending on the frequency, amplitude and the phase relationship of the two input signals, different patterns will appear on the CRO screen.

Let the two signals are:

$$x = A\sin\omega t$$

$$y = B\sin(\omega t + \phi)$$

Where,

A: amplitude of signal x, B: amplitude of signal y, Φ: phase angle.

Let us see different cases when both signals have equal frequencies.

i. $A \neq B$ and $\phi = 0$

Then two signals can be written as

$$x = A\sin\omega t$$
$$y = B\sin\omega t$$

We can write,

$$\frac{x}{A} = \frac{y}{B}$$

$$\Rightarrow \qquad\qquad y = \frac{B}{A}x$$

This equation represents a straight line passing through the origin.

ii. $A = B$ and $\phi = \Pi/2$

Then two signals can be written as

$$x = A\sin\omega t$$

$$y = A\sin\left(\omega t + \frac{\pi}{2}\right) = A\cos\omega t$$

We can write,

$$\frac{x^2}{A^2} + \frac{y^2}{A^2} = 1$$

$$\Rightarrow \qquad\qquad x^2 + y^2 = A^2$$

This equation represents a circle.

iii. $A \neq B$ and $\phi = \Pi/2$

Then two signals can be written as

$$x = A\sin\omega t$$

$$y = B\sin\left(\omega t + \frac{\pi}{2}\right) = B\cos\omega t$$

We can write,

$$\frac{x^2}{A^2} + \frac{y^2}{B^2} = 1$$

This equation represents a symmetrical ellipse.

iv. $A \neq B$ and $\phi = \Pi$

Then two signals can be written as

$$x = A\sin\omega t$$

$$y = B\sin(\omega t + \pi) = -B\sin\omega t$$

We can write,

$$\frac{x}{A} = -\frac{y}{B}$$

$$\Rightarrow \qquad\qquad y = -\frac{B}{A}x$$

This equation represents a straight line passing through the origin.

For signals with different phase angles, Lissajous patterns are shown in Fig. 1.6.1.

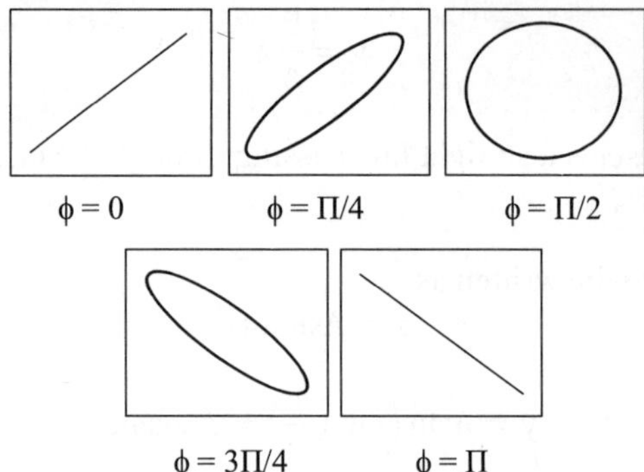

Fig 1.6.1 Lissajous patterns for signals with same frequencies

Frequency measurement

Lissajous figures can be used to measure the unknown frequency of a sinusoidal wave. This can be done by superimposing two ac voltages of different frequencies in mutually perpendicular directions. The sine wave signal with unknown frequency which is to be determined is applied to one set of deflection plates (say vertical) and sinusoidal voltage signal of known frequency is applied to another set of deflection plates (say horizontal). The frequency of this signal is varied till a suitable stationary lissajous figure pattern is observed on the oscilloscope.

The ratio of two frequencies is given by

$$\frac{f_y}{f_x} = \frac{f_{vertical}}{f_{horizontal}} = \frac{\text{Number of points touching the horizontal axis } (P_X)}{\text{Number of points touching the vertical axis } (P_Y)}$$

For signals with different frequencies, lissajous patterns are shown in Fig. 1.6.2.

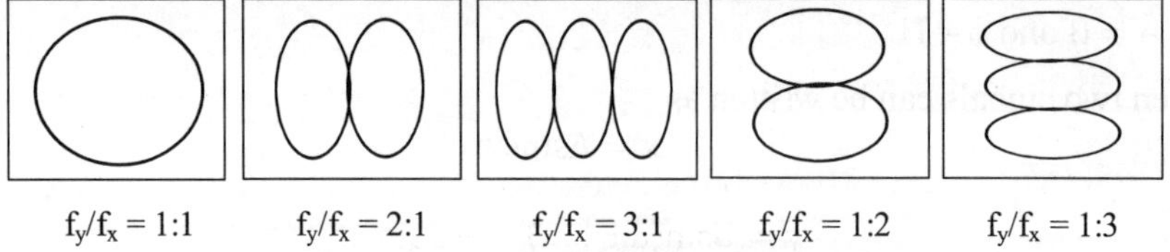

Fig 1.6.2 Lissajous pattern for signals with different frequencies

Phase measurement

The phase difference between two sinusoidal voltage signals of the same frequency can be measured either by measuring the time difference between the two waveforms or by obtaining the lissajous figure pattern on a cathode ray oscilloscope. The pattern can be observed by applying one of the signals to the horizontal deflection plates and the other to the vertical deflection plates. The resultant pattern will be an ellipse.

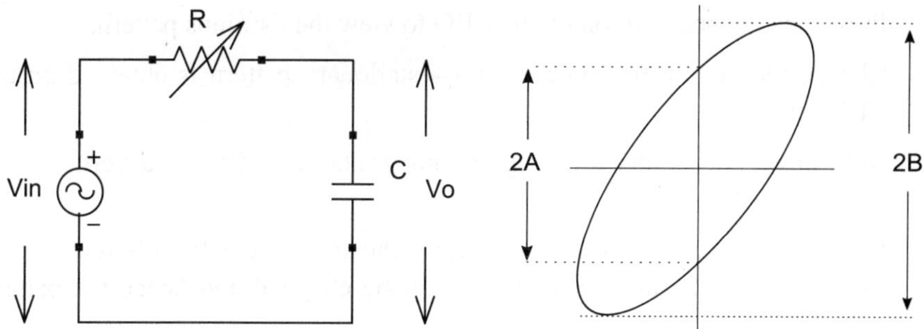

Fig. 1.6.3 Circuit for phase measurement   **Fig 1.6.4** Output at CRO in x-y mode

The input signal is applied across R and C and the output is taken across capacitor C. Because of the presence of capacitor, it produces a phase shift between input and output. Output voltage is given as

$$V_o = \frac{V_{in} \cdot (-jX_C)}{R - jX_C}$$

$\Rightarrow$

$$V_o = \frac{V_{in}}{1 - \dfrac{R}{jX_C}}$$

$\Rightarrow$

$$V_o = \frac{V_{in}}{1 + j2\pi fRC}$$

$\Rightarrow$

$$\frac{V_o}{V_{in}} = \frac{1}{1 + j\omega RC}$$

Phase difference is given as

$$\phi = \tan^{-1}(\omega RC)$$

By using Lissajous figure as shown in Fig. 1.6.4, phase difference between two signals is given as

$$\phi = \sin^{-1}\left(A/B\right)$$

PROCEDURE

Frequency Measurement

1. Connect the function generator with known frequency f_X to the X channel of CRO.

2. Connect the function generator with unknown frequency f_Y to the Y channel of CRO.

3. Press the x-y mode button on the CRO to view the lissajous pattern.

4. Change the known frequency f_X till a stationary pattern is obtained on the CRO screen.

5. Note down the points touching the horizontal axis (P_X) and vertical axis (P_Y).

6. For the same unknown frequency, repeat the above steps for different known frequency f_X. The number of loops will be changed and hence the pattern will also be changed.

7. Calculate the unknown frequency f_Y in each case using the formulae.

8. Repeat the above steps for different unknown frequency and take the readings again.

Note: On CRO, channel 1 and 2 are mentioned as X and Y channel for X-Y mode.

Phase Measurement

1. Connect the circuit as shown in Fig. 1.6.3 on the breadboard with $C = 0.1\mu F$.

2. Connect the function generator to give the input across RC network and set it to sine wave with frequency 1 kHz.

3. Connect channel 1 of the CRO across the input and channel 2 across the capacitor to view the input and output waveforms respectively.

4. Take out 1200Ω resistance from resistance box and observe the phase shift between two signals on CRO.

5. Press the x-y mode button on CRO to view the lissajous pattern *i.e.* ellipse in this case.

6. Press gnd button of both channel 1 and channel 2 of CRO to bring the electron gun dot at the center of the screen.

7. Press ac button of both channel 1 and channel 2 to view the ellipse again.

8. Measure number of divisions for 2A and 2B and take the trace of ellipse on trace paper marking the y and x axis.

9. Measure R, C, f, 2A and 2B and note them in a form of table.

10. Calculate the value of phase difference theoretically as well as practically and compare the two values.

11. Repeat the above steps for different values of resistances.

OBSERVATION TABLE

Frequency Measurement

For unknown frequency $f_Y = ...$ Hz

Table 1.6.1 Observation table for frequency measurement

S. No.	Lissajous figure	Known Frequency f_X (Hz)	No. of points on x-axis P_X	No. of points on y-axis P_Y	Unknown frequency, $f_Y = \left(\dfrac{P_X}{P_Y}\right) \cdot f_X$ (Hz)
1.					
2.					
3.					
4.					

Mean value of f_Y (practical) = ... Hz

f_Y (theoretical) = ... Hz

% error = ...

Phase Measurement

$C = 0.1\,\mu F$

Frequency $f = 1$ kHz

Table 1.6.2 Observation table for phase measurement

S. No.	Resistance R (kΩ)	2A (div)	2B (div)	Practical value $\phi = \sin^{-1}(A/B)$ (deg)	Theoretical value $\phi = \tan^{-1}(2\pi fRC)$ (deg)
1.	1400				41.32°
2.	1200				37°
3.	1000				32.13°
4.	800				26.67°
5.	600				20.64°
6.	400				14.1°

RESULT

The unknown frequency of an a.c. signal and the phase difference between the two a.c. signals have been measured successfully using Lissajous figure

DISCUSSION

Lissajous figures are used to compare the amplitude, frequency and phase between two signals. Just by looking at the shape of the pattern one can determine the difference between the two signals. If the signals are in phase, then the pattern is a straight line and if signals are out of phase by 90°, then the pattern will be of a circle.

Experiment 7

Z PARAMETERS

AIM

To Measure the Z parameters of a Two - Port Network.

APPARATUS REQUIRED

Breadboard, resistors - 1kΩ, 2.2kΩ, 10kΩ, 10V and 5V dc power supply, multimeter, connecting wires.

THEORY

Two Port Network

A two port network is a circuit or a device with two pair of terminals which can be connected to other external circuits. Each pair of terminal is called as a port only if the port condition is met *i.e.* current entering one terminal should be same as current leaving the other terminal. Port 1 is taken as input port and port 2 is taken as output port. It has four variables: input voltage V_1, input current I_1, output voltage V_2 and output current I_2.

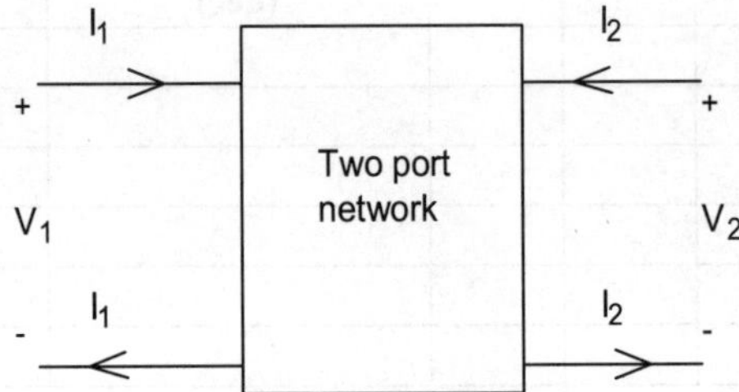

Fig. 1.7.1 Two port network

As shown in Fig. 1.7.1, the two port network can be seen as a black box which is described by the relationship between these four variables. This helps in calculating the response of the network to the signals applied at the input ports without the need of solving for internal voltages and currents in the network. Similar circuits or devices can be compared easily by using two port networks.

To establish the relation between input and output variables, different models are used such as Z parameters, Y parameters, h parameters, g parameters, ABCD parameters. Difference between various models depends on the variables that are taken as independent variables.

Z parameters or impedance parameters

Z parameters, also known as open circuit impedance parameters are used to represent the relation between input and output variables of a two port network. They are calculated under open circuit conditions *i.e.* current at either of the terminals is zero.

Consider a two port network as shown in Fig. 1.7.1. In this model, I_1 and I_2 are taken as independent variables and they are assumed to be the known variables and the voltages V_1 and V_2 are taken as dependent variables which can be calculated by using the following equations.

$$V_1 = Z_{11} * I_1 + Z_{12} * I_2$$

$$V_2 = Z_{21} * I_1 + Z_{22} * I_2$$

The above equations can be written in matrix form as

$$\begin{pmatrix} V_1 \\ V_2 \end{pmatrix} = \begin{pmatrix} Z_{11} & Z_{12} \\ Z_{21} & Z_{22} \end{pmatrix} \begin{pmatrix} I_1 \\ I_2 \end{pmatrix}$$

Where, Z_{11}, Z_{12}, Z_{21}, Z_{22} are impedances which are given as

$$Z_{11} = \left. \frac{V_1}{I_1} \right|_{I_2=0} = \text{ open circuit input impedance}$$

$$Z_{12} = \left. \frac{V_1}{I_2} \right|_{I_1=0} = \text{ open circuit transfer impedance from port1 to port2}$$

$$Z_{21} = \left. \frac{V_2}{I_1} \right|_{I_2=0} = \text{ open circuit transfer impedance from port2 to port1}$$

$$Z_{22} = \left. \frac{V_2}{I_2} \right|_{I_1=0} = \text{ open circuit output impedance}$$

If $Z_{11} = Z_{22}$, the network is said to be a symmetrical network and if $Z_{12} = Z_{21}$, the network is said to be a reciprocal network.

CIRCUIT DIAGRAM

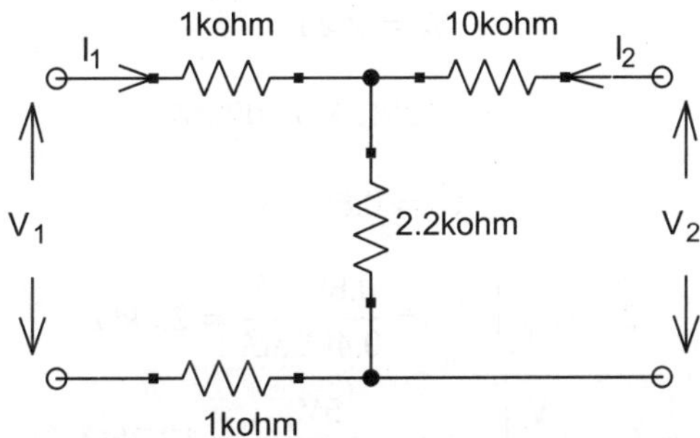

Fig. 1.7.2 Circuit diagram for Z parameters measurement

Case 1: When a voltage source V_1 (say 10V) is connected to the input terminals. To find Z_{11} and Z_{21}, I_2 is taken to be zero. Applying KVL in the mesh1 we get,

$$V_1 = I_1 + 2.2I_1 + I_1$$

$\Rightarrow$
$$V_1 = 4.2\,I_1$$

$\Rightarrow$
$$I_1 = \frac{10}{4.2k} = 2.38\text{mA}$$

The voltage between terminals 2 and 2' can be calculated as

$$V_2 = 2.2\,I_1$$

$\Rightarrow$
$$V_2 = 2.2\text{k}\Omega \times 2.38\text{mA}$$

$\Rightarrow$
$$V_2 = 5.236\text{V}$$

$$Z_{11} = \left.\frac{V_1}{I_1}\right|_{I_2=0} = \frac{10\text{V}}{2.38\text{mA}} = 4.2\text{k}\Omega$$

$$Z_{21} = \left.\frac{V_2}{I_1}\right|_{I_2=0} = \frac{5.236\text{V}}{2.38\text{mA}} = 2.2\text{k}\Omega$$

Case 2: When a voltage source V_2 (say 5V) is connected to the output terminals. To find Z_{12} and Z_{22}, I_1 is taken to be zero. Applying KVL in mesh2 we get,

$$V_2 = 10I_2 + 2.2I_2$$

$\Rightarrow$
$$V_2 = 12.2I_2$$

$\Rightarrow$
$$I_2 = \frac{5}{12.2k} = 0.409\text{mA}$$

The voltage between terminals 1 and 1' can be calculated as

$$V_1 - 2.2\,I_2$$

$\Rightarrow$
$$V_1 = 2.2\text{k}\Omega \times 0.409\text{mA}$$

$\Rightarrow$
$$V_1 = 0.8998\text{V}$$

$$Z_{12} = \left.\frac{V_1}{I_2}\right|_{I_1=0} = \frac{0.8998\text{V}}{0.409\text{mA}} = 2.2\text{k}\Omega$$

$$Z_{22} = \left.\frac{V_2}{I_2}\right|_{I_1=0} = \frac{5\text{V}}{0.409\text{mA}} = 12.2\text{k}\Omega$$

PROCEDURE

1. Connect the circuit as shown in Fig. 1.7.2 on the breadboard.

2. Keep terminals 2-2' open such that $I_2 = 0$.

3. Connect terminals 1-1' to 10V dc power supply and measure the corresponding current I_1 by connecting multimeter (as ammeter) in series with resistance 1kΩ.

4. Measure the voltage V_2 across 2.2kΩ resistance with the help of multimeter.

5. Calculate Z_{11} and Z_{21} by using the above values of V_1, I_1 and V_2 when $I_2 = 0$.

6. Now, keep terminals 1-1' open such that $I_1 = 0$.

7. Connect terminals 2-2' to 5V dc power supply and measure the corresponding current I_2 by connecting multimeter (as ammeter) in series with resistance 10kΩ.

8. Measure the voltage V_1 across 2.2kΩ resistance with the help of multimeter.

9. Calculate Z_{12} and Z_{22} by using above values of V_2, I_2 and V_1 when $I_2 = 0$.

OBSERVATION TABLE

Table 1.7.1 Observation table for Z parameter measurement

S.No.	Parameter	Theoretical value	Practical value
1.	Z_{11}	4.2kΩ	
2.	Z_{12}	2.2kΩ	
3.	Z_{21}	2.2kΩ	
4.	Z_{22}	12.2kΩ	

RESULT

Z parameters of the given network have been measured successfully. Z_{12} comes out to be same as Z_{21}. Hence, the given network is a reciprocal network.

DISCUSSION

A two port network is a network inside a black box with two pairs of accessible terminals known as input and output terminal. Such blocks are quite common in electronic systems, communication systems, and transmission and distribution systems.

Most of the amplifiers and other electrical networks have devices with three terminals. Since, one terminal is made common to both input and output therefore, two port network analyses can be applied on these circuits as well. Any linear circuit which has four terminals can be considered as a two port network only if it satisfies the port conditions.

Two port networks are also helpful in comparing similar circuits or devices, communication systems, control systems, power systems and electronic systems. They are also useful to simplify the complex circuits. Circuits such as filters, matching networks, transmission lines, transformers, small signal models for transistors, *etc.* can be analyzed as two port networks.

Experiment **8**

SERIES LCR

AIM

To Study the Frequency Response of a Series LCR Circuit and Determine its (a) Resonant Frequency (b) Impedance at Resonance (c) Quality Factor (d) Band Width.

APPARATUS REQUIRED

Breadboard, connecting wires, resistors, capacitor, inductor, milli ammeter, multimeter, function generator, probe.

THEORY

Series LCR

A series LCR circuit comprises of inductor, capacitor and resistor connected in series with an ac source as shown in Fig. 1.8.1.

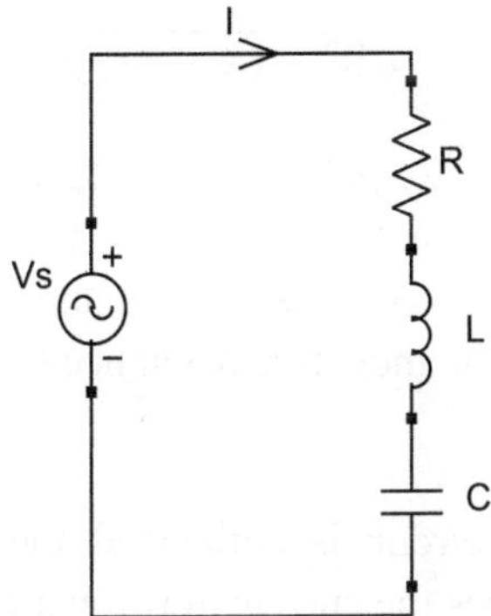

Fig. 1.8.1 Circuit for LCR series resonance

In ac circuits, the voltage and current are usually out of phase. In case of inductor, current lags the voltage by 90°, for capacitor current leads the voltage by 90° and for resistor voltage and current are in phase with each other.

In series LCR circuit, depending on the values of X_L and X_C, current will either lag behind or lead the applied input voltage. When $X_L > X_C$, circuit is inductive and when $X_C > X_L$, circuit is capacitive. If the current is made in phase with the applied input voltage *i.e.* $X_C = X_L$, circuit is said to be in resonance and it behaves as a pure resistor. The resonance can be produced either by changing the frequency of the ac signal keeping C and L constant or by changing either L or C keeping frequency constant.

Resonant frequency

It is the frequency at which the resonance occurs in a LCR circuit *i.e.* the inductive and capacitive reactance become equal. At resonance,

$$X_L = X_C$$

$$\Rightarrow \qquad 2\pi f_r L = \frac{1}{2\pi f_r C}$$

$$\Rightarrow \qquad f_r = \frac{1}{2\pi\sqrt{LC}} \qquad\qquad - \text{eqn 1}$$

Impedance at resonance

At resonant frequency, the series LCR circuit offers lowest impedance and thus, passes those frequency that are nearer to the resonant frequency. Total impedance of the circuit is given as

$$Z = R + j(X_L - X_C)$$

Magnitude is given as

$$|Z| = \sqrt{R^2 + (X_L - X_C)^2}$$

Series resonance occurs when $X_L = X_C$. Hence, at resonance, the net impedance in series LCR circuit becomes $Z = R$ *i.e.* circuit becomes purely resistive. Before the resonant frequency, the reactance will be capacitive while after the resonant frequency, the inductive reactance dominates.

Bandwidth

Bandwidth of a series LCR circuit is defined as the range of frequencies for which the current is 0.707 times the current at resonance frequency *i.e.* $I_r/\sqrt{2}$.

$$BW = f_2 - f_1 = \frac{R}{2\pi L} \qquad\qquad - \text{eqn 2}$$

Quality factor

It is a measure of the selectivity of a signal of a given frequency. Its value should be high for the better selectivity. Also, higher is the value of quality factor, narrower is the bandwidth. Quality factor is given as

$$Q = \frac{f_r}{f_2 - f_1}$$

$$\Rightarrow \qquad Q = \frac{f_r}{BW}$$

Using equation 2, we get

$$\Rightarrow \qquad Q = \frac{f_r \cdot 2\pi L}{R}$$

$$\Rightarrow \qquad Q = \frac{\omega_r L}{R}$$

Since at resonance, $X_L = X_C$

$$\therefore \qquad \omega_r L = \frac{1}{\omega_r C}$$

Hence, we can write quality factor as,

$$Q = \frac{1}{\omega_r CR} = \frac{\omega_r L}{R} = \frac{2\pi f_r L}{R} = \frac{1}{R}\sqrt{\frac{L}{C}} \qquad - \text{eqn 3}$$

GRAPH

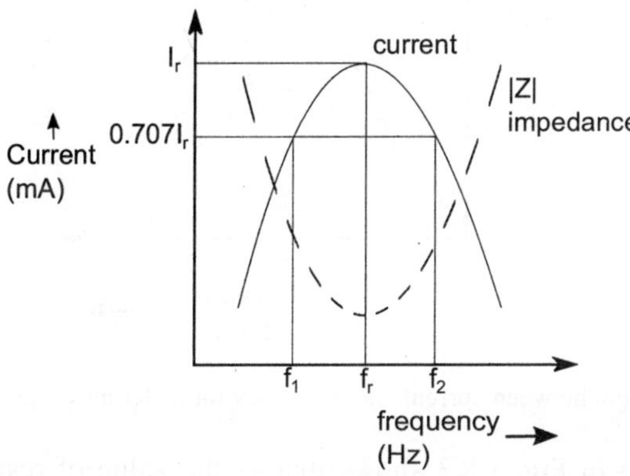

Fig. 1.8.2 Graph between current/impedance vs. frequency

Fig. 1.8.2 shows the variation between current/impedance with the frequency. At resonance, current flowing through the circuit is the maximum and impedance offered is the minimum.

Explanation

a. Dc signal

Capacitor acts as an open switch and inductor acts as a closed switch. Therefore, X_C is infinite and X_L is zero. Hence, total impedance of the circuit is infinite and hence no current flows through the circuit.

b. $0\text{Hz} < \text{frequency} < f_r$

For the frequencies less than resonance frequency, X_C is greater than X_L and hence total impedance is dominated by X_C. As frequency increases, X_C decreases therefore net impedance also decreases. Thus, current flowing in the circuit increases.

c. Frequency = f_r

At resonant frequency, $X_C = X_L$. Therefore, net impedance is purely resistive *i.e.* $Z = R$. This is the minimum impedance offered by the circuit and hence maximum current flows in the circuit.

d. Frequency > f_r

For the frequencies greater than resonance frequency, X_L is greater than X_C and hence total impedance is dominated by X_L. As frequency increases, X_L increases therefore net impedance also increases. Thus, current flowing in the circuit decreases.

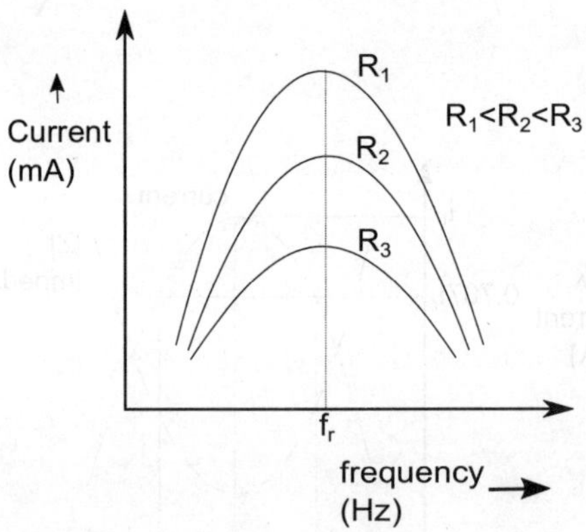

Fig. 1.8.3 Graph between current and frequency for different values of resistances

Graph shown in Fig. 1.8.3 shows that as the value of resistance increases, current flowing in the circuit decreases.

PROCEDURE

1.   Connect the circuit as shown in Fig. 1.8.1 on the breadboard.

2.   Choose the value of R, L and C such that resonance frequency lies within the range of 1000-3000Hz.

3.   Connect the function generator using a probe to give the input signal.

4.   Connect milli-ammeter in series with R, L and C to measure the current flowing in the circuit.

5.   Connect multimeter across R, L and C combination to measure the total impedance offered by the circuit.

6. Adjust the voltage of input signal such that current comes within the range of milli-ammeter.

7. Change the frequency in steps of 100Hz starting from value less than resonance frequency to above resonance frequency and note down the corresponding current in milli-ammeter and impedance in multimeter.

8. Now, change the value of R and repeat step 6 and 7.

9. Plot the graph between current/impedance and frequency for different value of resistances.

OBSERVATION TABLE

Table 1.8.1 Observation table for series LCR

S.No.	Frequency f (Hz)	R_1		R_2		R_3	
		I (mA)	Z(Ω)	I (mA)	Z(Ω)	I (mA)	Z(Ω)
1.	...						
2.	...						
3.	f_r						
4.	...						
5.	...						

$C = ...\ \mu F,\ L = ...\ mH,\ R_1 = ...\ \Omega,\ R_2 = ...\ \Omega,\ R_3 = ...\ \Omega$

Table 1.8.2 Comparison of theoretical and practical values

Resistance	Resonance frequency f_r (Hz)		Bandwidth (Hz)		Quality factor $\dfrac{f_r}{f_2 - f_1}$	Impedance at Resonance (Ω)	
	Th. $\dfrac{1}{2\pi\sqrt{LC}}$	Pr. (graph)	Th. $\dfrac{R}{2\pi L}$	Pr. (graph) $f_2 - f_1$		Th. Z = R	Pr. (graph)
R1 = ...							
R2 = ...							
R3 = ...							

RESULT

Series LCR circuit has been studied successfully and graph between current/impedance and frequency has been plotted for different values of resistances. The values of below mentioned parameters are calculated theoretically and graphically.

Table 1.8.3 Result of series LCR

S.No.	Parameter	Theoretical value	Graphical value	Percentage error
1.	Resonant frequency f_r			
2.	Bandwidth, BW			
3.	Quality factor, Q			

DISCUSSION

A series LCR circuit is also known as an acceptor circuit because it allows maximum current to flow through the circuit at a particular frequency while rejecting others. It is used in oscillating and tuner circuits such as in radio receivers and TV sets to select a particular frequency. The ability of a radio receiver to select a certain frequency that has been transmitted by a station and to eliminate frequencies from other stations is based on the principle of resonance.

Experiment 9

PARALLEL LCR

AIM

To Study the Frequency Response of a Parallel LCR Circuit and Determine its (a) Resonant Frequency (b) Impedance at Resonance (c) Quality Factor (d) Band Width.

APPARATUS REQUIRED

Breadboard, connecting wires, resistors, capacitor, inductor, milli ammeter, multimeter, function generator, probe.

THEORY

Parallel LCR

A parallel LCR circuit comprises of series combination of inductor and resistor connected in parallel with a capacitor and an ac source as shown in Fig. 1.9.1.

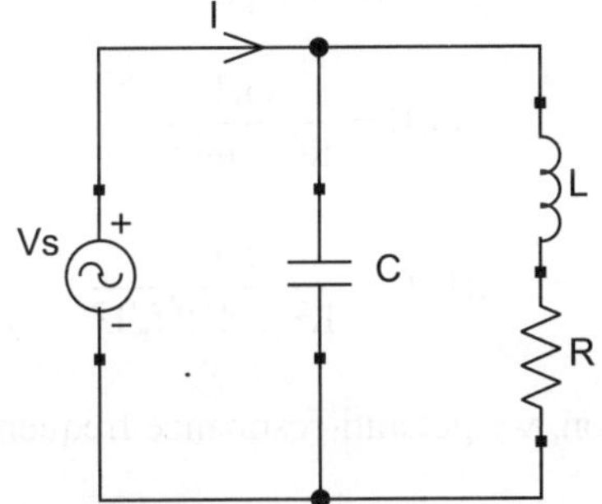

Fig. 1.9.1 Circuit for LCR parallel resonance

A parallel resonant circuit is also called as a tank circuit which stores the energy in the form of magnetic field of the coil and electric field of the capacitor. The circuit is said to be in resonant condition when the voltage applied and current flowing in the circuit are in phase with each other *i.e.* when the susceptance part of admittance is zero.

Admittance

Total admittance of the circuit is given by

$$\frac{1}{Z} = \frac{1}{X_C} + \frac{1}{X_L + R}$$

Since,

$$X_C = \frac{1}{j\omega C} \quad \text{and} \quad X_L = j\omega L$$

Therefore, admittance can be written as

$$\frac{1}{Z} = \frac{1}{1/j\omega C} + \frac{1}{j\omega L + R}$$

$$\Rightarrow \qquad \frac{1}{Z} = j\omega C + \frac{R - j\omega L}{R^2 + \omega^2 L^2}$$

Separating the real and imaginary terms,

$$\Rightarrow \qquad \frac{1}{Z} = \frac{R}{R^2 + \omega^2 L^2} + j\left[\omega C - \frac{\omega L}{R^2 + \omega^2 L^2}\right] \qquad - \text{eqn 1}$$

Anti-resonance frequency

At resonance, current and voltage are in phase with each other. Therefore, the reactive term which gives the phase change is zero at anti-resonance frequency. Equating susceptance part of equation 1 to zero we get,

$$\omega_r C - \frac{\omega_r L}{R^2 + \omega_r^2 L^2} = 0$$

$$\Rightarrow \qquad \omega_r C = \frac{\omega_r L}{R^2 + \omega_r^2 L^2}$$

$$\Rightarrow \qquad 2\Pi f_r C = \frac{2\Pi f_r L}{R^2 + 4\Pi^2 f_r^2 L^2}$$

Solving the above equation, we get anti-resonance frequency as

$$f_r = \frac{1}{2\Pi}\sqrt{\frac{1}{LC} - \frac{R^2}{L^2}} \qquad - \text{eqn 2}$$

If R is very small, the anti-resonance frequency becomes

$$f_r = \frac{1}{2\Pi\sqrt{LC}}$$

Impedance at resonance

Total admittance of the circuit is given as

$$\frac{1}{Z} = \frac{R}{R^2 + \omega^2 L^2} + j\left[\omega C - \frac{\omega L}{R^2 + \omega^2 L^2}\right]$$

At resonance, the susceptance part becomes zero. Hence, we get admittance as

$$\frac{1}{Z} = \frac{R}{R^2 + \omega_r^2 L^2}$$

Hence, impedance offered by the circuit at resonance becomes

$$Z = \frac{R^2 + \omega_r^2 L^2}{R}$$

Putting the value of anti-resonance frequency from equation 2 in the above equation, we get

$$Z = R + \frac{L^2}{R}\left(\frac{1}{LC} - \frac{R^2}{L^2}\right)$$

$$\Rightarrow \qquad Z = {}^{L}\!/_{RC}$$

This is the maximum impedance offered by the circuit and is also known as dynamic resistance of the parallel LCR circuit. The current flowing in the circuit will be the minimum at resonance and therefore, the condition of resonance for parallel LCR circuit is known as anti-resonance and the corresponding frequency is called as anti-resonance frequency.

Bandwidth

Bandwidth of a parallel LCR circuit is defined as the range of frequencies for which impedance offered by the circuit is 0.707 times the impedance at resonance frequency *i.e.* $Z_r/\sqrt{2}$.

Quality factor

It is a measure of the selectivity of a signal. The higher the value of impedance at resonance, the better is the selectivity. It is given as

$$Q = \frac{1}{\omega_r CR} = \frac{\omega_r L}{R} = \frac{2\pi f_r L}{R} = \frac{1}{R}\sqrt{L/C}$$

GRAPH

The graph shown in Fig. 1.9.2 shows the variation between current/impedance with the frequency. Impedance offered by the circuit is the maximum at the anti-resonant frequency and decreases as frequency increases or decreases.

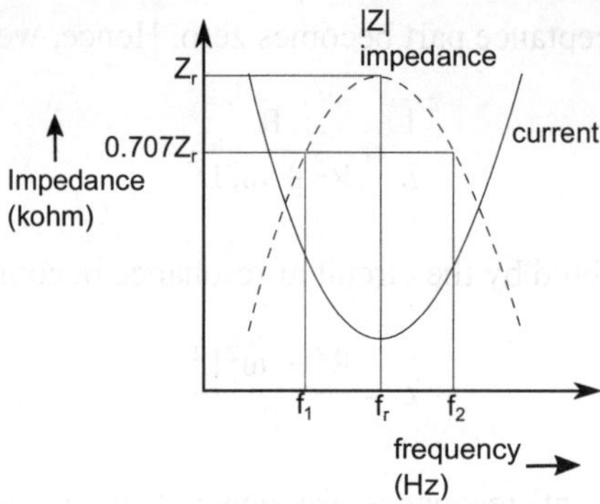

Fig. 1.9.2 Graph between impedance/current vs. frequency

Explanation

a. Dc signal

Capacitor acts as an open switch and inductor acts as a closed switch. Therefore, X_C is infinite and X_L is zero. Since, inductor and capacitor are in parallel with each other therefore, net impedance is effectively inductive only which is zero for zero frequency. Hence maximum current flows through the circuit.

b. $0Hz <$ frequency $< f_r$

For the frequencies lesser than resonance frequency, X_C is greater than X_L and hence total impedance is dominated by X_L. As the frequency increases, X_L increases therefore net impedance also increases. Thus, current flowing in the circuit decreases.

c. Frequency $= f_r$

At resonant frequency, $X_C = X_L$. Therefore, net impedance is the maximum and hence minimum current flows in the circuit.

d. Frequency $> f_r$

For the frequencies greater than resonance frequency, X_L is greater than X_C and hence total impedance is dominated by X_C. As the frequency increases, X_C decreases therefore net impedance also decreases. Thus, current flowing in the circuit increases.

The graph shown in Fig. 1.9.3 shows the variation of current with frequency for different values of resistances. Expression for anti-resonant frequency for a parallel LCR circuit (refer equation 2) shows that as value of R increases, anti-resonant frequency decreases.

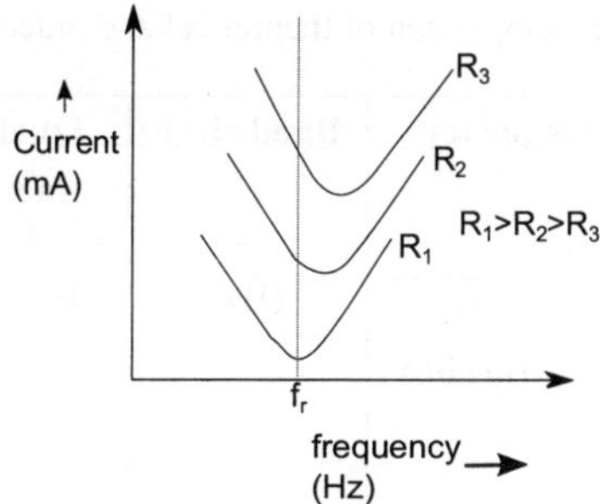

Fig. 1.9.3 Graph between current and frequency for different values of resistances

PROCEDURE

1. Connect the circuit as shown in Fig. 1.9.1 on the breadboard.
2. Choose the value of R, L and C such that resonance frequency lies within the range of 1000-3000Hz.
3. Connect the function generator using a probe to give the input signal.
4. Connect milli-ammeter in series with the voltage source to measure the current flowing in the circuit.
5. Connect multimeter across R-L combination to measure the total impedance offered by the circuit.
6. Adjust the voltage of input signal such that current comes within the range of milli-ammeter.
7. Change the frequency in steps of 100Hz starting from value less than anti-resonance frequency to above anti-resonance frequency and note down the corresponding current in milli-ammeter and impedance in multimeter.
8. Now, change the value of R and repeat step 6 and 7.
9. Plot the graph between current/impedance and frequency for different value of resistances.

OBSERVATION TABLE

Table 1.9.1 Observation table for parallel LCR

S.No.	Frequency	R_1		R_2		R_3	
	f (Hz)	I (mA)	Z(Ω)	I (mA)	Z(Ω)	I (mA)	Z(Ω)
1.	...						
2.	...						
3.	...						
4.	f_r						
5.	...						
6.	...						
...	...						

$C = \ldots\ \mu F,\ L = \ldots\ mH,\ R_1 = \ldots\ \Omega,\ R_2 = \ldots\ \Omega,\ R_3 = \ldots\ \Omega$

Table 1.9.2 Comparison of theoretical and practical values

Resistance	Resonance frequency f_r (Hz)		Bandwidth $f_2 - f_1$ (Hz)	Quality factor $\dfrac{f_r}{f_2 - f_1}$	Impedance at Resonance (Ω)	
	Th. $\dfrac{1}{2\Pi}\sqrt{\dfrac{1}{LC} - \dfrac{R^2}{L^2}}$	Pr. (graph)			Th. $^L\!/_{RC}$	Pr. (graph)
R1 = ...						
R2 = ...						
R3 = ...						

RESULT

Parallel LCR circuit has been studied successfully and graph between current/impedance and frequency has been plotted for different values of resistances. The values of below mentioned parameters are calculated theoretically and graphically.

Table 1.9.3 Result of parallel LCR

S.No.	Parameter	Theoretical value	Graphical value	Percentage error
1.	Resonant frequency f_r			
2.	Bandwidth, BW			
3.	Quality factor, Q			

DISCUSSION

A parallel LCR circuit is also known as a rejecter circuit because at anti-resonance frequency, minimum current flows through the circuit. Hence, it rejects a particular frequency while allowing others to pass through it. It can be used in circuits such as voltage multiplier, oscillator, pulse discharge circuits, filters, radio tuners, audio receivers *etc*.

<table><tr><td>

2

</td><td>

Experiments on Semiconductor Devices

</td></tr></table>

LIST OF EXPERIMENTS

1. To Study the I-V Characteristics of Diode – Ordinary and Zener Diode.

2. To Study the I-V Characteristics of the Common Emitter and Common Base Configurations of BJT and Obtain the H-Parameters.

3. To Study the I-V Characteristics of the UJT.

4. To Study the I-V Characteristics of the SCR.

5. To Study the I-V Characteristics of the Common Source FET Configuration.

Experiment 1

ORDINARY and ZENER DIODE

AIM

To Study the I-V Characteristics of Diode – Ordinary and Zener Diode.

APPARATUS REQUIRED

Breadboard, ordinary diode – IN4007, zener diode, variable dc power supply – 0-2V, 0-20V, milli ammeter – 0-100mA, micro ammeter – 0-500μA, voltmeter – 0-1V, 0-20V, resistor – 100Ω, multimeter, connecting wires.

THEORY

P-N Junction Diode

It is a two terminal p-n junction device that consists of an n-type semiconductor material bonded to a p-type semiconductor material forming a p-n junction as shown in Fig. 2.1.2. The p-type material of diode is the positive electrode called as anode and the n-type material is the negative electrode called as cathode.

The diode offers very low resistance when forward biased and very high resistance when reverse biased *i.e.* it conducts only in one direction when it is forward biased. When the diode is reverse biased, very small current, known as reverse leakage current flows through the device which is due to the presence of minority carriers. As the reverse voltage is increased beyond the breakdown voltage, a large current starts flowing through the device which can damage the simple p-n junction diode by overheating it.

Diodes that are mostly used today are made up of Silicon because of their higher peak inverse voltage, current ratings and wider temperature ranges.

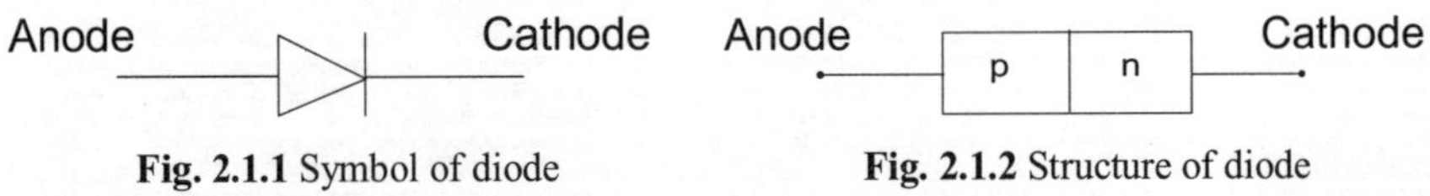

Fig. 2.1.1 Symbol of diode   **Fig. 2.1.2** Structure of diode

Unbiased diode

The word unbiased means that no external voltage is applied across the p-n junction diode. The p-type semiconductor has majority of holes and n-type semiconductor has majority of electrons. Hence, due to the density gradient, electrons diffuse from n to p side and holes diffuse from p to n side where they recombine with each other leaving behind the positive and negative immobile ions on n and p side respectively. These immobile ions form a region which is depleted of mobile charge carriers; hence that region is called as depletion region. An electric field is created across the junction because of these positive and negative immobile charges and a potential is developed across it known as built-in-potential. This field opposes further diffusion of electrons and holes. Once the equilibrium is established, a barrier is set up called as potential barrier which does not allow any further movement of holes and electrons across the junction. To overcome this barrier, an external voltage called as knee voltage is applied and its value depends upon the type of semiconductor material used. For Silicon diodes, it is 0.6 - 0.7V and for Germanium diodes, it is 0.3 - 0.35V.

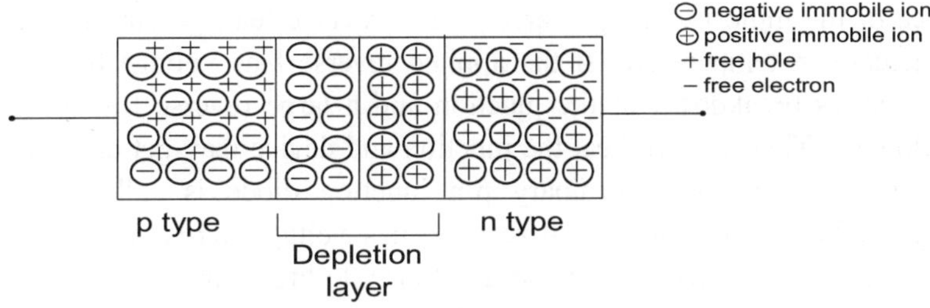

Fig. 2.1.3 Unbiased p-n junction diode

Biased diode

Depending upon the polarity of an external voltage applied to the p and n side of the diode, the potential barrier around the junction can be changed. The potential difference across the p-n junction can be applied in the following two ways.

a. Forward biasing

The diode is said to be forward biased when p-side and n-side of the diode are connected to the positive and negative terminal of the battery respectively as shown in Fig. 2.1.4.

This external bias exerts a force on the holes on p-side and electrons on the n-side such that they start moving towards the junction thereby, reducing the width of uncovered charges and the potential barrier. Thus, a large current flows in the external circuit due to the movement of charge carriers. The external applied voltage at which the potential barrier reduces and the current flows in the circuit is called knee voltage.

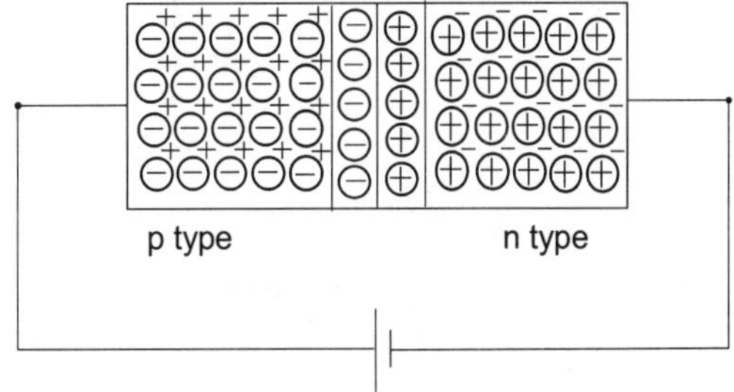

Fig. 2.1.4 Forward biased p-n junction diode

b. Reverse biasing

The diode is said to be reverse biased when p-side and n-side of the diode are connected to the negative and positive terminal of the battery respectively as shown in Fig. 2.1.5.

This external bias attracts holes on the p-side and electrons on the n-side towards the negative and positive terminal of the external voltage respectively which causes the charge carriers to move away from the junction. Hence, width of the depletion region and the barrier height increases. As the reverse bias voltage is increased, the potential barrier also increases which further restricts the flow of charge carriers. A negligible amount of current called as reverse saturation current or leakage current flows through the circuit because of the presence of the minority charge carriers. The reverse leakage current increases with increase in temperature. If the applied reverse bias voltage becomes too high, it causes breakdown of the junction and a large reverse leakage current starts flowing. This can overheat the junction and device may get damaged. The breakdown that occurs in ordinary p-n junction diode is called avalanche breakdown. When there is a high reverse bias voltage across the device, the thermally generated carriers gain enough energy to break the covalent bond and new electron-hole pairs are generated which again disturb the other bonds. So, there is a cumulative effect of generation of avalanche carriers which results in large current flow through the circuit. A diode's maximum reverse-bias voltage rating is known as the peak-inverse voltage (PIV).

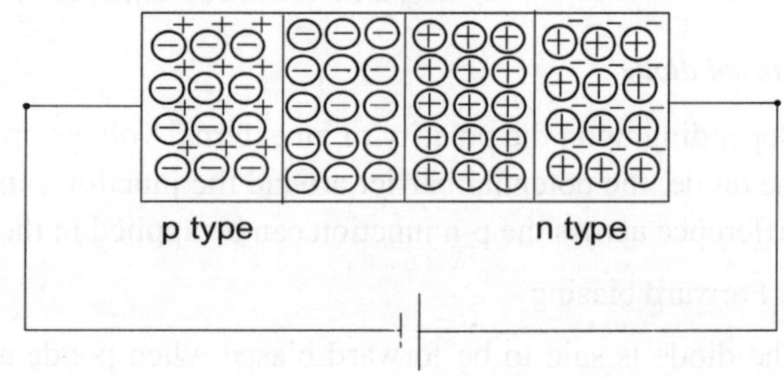

Fig. 2.1.5 Reverse biased p-n junction diode

I-V characteristics of p-n junction diode

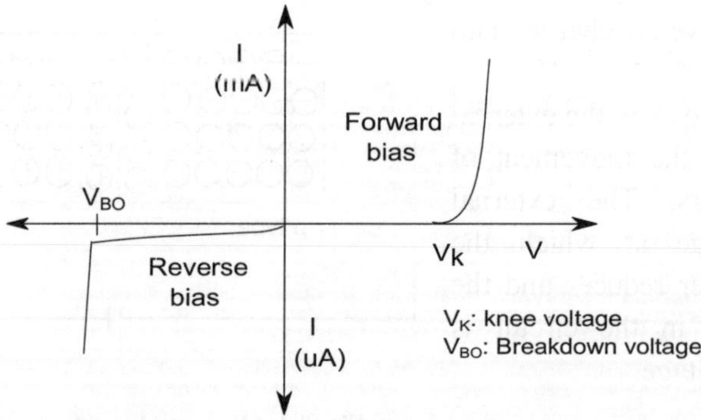

Fig. 2.1.6 I-V characteristics of p-n junction diode

I-V characteristics of a simple p-n junction diode as shown in Fig. 2.1.6 show that the diode is a non-linear or non-ohmic device.

Zener Diode

A zener diode is a highly doped p-n junction device with a very narrow junction that allows flow of current in both the directions *i.e.* forward as well as reverse. They behave same as ordinary p-n junction diode when forward biased.

When reverse biased, initially small leakage current flows through the device but as the reverse voltage is increased beyond the breakdown voltage known as zener voltage, zener breakdown occurs. This is due to the fact that the zener diode is heavily doped. As a result, the potential barrier and depletion width are reduced and high electric field is set up. This results in tunneling of carriers which give rise to sudden increase in reverse current.

The zener diode may be of silicon or germanium but silicon is preferred because of higher operating temperature and current capability. Moreover, the knee point in case of silicon diodes is sharper as compared to germanium diodes.

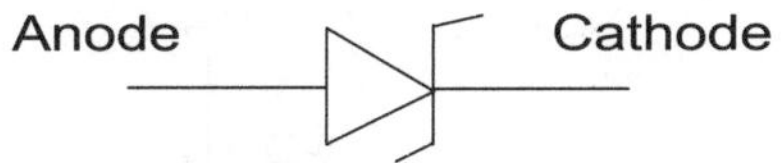

Fig. 2.1.7 Symbol of zener diode

I-V characteristics of zener diode

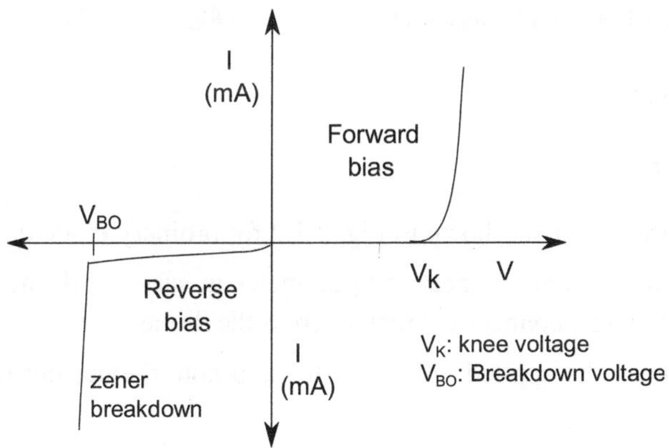

Fig. 2.1.8 I-V characteristics of zener diode

CIRCUIT DIAGRAM

P-n Junction Diode

Circuit diagram of p-n junction diodes are shown in Figs. 2.1.9 & 2.1.10.

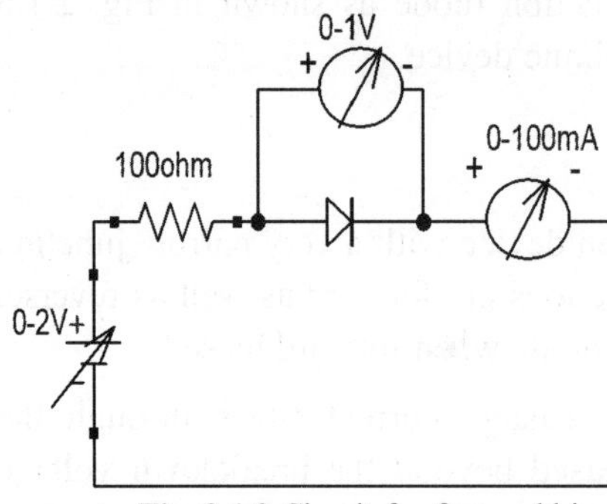

Fig. 2.1.9 Circuit for forward bias

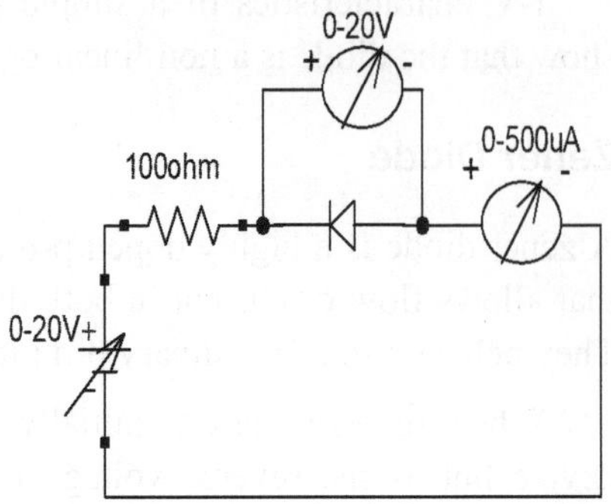

Fig. 2.1.10 Circuit for reverse bias

Zener Diode

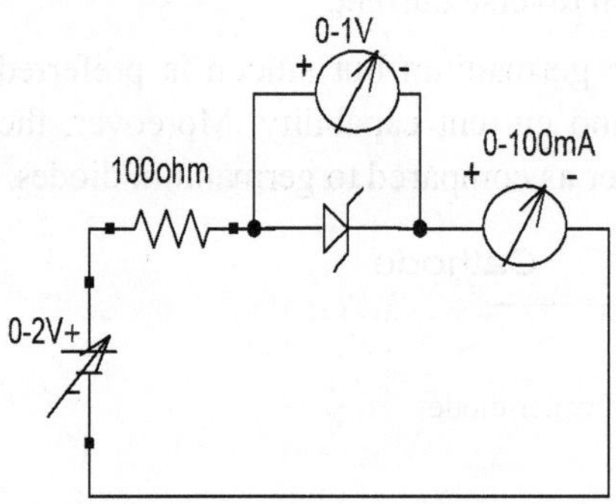

Fig. 2.1.11 Circuit for forward bias

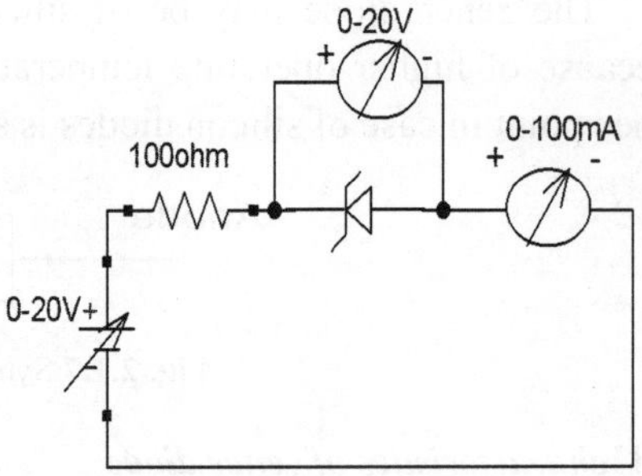

Fig. 2.1.12 Circuit for reverse bias

PROCEDURE

Forward Bias

1. Connect the circuit as shown in Fig. 2.1.9 for ordinary diode on breadboard.

2. To measure current, connect milli ammeter in series with the diode and to measure voltage, connect voltmeter across the diode.

3. Initially keep the applied voltage at 0V and note the current flowing in the circuit.

4. Increase the voltage in steps of 0.1V and note the corresponding voltage and current in voltmeter and milli-ammeter respectively.

5. Take the observations and note the value of knee voltage.

6. Plot the graph between voltage and current.

7. Now, connect the circuit for zener diode as shown in Fig. 2.1.11 and repeat the above steps.

Reverse Bias

1. Connect the circuit as shown in Fig. 2.1.10 for ordinary diode on breadboard.

2. To measure current, connect micro-ammeter in series with the diode.

3. To measure voltage, connect voltmeter across the diode.

4. Initially keep the applied voltage at 0V and note the current flowing in the circuit.

5. Increase the voltage in steps of 1V and note the corresponding voltage and current in voltmeter and micro-ammeter.

6. Keep the applied reverse voltage below the breakdown voltage of the diode which can be noted from manufacturer's datasheet.

7. Take the observations and note the readings in a tabular form.

8. Plot the graph between voltage and current.

9. Now, connect the circuit for zener diode as shown in Fig. 2.1.12.

10. Connect milli-ammeter for measuring the current in the circuit.

11. Repeat the above steps (except step 6).

12. Take the readings for zener diode till the breakdown occurs.

13. Measure the zener breakdown voltage and plot the graph between voltage and current.

OBSERVATION TABLE

P-n Junction Diode

Table 2.1.1 Observation table for ordinary p-n junction diode

Forward Bias		Reverse Bias	
Voltage (V)	**Current (mA)**	**Voltage (V)**	**Current (µA)**
0		1	
0.1		2	
0.2		3	
0.3		4	
0.4		5	
0.5		6	
0.6		7	
0.7		8	

Zener Diode

Table 2.1.2 Observation table for zener diode

Forward Bias		Reverse Bias	
Voltage (V)	Current (mA)	Voltage (V)	Current (mA)
0		1	
0.1		2	
0.2		3	
0.3		4	
0.4		5	
0.5		6	
0.6		7	
0.7		8	

RESULT

I-V characteristics of ordinary p-n junction diode and zener diode in forward bias mode and reverse bias mode have been studied and drawn successfully. Following parameters are calculated through the graph.

Knee voltage, V_k = ...

Zener breakdown voltage, V_{BO} = ...

DISCUSSION

Diodes are widely used as:

- Switch in digital logic circuits.

- Rectifier to convert ac to dc in power supplies.

- Clipper circuits which are used as wave shaping circuits in computers, radars, radios and TV receivers.

- Clampers which are used as dc restorers in TV receivers and voltage multipliers.

- Over-voltage protection circuits.

Zener diodes are widely used as:

- Surge protectors.

- Meter protection circuits.

- Voltage regulators.

- Peak clippers.

Experiment 2

COMMON EMITTER and COMMON BASE CONFIGURATIONS of BJT

AIM

To Study the I-V Characteristics of the Common Emitter and Common Base Configurations of BJT and Obtain the h-Parameters.

APPARATUS REQUIRED

BJT - BC108, resistors – 100Ω, 100kΩ, variable dc power supply – 0-5V, 0-20V, voltmeter – 0-1V, 0-20V, ammeter – 0-300µA, 0-100mA, 0-200mA, breadboard, connecting wires.

THEORY

Bipolar Junction Transistor

It is a three terminal device with terminals named as emitter (E), base (B) and collector (C). The term 'bipolar' signifies that two types of carriers *i.e.* electrons and holes are involved in the flow of current through the transistor, the term 'junction' indicates that it has two junctions known as emitter base junction and base collector junction and the term 'transistor' is a combination of two words *i.e.* 'transfer + resistor' which means current is transferred from region of 'low resistance' to 'high resistance'.

It has three regions of operation: cut off, saturation and active region. When the transistor is made to operate between cut off and saturation, it acts as a switch which is the basis of digital electronics and when the transistor is made to operate in active region, it works as an amplifier which is the basis of analog electronics.

Based on construction, there are two basic types of bipolar transistor: n-p-n and p-n-p transistor.

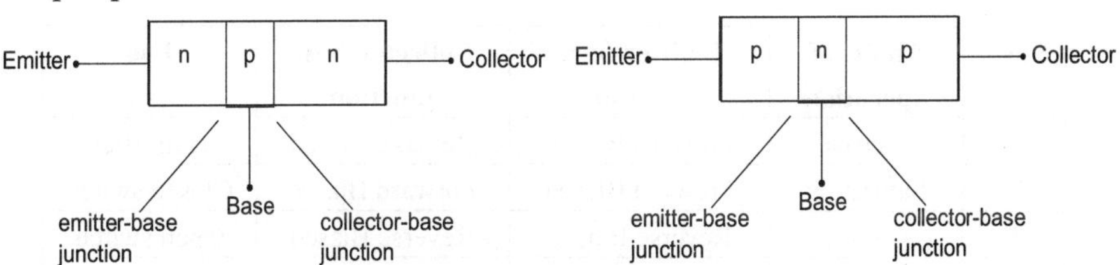

Fig. 2.2.1 Structure for n-p-n and p-n-p transistor

As shown in Fig. 2.2.1, in p-n-p transistor, two p-type semiconductor regions are separated by a thin n-type region and the conduction is by holes whereas in npn transistor, two n-type semiconductor regions are separated by a thin p-type region and the conduction is by electrons. The thin lightly doped central region is called as base and on the one side, there is highly doped region called as emitter which is a source of majority carriers and on the other side, there is a collector region which is comparatively lightly doped and collects the carriers.

The input circuit *i.e.* emitter-base junction is forward biased and offers low resistance whereas the output circuit *i.e.* collector-base junction is reverse biased and offers high resistance. Therefore, a transistor transfers the input signal current from a low resistance circuit to high resistance circuit which is responsible for the amplifying action of a transistor.

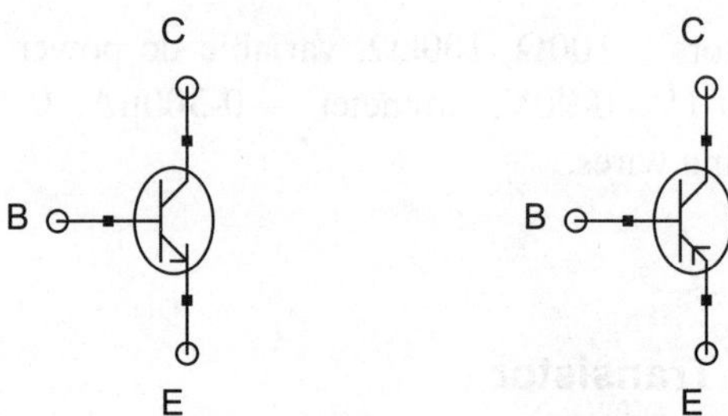

Fig. 2.2.2 Symbol for npn and pnp transistor

In the symbol shown in Fig. 2.2.2, the arrow indicates the direction of flow of conventional current between the base and emitter terminals.

n-p-n transistors are preferred over pnp transistors because of the following reasons:

- Electrons are the majority carriers in case of npn and they have better mobility as compared to holes.

- High frequency response is better in case of npn transistors.

Different modes of operation of a transistor

Depending upon the different biasing of two junctions (Emitter-base and Collector-base junctions), transistor may operate in different modes as listed below.

Table 2.2.1 Different modes of operation of transistor

S.No.	Mode of operation	Emitter-base junction	Collector-base junction	Use
1.	Active	Forward Biased	Reverse Biased	Amplifier
2.	Saturation	Forward Biased	Forward Biased	Closed switch
3.	Cut off	Reverse Biased	Reverse Biased	Open switch
4.	Inverted	Reverse Biased	Forward Biased	Not used

Transistor configurations

Transistor can operate in three configurations depending on which electrode is made common to both the input and output sections: Common Base (CB), Common Emitter (CE), and Common Collector (CC).

Transistor as a two port model

Transistor is a three terminal device with one of its terminals as grounded which is made common between input and output section. Hence, while studying the characteristics of transistor, it can be considered as a two port network with four terminals - two input and two output terminals.

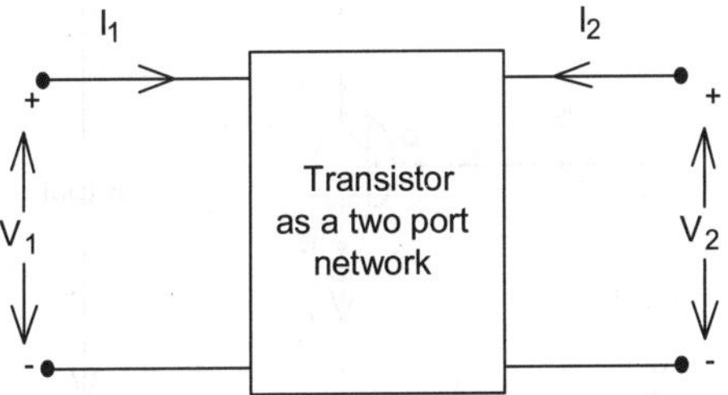

Fig. 2.2.3 Transistor as a two port network

Transistor as a two port network is shown in Fig. 2.2.3. The input voltage and current are denoted as V_1 and I_1 and output voltage and current as V_2 and I_2.

The current-voltage equations for two port network in terms of h-parameters are given as

$$V_1 = h_{11}I_1 + h_{12}V_2$$

$$I_2 = h_{21}I_1 + h_{22}V_2$$

Where,

V_1 and I_1: R.M.S. value of input voltage and input current

V_2 and I_2: R.M.S. value of output voltage and output current

h_{11}, h_{12}, h_{21} and h_{22}: hybrid or h-parameters which are defined as

$h_{11} = h_i = \left(V_1/I_1 \right)_{V_2=0}$ = short circuit input impedance

$h_{21} = h_f = \left(I_2/I_1 \right)_{V_2=0}$ = short circuit forward current transfer ratio

$h_{12} = h_r = \left(V_1/V_2 \right)_{I_1=0}$ = open circuit reverse voltage transfer ratio

$h_{22} = h_o = \left(I_2/V_2 \right)_{I_1=0}$ = open circuit output admittance

Common Emitter Configuration (CE)

In CE configuration as shown in Fig. 2.2.4, emitter is made common to both the input and output signals with the input signal being applied between the base and emitter terminals and the output signal is taken between the collector and emitter terminals. Bias voltages V_{BB} and V_{CC} are connected between base-emitter and collector-emitter junctions respectively. When the bias voltage V_{BB} is applied, base current I_B flows and when the bias voltage V_{CC} is applied such that collector is made more positive than emitter then the collector current I_C flows.

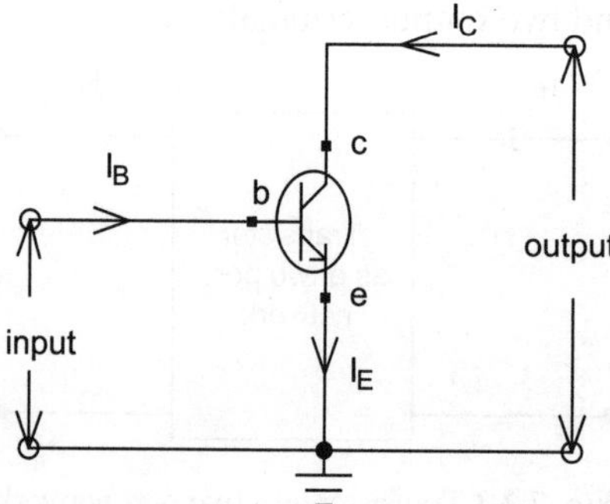

Fig. 2.2.4 NPN transistor in common emitter configuration

Basic current equation,

$$I_E = I_C + I_B$$

Current gain,

$$\beta = I_c / I_B$$

Total collector current,

$$I_C = \beta I_B + I_{CEO}$$

Where, βI_B is the current component produced by majority carriers and I_{CEO} is the leakage current due to the movement of minority carriers across emitter-collector junction due to reverse biasing.

Determination of h-parameters in common emitter configuration

1. h_{11} or h_{ie}

- Find Quiescent Q point of the input characteristic curve for $V_{CE} = V_{C2}$.

- Draw tangent at Q point mark two points A and B on either side of Q as shown in the Fig. 2.2.5.

- Draw perpendiculars on x and y axis.

- Inverse of the slope gives the value of h_{ie} as

$$h_{ie} = \frac{(V_{B2} - V_{B1})}{(I_{B2} - I_{B1})}$$

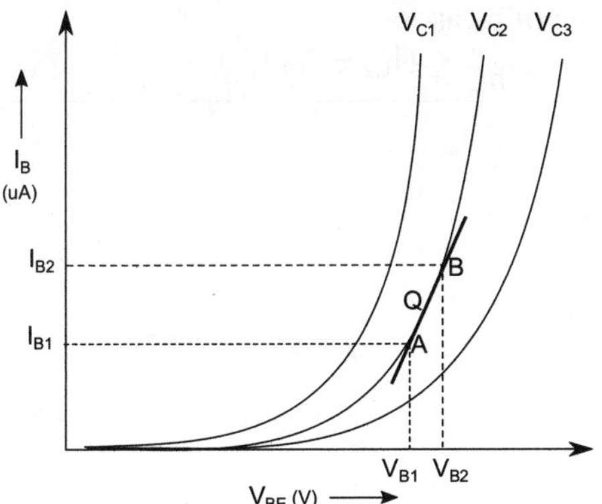

Fig. 2.2.5 Calculation of h_{11} for CE configuration

2. h_{12} or h_{re}

- Find Q point and draw a horizontal line parallel to x-axis through Quiescent Q point cutting the three input characteristic curves as shown in the Fig. 2.2.6.

- Draw perpendiculars from points A and B on x-axis.

- Calculate the value of h_{re} as

$$h_{re} = \left.(V_{B2} - V_{B1})\right/(V_{C2} - V_{C1})$$

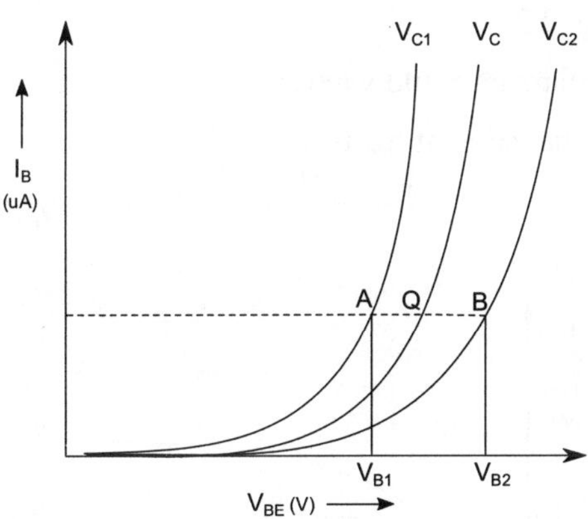

Fig. 2.2.6 Calculation of h_{12} for CE configuration

3. h_{21} or h_{fe}

- Find Quiescent Q point of the output characteristic curve.

- Draw a vertical line parallel to y-axis through Q point cutting the three output characteristic curves as shown in the Fig. 2.2.7.

- Draw perpendiculars from points A and B on y-axis.

- Calculate the value of h_{fe} as

$$h_{fe} = \frac{(I_{C2} - I_{C1})}{(I_{B2} - I_{B1})}$$

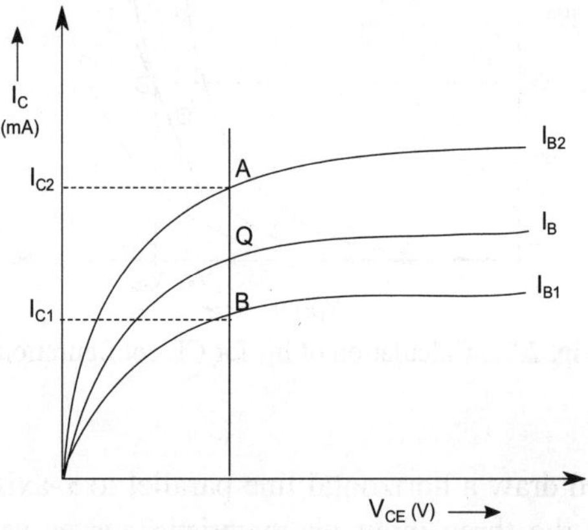

Fig. 2.2.7 Calculation of h21 for CE configuration

4. h_{22} or h_{oe}

- Find Quiescent Q point of the output characteristic curve.

- Draw tangent at Q and mark two points A and B on either side of Q as shown in the Fig. 2.2.8.

- Draw perpendiculars on x and y axis.

- The slope gives the value of h_{oe} as

$$h_{oe} = \frac{(I_{C2} - I_{C1})}{(V_{C2} - V_{C1})}$$

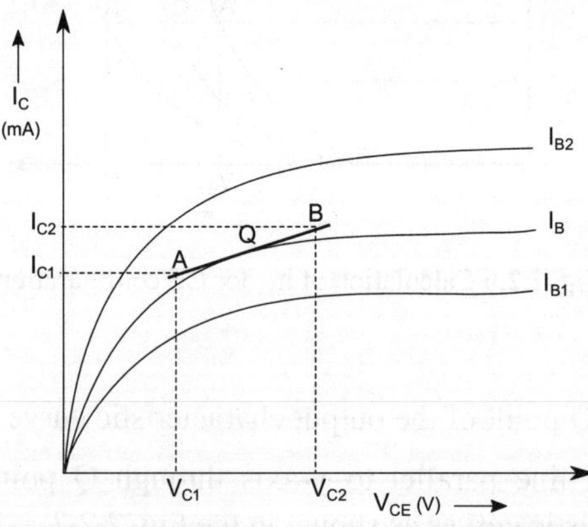

Fig. 2.2.8 Calculation of h22 for CE configuration

Hybrid model for common emitter

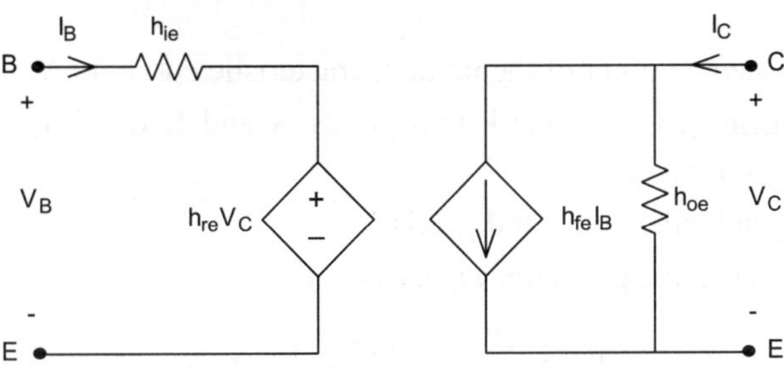

Fig. 2.2.9 Hybrid model for npn transistor in CE configuration

Common Base Configuration (CB)

In CB configuration as shown in Fig. 2.2.10, base is made common to both the input and output signals with the input signal being applied between the emitter and base terminals and output signal is taken between the collector and base terminals. Emitter-base junction is forward biased and collector-base junction is reverse biased. When bias voltage V_{EE} is applied across emitter-base junction, the emitter current I_E flows in the input circuit. As the base region is very thin, most of the electrons from the emitter region move towards the collector region and thus contributes to collector current I_C in the output circuit. Few electrons that recombine with the holes in the base give rise to the base current I_B.

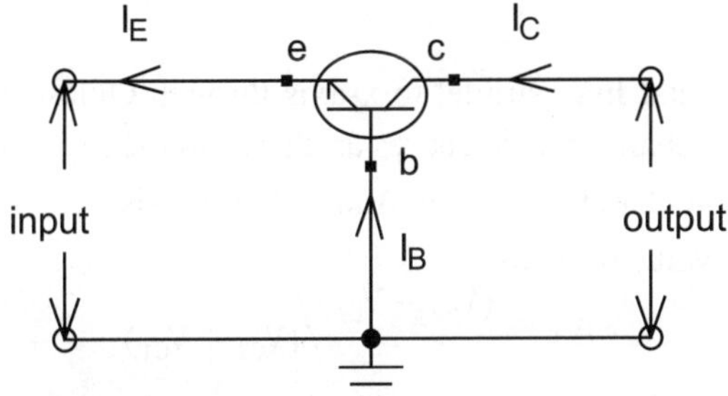

Fig. 2.2.10 Npn transistor in common base configuration

Basic current equation,

$$I_E = I_C + I_B$$

DC current gain or current amplification factor,

$$\alpha = I_C / I_E$$

Total collector current,

$$I_C = \alpha I_E + I_{CBO}$$

Where, αI_E is the current component produced by the majority carriers and I_{CBO} is the leakage current due to the movement of minority carriers across the base-collector junction due to reverse biasing.

Determination of h-parameters in common base configuration

1. h_{11} or h_{ib}

- Find Quiescent Q point of the input characteristic curve for $V_{CB} = V_{C2}$.
- Draw tangent at Q and mark two points A and B on either side of Q as shown in the Fig. 2.2.11.
- Draw perpendiculars on x and y axis.
- Inverse of the slope gives the value of h_{ib} as

$$h_{ib} = \left.(V_{E2} - V_{E1})\middle/(I_{E2} - I_{E1})\right.$$

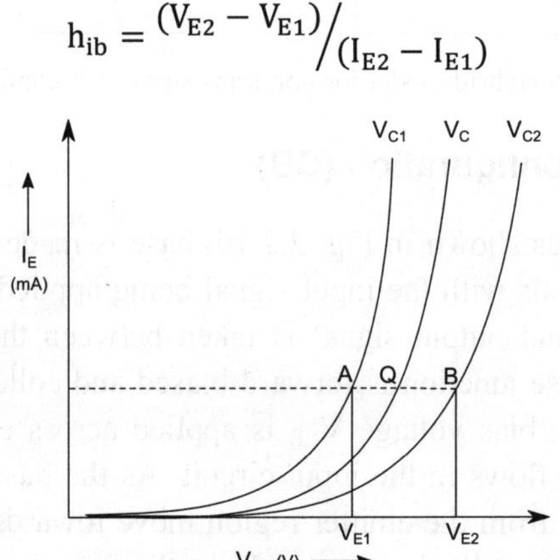

Fig. 2.2.11 Calculation of h_{11} for CB configuration

2. h_{12} or h_{rb}

- Draw a horizontal line parallel to x-axis through Quiescent Q point cutting the three input characteristic curves as shown in the Fig. 2.2.12.
- Draw perpendiculars from points A and B on x-axis.
- Calculate the value of h_{rb} as

$$h_{rb} = \left.(V_{E2} - V_{E1})\middle/(V_{C2} - V_{C1})\right.$$

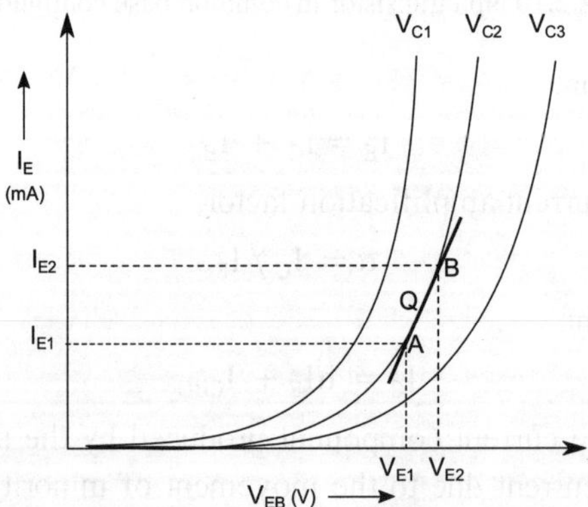

Fig. 2.2.12 Calculation of h_{12} for CB configuration

3. h_{21} or h_{fb}

- Find Q point of the output characteristic curve for I_E.
- Draw a vertical line parallel to y-axis through Q point cutting the three output characteristic curves as shown in the Fig. 2.2.13.
- Draw perpendiculars from points A and B on y-axis.
- Calculate the value of h_{fb} as

$$h_{fb} = (I_{C2} - I_{C1})/(I_{E2} - I_{E1})$$

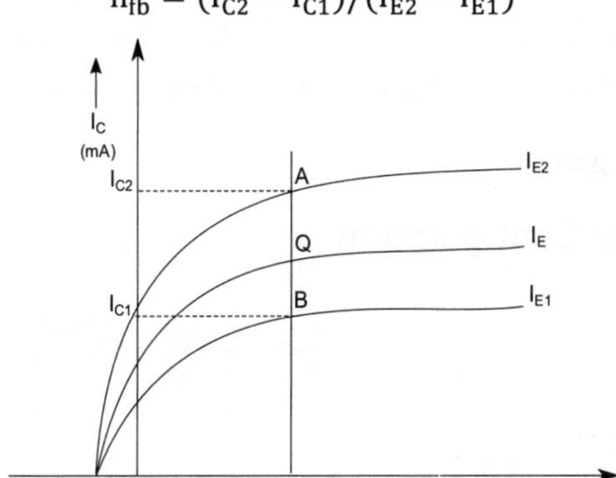

Fig. 2.2.13 Calculation of h_{21} for CB configuration

4. h_{22} or h_{ob}

- Find Quiescent Q point of the output characteristic curve for I_E.
- Draw tangent at Q and mark two points A and B on either side of Q as shown in the Fig. 2.2.14.
- Draw perpendiculars on x and y axis.
- The slope gives the value of h_{ob} as

$$h_{ob} = (I_{C2} - I_{C1})/(V_{C2} - V_{C1})$$

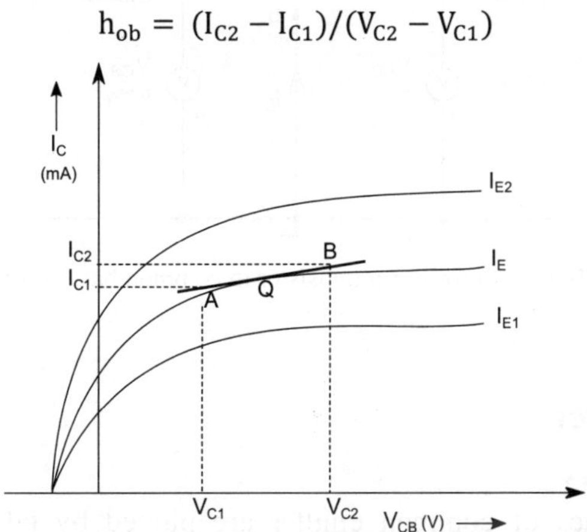

Fig. 2.2.14 Calculation of h_{22} for CB configuration

Hybrid model for common base

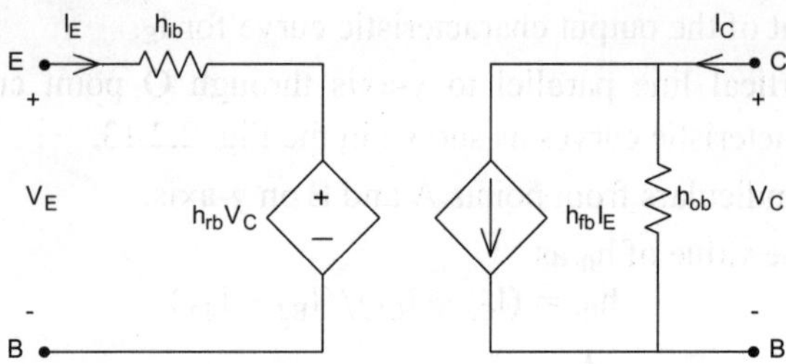

Fig. 2.2.15 Hybrid model for CB n-p-n transistor

CIRCUIT DIAGRAM

Common Emitter Configuration

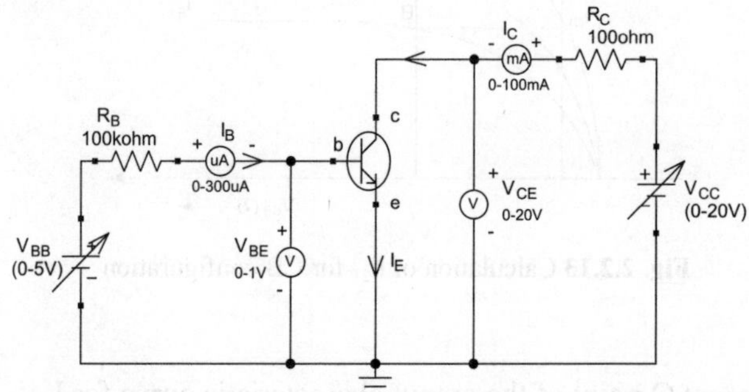

Fig. 2.2.16 Circuit for n-p-n transistor in common emitter configuration

Common Base Configuration

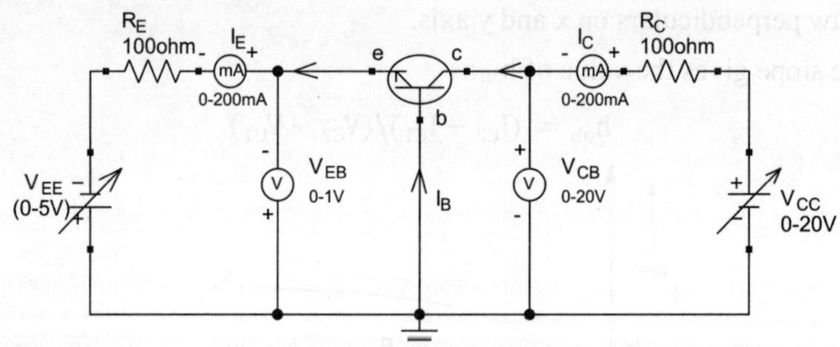

Fig. 2.2.17 Circuit for n-p-n transistor in common base configuration

PROCEDURE

Common Emitter

Input characteristics

Input characteristics of common emitter are plotted by taking the variation between the input voltage (V_{BE}) and input current (I_B) keeping the output voltage (V_{CE}) constant.

1. Connect the circuit as shown in Fig. 2.2.16 on breadboard.
2. Initially fix $V_{CE} = 0V$ by using Vcc.
3. Vary V_{BB} starting from 0V and measure the corresponding base emitter voltage V_{BE} and base current I_B.
4. Repeat the above step for different values of V_{CE}.
5. Plot the graph and calculate the following parameters:
    - Short circuit input impedance, $h_{ie} =$________
    - Open circuit reverse voltage transfer ratio, $h_{re} =$______

Output characteristic

Output characteristics of common emitter are plotted by taking the variation between the output voltage (V_{CE}) and output current (I_C) keeping the input current (I_B) constant.

1. Initially fix $I_B = 10\mu A$.
2. Vary V_{CC} starting from 0V and measure the corresponding collector to emitter voltage V_{CE} and collector current I_C.
3. Repeat the above step for different values of I_B.
4. Plot the graph and calculate the following parameters:
    - Short circuit forward current transfer ratio, $h_{fe} =$________
    - Open circuit output admittance, $h_{oe} =$________

Common Base

Input characteristics

Input characteristics of common base are plotted by taking the variation between the input voltage (V_{EB}) and input current (I_E) keeping the output voltage (V_{CB}) constant.

1. Connect the circuit as shown in the Fig. 2.2.17.
2. Initially fix $V_{CB} = 0V$ by using V_{CC}.
3. Vary V_{EE} starting from 0V and measure the corresponding emitter base voltage V_{EB} and emitter current I_E.
4. Repeat the above step for different values of V_{CB}.
5. Plot the graph and calculate the following parameters:
    - Short circuit input impedance, $h_{ib} =$________
    - Open circuit reverse voltage transfer ratio, $h_{rb} =$______

Output characteristics

Output characteristics of common base are plotted by taking the variation between the output voltage (V_{CB}) and output current (I_C) keeping the input current (I_E) constant.

1.   Initially fix $I_E = 1mA$ by varying V_{EE}.
2.   Vary V_{CC} starting from 0V and measure the corresponding collector to base voltage V_{CB} and collector current I_C.
3.   Repeat the above step for different values of I_E.
4.   Plot the graph and calculate the following parameters:
     - Short circuit forward current transfer ratio, h_{fb} =________
     - Open circuit output admittance, h_{ob} =________

OBSERVATION TABLE

Common Emitter

Input characteristics

Table 2.2.2 Observation table for common emitter input characteristics

V_{CE1}(V)		V_{CE2}(V)		V_{CE3}(V)	
V_{BE}(V)	I_B(mA)	V_{BE}(V)	I_B(mA)	V_{BE}(V)	I_B(mA)
0		0		0	
0.1		0.1		0.1	
0.2		0.2		0.2	
...		...		...	
0.7		0.7		0.7	

Output characteristics

Table 2.2.3 Observation table for common emitter output characteristics

I_{B1}(μA)		I_{B2}(μA)		I_{B3}(μA)	
V_{CE}(V)	I_C(mA)	V_{CE}(V)	I_C(mA)	V_{CE}(V)	I_C(mA)
0		0		0	
1		1		1	
2		2		2	
...		...		...	
20		20		20	

Common Base

Input characteristics

Table 2.2.4 Observation table for common base input characteristics

V_{CB1}(V)		V_{CB2}(V)		V_{CB3}(V)	
V_{EB}(V)	I_E(mA)	V_{EB}(V)	I_E(mA)	V_{EB}(V)	I_E(mA)
0		0		0	
0.1		0.1		0.1	
0.2		0.2		0.2	
...		...		...	
0.7		0.7		0.7	

Output characteristics

Table 2.2.5 Observation table for common base output characteristics

I_{E1}(mA)		I_{E2}(mA)		I_{E3}(mA)	
V_{CB}(V)	I_C(mA)	V_{CB}(V)	I_C(mA)	V_{CB}(V)	I_C(mA)
0		0		0	
1		1		1	
2		2		2	
...		...		...	
20		20		20	

GRAPH

Common Base

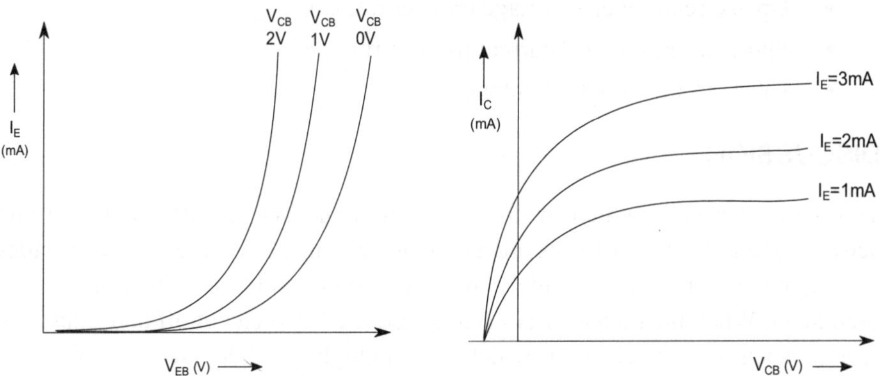

Fig. 2.2.18 Input and output characteristics

Common Emitter

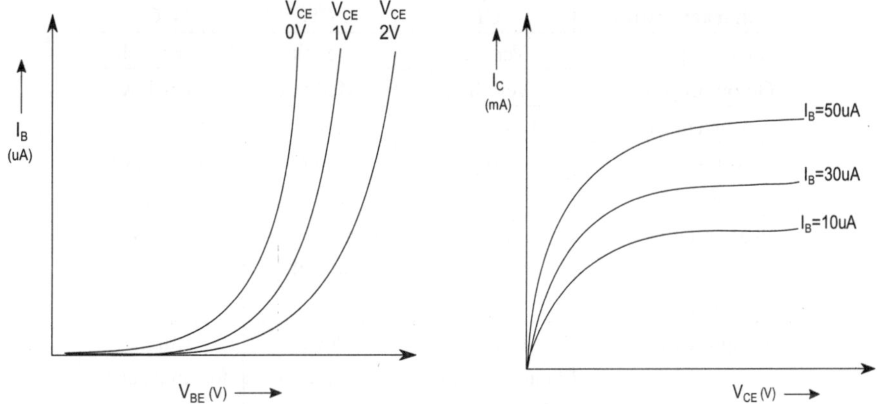

Fig. 2.2.19 Input and output characteristics

RESULT

The input and output characteristics of the transistor in common emitter and common base configurations have been studied and plotted successfully. The following parameters are calculated.

CE configuration

- Short circuit input impedance, h_{ie} = ____
- Open circuit reverse voltage transfer ratio, h_{re} = ____
- Short circuit forward current transfer ratio, h_{fe} = ____
- Open circuit output admittance, h_{oe} = ____

CB configuration

- Short circuit input impedance, h_{ib} = ____
- Open circuit reverse voltage transfer ratio, h_{rb} = ____
- Short circuit forward current transfer ratio, h_{fb} = ____
- Open circuit output admittance, h_{ob} = ____

DISCUSSION

Transistors can be operated in different region such as cut off, saturation and active. When the transistor is made to operate in active region, it provides amplifying action and is widely used in audio and video amplifiers and oscillators. When the transistor is made to operate between cut off and saturation region, it acts as a switch and is widely used in high speed digital electronics.

Common base, common emitter and common collector configurations have different characteristics which are listed in the table shown below.

Table 2.2.6 Comparison of CB, CE and CC configurations

Characteristics	CB	CE	CC
Input impedance	Very low	Medium	Very high
Output impedance	Very high	Medium	Very low
Current gain	Less than 1	High	Very high
Voltage gain	High	High	Less than 1
Power gain	Medium	Highest	Medium
Phase relationship between input and output	In phase	Out of phase by 180°	In phase
Applications	Constant current source	Power amplifier	Buffer for impedance matching applications

Experiment 3

UNI - JUNCTION TRANSISTOR

AIM

To Study the I-V Characteristics of the UJT.

APPARATUS REQUIRED

UJT 2N2646, bread board, connecting wires, resistor – 10kΩ, two variable dc power supply – 0-30V, voltmeter – 0-30V, ammeter – 0-5mA.

THEORY

Uni-Junction Transistor

A UJT or Uni-junction transistor is a semiconductor device that has a single junction with three terminals named as Emitter (E) and two bases (B_1 and B_2). It acts as a triggering device which changes its state abruptly from high impedance to low impedance state *i.e.* from non-conducting/OFF state to conducting/ON state.

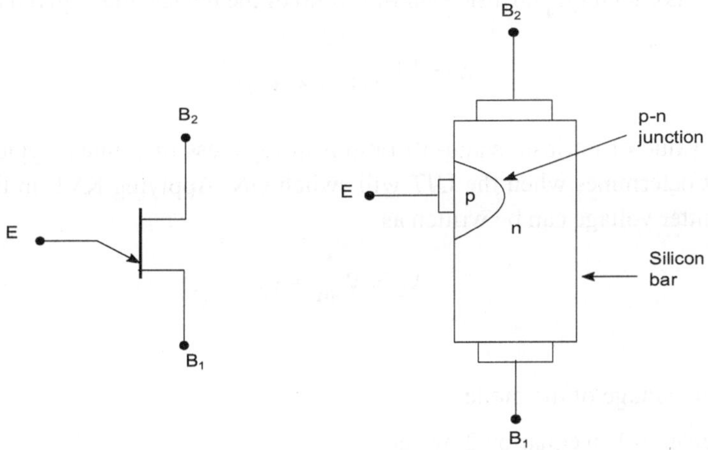

Fig. 2.3.1 Circuit symbol and basic structure of UJT

A UJT consists of uniformly lightly doped n-type Silicon bar with two contacts on either side designated as two bases: Base 1 (B_1) and Base 2 (B_2) and a p-type heavily doped region termed as Emitter (E) which lies closer to B_2 than B_1 region as shown in Fig. 2.3.1.

The equivalent circuit for UJT is shown in Fig. 2.3.2. The diode depicts p-n junction in the emitter and the two series resistors R_{B1} and R_{B2} represent the high impedance of lightly doped n-type Si bar.

Where,

R_{B1}: Internal resistance of n-type bar from base B_1.

R_{B2}: Internal resistance of n-type bar from base B_2.

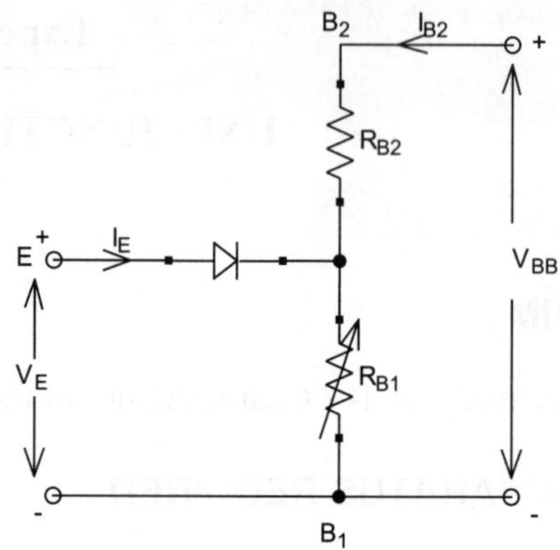

Fig. 2.3.2 Equivalent circuit for UJT

R_{BB} is the inter-base resistance and is equal to the sum of R_{B1} and R_{B2} with emitter terminal E open *i.e.* it is the total resistance of Si bar This inter-base resistance is relatively high, typically $5K\Omega$ to $10K\Omega$. Since, the n-type Si bar has a high resistance and emitter is closer to B_2 than B_1 terminal, therefore R_{B1} is greater than R_{B2}. When voltage V_{BB} is applied between B_2 and B_1, the voltage appears across R_{B1} which is given as

$$V_{RB1} = \eta \cdot V_{BB} \qquad - \text{ eqn 1}$$

Where, $\square$ is called the intrinsic stand-off ratio of the device and is given as

$$\eta = {R_{B1}}/{(R_{B1} + R_{B2})}$$

The value of intrinsic stand-off ratio is always less than unity, typically 0.5 to 0.8. It determines when the UJT will switch ON. Applying KVL in the input loop, emitter voltage can be written as

$$V_E = V_{RB_1} + V_K \qquad - \text{ eqn 2}$$

Where,

V_K: cut-in voltage of the diode.

Using equation 1 in equation 2, we get

$$V_E = \eta \cdot V_{BB} + V_K \qquad - \text{ eqn 3}$$

In Fig. 2.3.2, anode of diode is at voltage V_E and cathode of diode is at voltage V_{RB1}. Initially, when V_{BB} is 0V, V_{RB1} will also be zero and hence UJT will behave like an ordinary diode and it will remain forward biased for all values of $V_E > V_K$. As V_{BB} is increased to some positive value, there is a voltage drop across the resistance R_{B1} *i.e.* V_{RB1} (refer eqn1). Voltages V_E and V_{RB1} will decide whether the diode is forward biased or reverse biased. If the emitter voltage is such that

- $V_E < V_{RB1}$

  The diode becomes reverse biased which results in flow of small reverse saturation current.

- $V_E = V_{RB1}$

  Since, voltage at both p and n terminals of diode is equal, neither reverse nor forward current will flow.

- $V_E > V_{RB1} + V_K$

  Since, voltage at p terminal of diode is greater than voltage at n terminal, the diode becomes forward biased and it starts conducting. The emitter voltage at which diode starts conducting is termed as peak voltage V_P and is given as

$$V_P = V_{R_{B1}} + V_K \qquad\qquad - \text{eqn 4}$$

When the diode becomes forward biased, holes are swept by the electric field towards B_1 which increases emitter current I_E. As the conductivity increases, the base resistance R_{B1} reduces which results in reduction of V_{RB1}. This implies that if more number of holes is injected, it results in further reduction of V_{RB1}. Hence, emitter voltage V_E also decreases (refer eqn2). This decrease in voltage with increase in current gives the negative resistance region. As the emitter current I_E further increases and becomes equal to valley current I_V, the device goes into saturation. The voltage at this point is called as valley point voltage V_V. At this point, there is no further decrease in the voltage and this characteristic curve is similar to that for a simple diode.

If the value of V_{BB} is increased, the drop across resistance R_{B1} *i.e.* V_{RB1} also increases (refer eqn1). As a result, the voltage required to turn on the diode *i.e.* peak voltage V_P also increases (refer eqn4). Hence, switching voltage *i.e.* the voltage at which the UJT is turned ON can be changed by changing the external applied voltage between base B_1 and base B_2 terminals *i.e.* V_{BB}. The switching time from peak to valley points depends upon the device geometry and biasing conditions.

CHARACTERISTICS OF UJT

Point A: valley point, Point B: peak point

V_V (valley Voltage): emitter voltage at the valley point and increases with increase in V_{BB}.

I_V (valley Current): emitter current at the valley point and increases with increase in V_{BB}.

V_P (peak Voltage): emitter voltage at the peak point and increases with increase in V_{BB}.

I_P (peak Current): emitter current at the peak point; minimum current required to trigger the UJT.

Region between A and B is called negative resistance region.

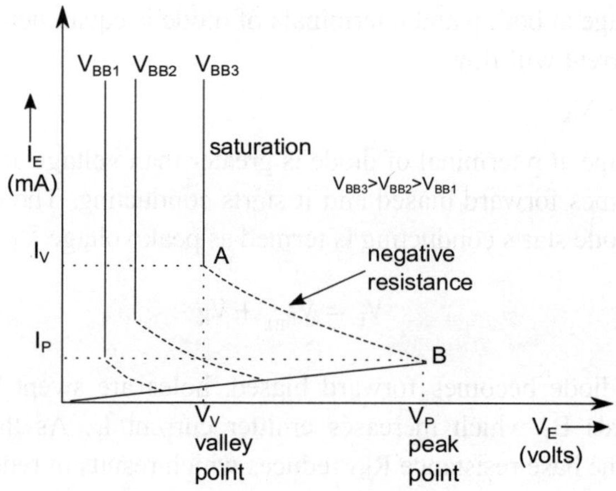

Fig. 2.3.3 I-V characteristics of UJT

CIRCUIT DIAGRAM

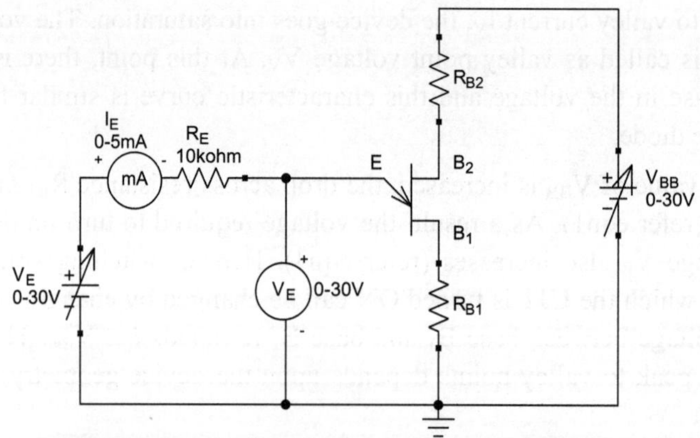

Fig. 2.3.4 Circuit to study I-V characteristics of UJT

PROCEDURE

1. Connect the circuit as shown in Fig. 2.3.4 on breadboard.

2. Connect milli-ammeter and voltmeter to measure I_E and V_E.

3. Set the inter-base voltage V_{BB} to 0V.

4. Increase the emitter voltage V_E slowly from 0V and measure the corresponding voltage V_E and emitter current I_E.

5. Note the readings in tabular form.

6. Repeat steps 4 and 5 for V_{BB} equal to 5V and 10V.

7. Plot the graph between V_E and I_E for different V_{BB} values.

8. Calculate the value of intrinsic stand-off ratio η.

OBSERVATION TABLE

Table 2.3.1 Observation table for UJT characteristics

$V_{BB1}(V) = 0V$		$V_{BB2}(V) = 5V$		$V_{BB3}(V) = 10V$	
$V_E(V)$	$I_E(mA)$	$V_E(V)$	$I_E(mA)$	$V_E(V)$	$I_E(mA)$
0		0		0	
1		1		1	
..		..		..	
..		..		..	

Table 2.3.2 Calculation of intrinsic stand-off ratio

S.No.	Inter base voltage V_{BB} (volts)	Peak point voltage V_P (volts) (From graph)	Intrinsic stand-off ratio η $V_P = \eta V_{BB} + V_K$
1.	0V		
2.	5V		
3.	10V		

Where, V_K is 0.7V for Si.

RESULT

I-V characteristics of UJT have been studied and drawn successfully. Intrinsic stand-off ratio for the given UJT is found to be …..

DISCUSSION

UJT is used as a triggering device for SCR. Since, UJT has a low output impedance of around 3Ω - 9Ω and SCR has low input impedance, therefore, by using UJT as a triggering device for SCR, maximum power can be transferred (because of impedance matching).

UJT is most widely used as a relaxation oscillator which provides saw-tooth waveform at the output. Other applications of UJT include voltage or current limiters, voltage or current regulators, phase control circuits and switching devices.

Experiment 4

SEMICONDUCTOR CONTROLLED RECTIFER

AIM

To Study the I-V Characteristics of the SCR.

APPARATUS REQUIRED

SCR 2P4M, power resistors – 1kΩ, 470Ω, variable dc power supply – 0-35V, 0-15V, voltmeter – 0-35V, 0-15V, two ammeters – 0-25mA, breadboard, connecting wires.

THEORY

Semiconductor Controlled Rectifier

The Semiconductor Controlled Rectifier, also called as Silicon Controlled Rectifier, abbreviated as SCR, belongs to the family of semiconductor devices called as thyristors. It is a unidirectional device which conducts only in one direction.

It has four alternate layers of p and n materials with three p-n junctions (J_1, J_2, and J_3) and three terminals named as anode, cathode and gate as shown in Fig. 2.4.2. The gate terminal is used to control the firing of SCR and thus, the device acts as a controlled rectifier. It is widely used as a switch as it can go from blocking *i.e.* OFF state to conducting *i.e.* ON state and vice-versa. Both the states are stable and reversible.

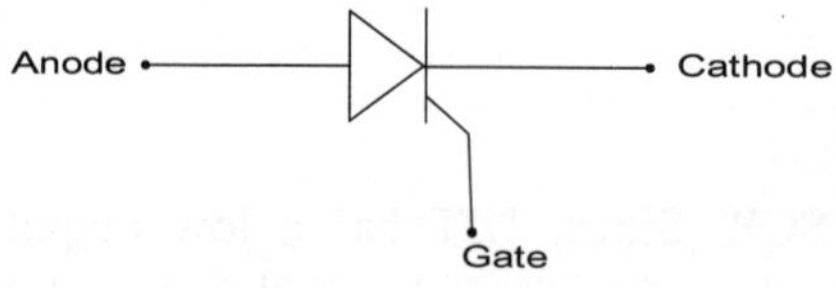

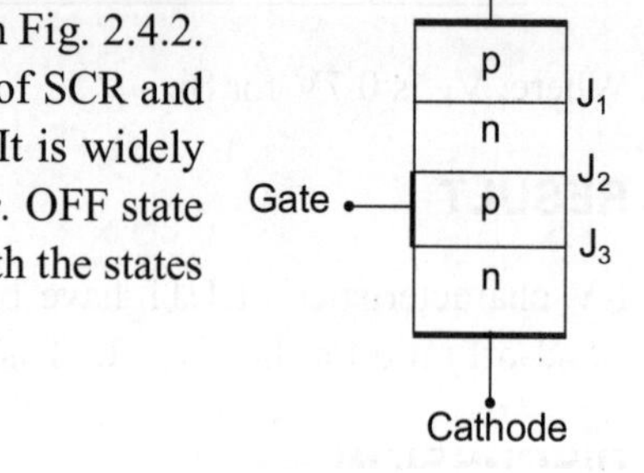

Fig. 2.4.2 SCR structure

Fig. 2.4.1 SCR symbol

The doping concentration varies between different layers of SCR. First and the fourth p and n layers are highly doped. The inner n layer is very wide and lightly doped which enables wide space charge layer and thus provides a high forward and reverse voltage blocking capability of SCR.

The material used for SCR is Silicon because of the following reasons:

- Si has high intrinsic resistivity (around 2,30,000 Ω-cm). Hence, it can sustain high breakdown voltages and has high power handling capability.
- Si devices can operate at high temperatures of about 150°C.
- Si has less leakage current and hence it can be used as a switch.

Different modes of operation of SCR

1. Forward Blocking or OFF state

When anode is made positive with respect to cathode, junctions J_1 and J_3 are forward biased and junction J_2 is reverse biased. Therefore, small leakage current, also known as OFF state current, flows through anode to cathode and SCR is said to be operated in forward blocking state or OFF state.

2. Forward Conducting or ON state

When anode to cathode voltage V_{AK} is increased to a sufficiently large value, breakdown occurs in reverse biased junction J_2. Junctions J_1 and J_3 are forward biased hence, there will be free movement of carriers across all the three junctions. This result in large anode to cathode forward current and ohmic drop across four layers will become very small. The voltage at which avalanche breakdown occurs is called forward break-over voltage V_{FBO} and the device is said to be operated in conducting or ON state. An external resistance is used to limit the current flowing through the device. In order to maintain the required amount of carrier flow across the junctions *i.e.* to keep the device in ON state, the forward or anode current has to be more than the latching current I_L. If forward anode current is reduced below a level known as holding current I_H, the depletion region will begin to develop around junction J_2 due to reduced number of carriers and the device will go to blocking *i.e.* OFF state.

3. Reverse blocking or OFF state

When voltage at the cathode terminal is made positive with respect to anode, the junctions J_1 and J_3 become reverse biased and J_2 becomes forward biased, then SCR is said to be operated in reverse blocking state. The motion of carriers will be restricted due to the reverse biased junctions J_1 and J_3 and only reverse current will flow through the device. If the reverse voltage is gradually increased beyond its breakdown voltage, avalanche breakdown occurs in the device and SCR starts conducting heavily in the reverse direction. The corresponding maximum reverse voltage is called reverse breakdown voltage.

Gate control triggering

SCR can be turned ON by applying the anode to cathode voltage beyond the forward break-over voltage V_{FBO}. But this process can be destructive. Therefore, practically, forward voltage V_A is maintained below V_{FBO} only and the device is turned ON by applying a positive voltage between gate and cathode terminal. As the signal is applied on the gate terminal, the avalanche breakdown occurs in reverse biased junction J_2 at a lower anode to cathode voltage and the anode current becomes greater than the holding current and the device is turned ON. Even if the signal at gate terminal is removed, the device continues to conduct due to the positive biasing. This method of turning ON the SCR by applying the positive voltage between gate and cathode is called gate control method.

CIRCUIT DIAGRAM

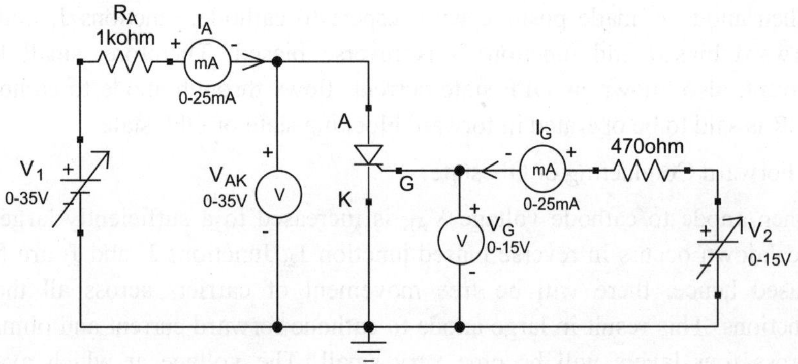

Fig. 2.4.3 Circuit to study I-V characteristics of SCR

CHARACTERISTICS OF SCR

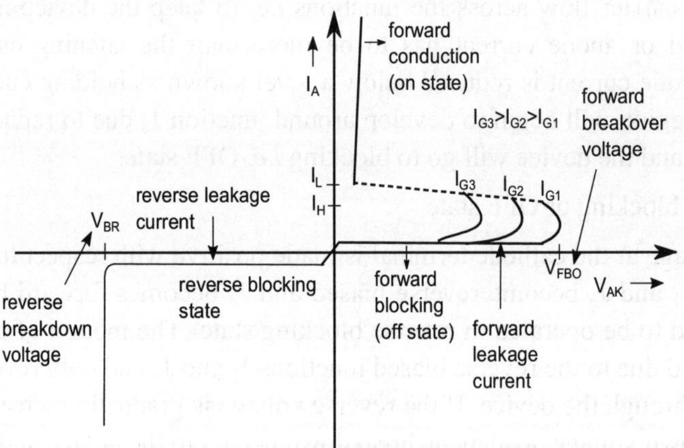

Fig. 2.4.4 I-V characteristics of SCR

Where, I_H (holding current): minimum anode current required to bring the SCR in OFF state, I_L (latching current): minimum anode current required to maintain the SCR in the ON state.

PROCEDURE

1. Connect the circuit as shown in Fig. 2.4.3 on breadboard.

2. To determine the value of gate current I_G at which given SCR would be turned ON, slowly increase the voltage between anode and cathode terminals V_1 till the voltmeter V_{AK} reaches to its maximum value, say if voltmeter is of the range (0-35V), increase V_1 such that voltmeter reads 35V.

3. Now, slowly increase the gate voltage V_G by increasing V_2 till the breakdown occurs *i.e.* when anode current I_A shoots up and anode to cathode voltage V_{AK} is reduced.

4. This will give the minimum value of gate current I_{Gmin} at which SCR is switched ON. The corresponding voltage is referred as V_{Gmin}.

5. Set all the voltages at 0V and turn OFF the power supply for few minutes so that SCR come back to forward blocking state.

6. Turn ON the power supply and keep the gate voltage $V_{G1} = V_{Gmin}$. Note the gate current I_G as I_{Gmin} through the ammeter connected at the gate terminal.

7. Increase V_1 such that anode to cathode voltage V_{AK} increases slowly from 0V and measure the corresponding anode current I_A.

8. Take the readings till the breakdown occurs and note them in a tabular form.

9. For different set of readings, keep the gate voltage V_G greater than V_{Gmin} and I_G greater than I_{Gmin} and repeat step 7 and 8.

10. Plot the graph between V_{AK} and I_A for different values of I_G.

11. Note the forward break-over voltage V_{FBO} for different gate current I_G.

OBSERVATION TABLE

Table 2.4.1 Observation table for I-V characteristics of SCR

I_{G1}(mA)		I_{G2}(mA)		I_{G3}(mA)	
V_{AK}(V)	I_A(mA)	V_{AK}(V)	I_A(mA)	V_{AK}(V)	I_A(mA)
0		0		0	
1		1		1	
..		..		..	
..		..		..	

RESULT

I-V characteristics of SCR have been studied and drawn successfully. Forward break over voltages and holding current for different gate currents are found to be:

$$I_{G1} = \rule{2cm}{0.4pt}, \quad V_{FBO1} = \rule{2cm}{0.4pt}, \quad I_{H1} = \rule{2cm}{0.4pt}$$

$$I_{G2} = \rule{2cm}{0.4pt}, \quad V_{FBO2} = \rule{2cm}{0.4pt}, \quad I_{H2} = \rule{2cm}{0.4pt}$$

$$I_{G3} = \rule{2cm}{0.4pt}, \quad V_{FBO3} = \rule{2cm}{0.4pt}, \quad I_{H3} = \rule{2cm}{0.4pt}$$

DISCUSSION

SCR is widely used as:

- Rectifier for conversion of ac into dc.
- Inverter for conversion of dc into ac.
- Speed control of ac and dc motors.
- Dc chopper (dc to dc converter).
- HVDC transmission lines.
- Over light detector.
- Over voltage protection.
- Power supplies for computers.
- In switching and control applications like power switches, static switches, relay control, phase control, power control and control of welding machines.

Experiment 5

FIELD EFFECT TRANSISTOR

AIM

To Study the I-V Characteristics of the Common Source FET Configuration.

APPARATUS REQUIRED

JFET BFW10, variable dc power supply – 0-10V, 0-20V, voltmeter – 0-10V, 0-20V, ammeter – 20mA, resistor – 500Ω, 1kΩ breadboard, connecting wires.

THEORY

Field Effect Transistor

A Field effect transistor abbreviated as FET is a semiconductor device with three terminals named as Drain, Source and Gate. It is a unipolar device because current conduction takes place by only one type of carriers *i.e.* majority carriers.

FETs are classified into two types

- N-channel FETs with electrons as majority carriers.
- P-channel FETs with holes as majority carriers.

  N-channel FETs are preferred over p-channel FETs because of higher electron mobility as compared to the holes.

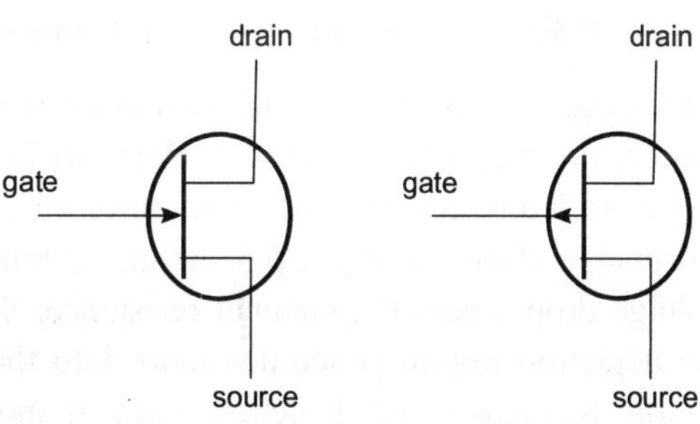

Fig. 2.5.1 Symbol for n-channel and p-channel JFET

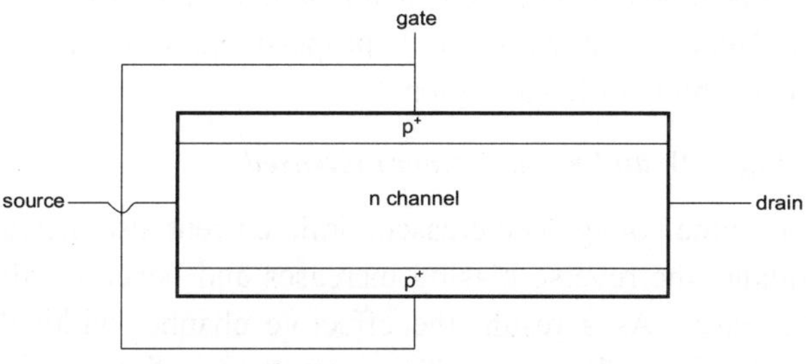

Fig. 2.5.2 Structure of n-channel JFET

FET consists of uniformly doped semiconductor bar called as channel with two ohmic contacts at both the ends, known as Source S and Drain D. The bar has two junctions at its both sides which are connected internally to form a common terminal known as gate. In n-channel FET, the semiconductor bar is of n-type and junctions on both sides of the channel are highly doped p^+ region as shown in Fig. 2.5.2. In case of p-channel FET, the semiconductor bar is of p-type and junctions on both sides of the channel are highly doped n^+ region.

Notations in FET

1. Source: The terminal through which the majority carriers enter the channel.

2. Drain: The terminal through which the majority carriers leave the channel.

3. Gate: The terminal obtained by connecting two heavily doped impurity regions which control the flow of carriers from source to drain.

4. Channel: The region between the source and drain sandwiched between the gate and majority carriers flow from source to drain through the channel.

Let us consider different cases when biasing voltages are applied between gate and source terminal and drain and source terminal.

Case 1: When $V_{GS} = 0V$ and $V_{DS} = 0V$

With zero voltages, no current flows through the device. The depletion regions around the p-n junctions are of equal thickness and they are symmetrical.

Case 2: When $V_{GS} = 0V$ and $V_{DS} =$ small positive voltage

As voltage V_{DS} is applied such that drain is made more positive with respect to source, the majority carriers *i.e.* electrons in n-channel FET start flowing from source to drain terminal and drain current I_D starts flowing from drain to source terminal as shown in Fig. 2.5.3. As the current I_D flows through the bar, it causes voltage drop across the channel resistance. Since, gate region is heavily doped, the depletion region penetrates more into the channel as compared to the gate region. Moreover, the depletion width is more towards the drain side than the source side because the drain is at positive potential with respect to the source *i.e.* drain end is more reverse biased. For small values of V_{DD}, V_{DS} is small and depletion region is almost negligible which results in free flow of electrons from source to drain. Hence, drain current I_D is proportional to V_{DS} and therefore, this region is called as ohmic or linear region.

Case 3: When $V_{GS} = 0V$ and V_{DS} is further increased

As the drain to source voltage is increased, drain current also increases. But with increase in voltage, the reverse biasing increases and hence width of depletion region also increases. As a result, the effective channel width decreases and hence, its conductivity decreases. This results in slight deviation of I-V characteristics from linearity and hence, there is lesser increase in I_D with increase in V_{DS}. As V_{DS} is further increased, the width of depletion region increases further and the channel region becomes more restricted near the drain end. For some value of drain to source voltage, known as pinch-off voltage V_P, the depletion region on both sides of the channel becomes so wide that they start touching each other as shown in Fig. 2.5.4. This result in saturation of drain current (drain current is denoted as I_{DSS}) and this point is called as pinch-off point. As the voltage V_{DS} is further increased, there is no further increase in drain current and it becomes constant. On further increase of voltage V_{DS}, breakdown takes place which results in sudden increase of the current. This should be avoided as it may damage the device.

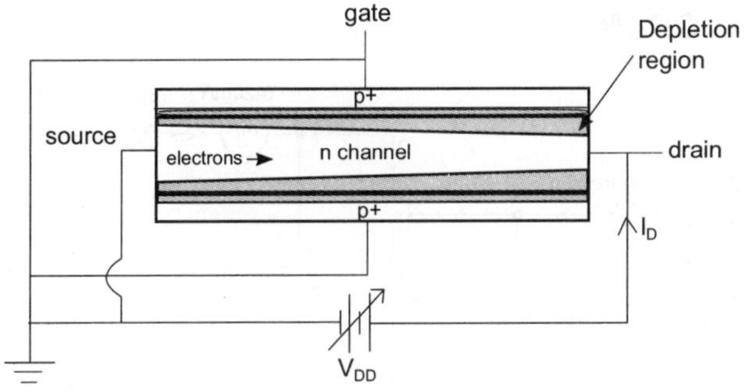

Fig. 2.5.3

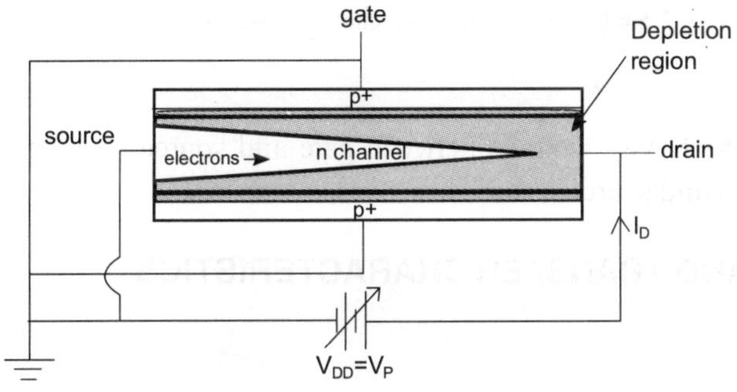

Fig. 2.5.4

Case 4: When V_{GS} = negative voltage

As the gate terminal is made negative with respect to the source terminal, the gate to source junction becomes reverse biased as shown in Fig. 2.5.5. This results in larger depletion width which reduces the effective channel width and hence, the conductivity decreases. If V_{GS} is further increased, the channel becomes narrower at the drain end and the drain current further reduces. Hence, gate voltage is used to control the drain current flowing through the channel. Thus, as the gate to source voltage increases for a fixed value of drain to source voltage, the drain current decreases and becomes zero.

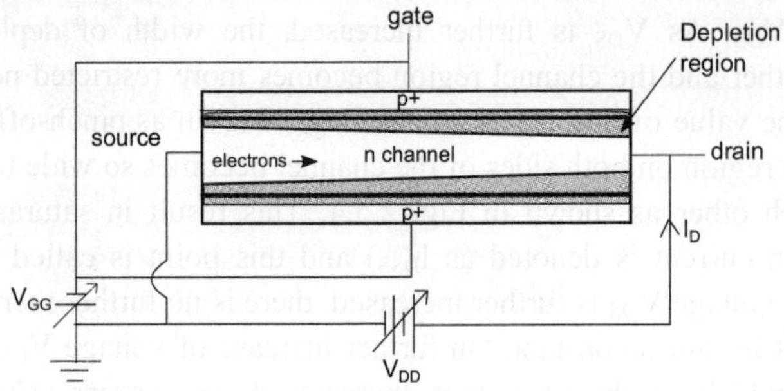

Fig. 2.5.5

CIRCUIT DIAGRAM

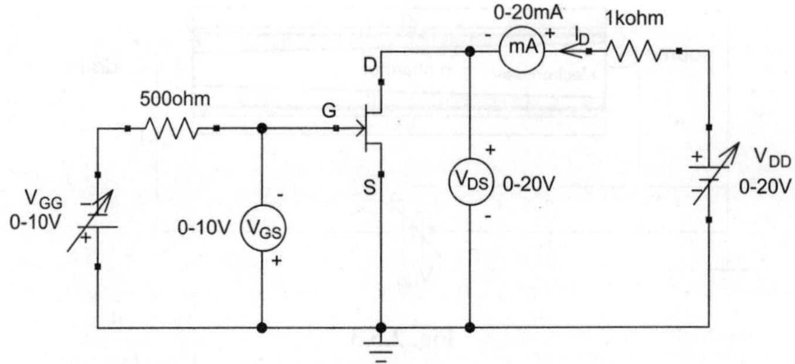

Fig. 2.5.6 Circuit to study I-V characteristics of n-channel JFET

Where,

I_D: drain current, V_{GS}: voltage between gate and source terminal, V_{DS}: voltage between drain and source terminal.

OUTPUT AND TRANSFER CHARACTERISTICS

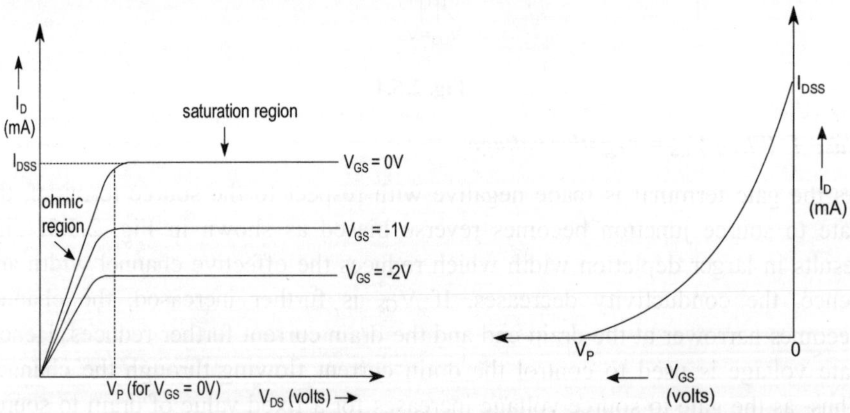

Fig. 2.5.7 Output characteristics   **Fig. 2.5.8** Transfer characteristics

Where,

V_P: pinch off voltage, I_{DSS}: maximum I_D defined for $V_{GS} = 0V$ and $V_{DS} > V_P$.

Output or drain characteristics

The Drain characteristics of FET is plotted by keeping the gate to source voltage (V_{GS}) constant and varying the drain to source voltage (V_{DS}) and measuring the corresponding drain current (I_D) as shown in Fig. 2.5.7.

Transfer characteristics

The transfer characteristics of FET is plotted by keeping the drain to source voltage (V_{DS}) constant and varying the gate to source voltage (V_{GS}) and measuring the corresponding drain current (I_D) as shown in Fig. 2.5.8.

Drain resistance

It is defined as the ratio of the change in drain to source voltage to change in drain current at a constant gate to source voltage. (Refer Fig. 2.5.9)

$$r_D = \left.\frac{\Delta V_{DS}}{\Delta I_D}\right|_{V_{GS}}$$

Transconductance

It is defined as the ratio of the change in drain current to the change in gate to source voltage at a constant drain to source voltage. (Refer Fig. 2.5.10)

$$g_m = \left.\frac{\Delta I_D}{\Delta V_{GS}}\right|_{V_{DS}}$$

Amplification factor

It is defined as the ratio of the change in drain to source voltage to the change in gate to source voltage at a constant drain current.

$$\mu = \left.\frac{\Delta V_{Ds}}{\Delta V_{GS}}\right|_{I_D} = r_D g_m$$

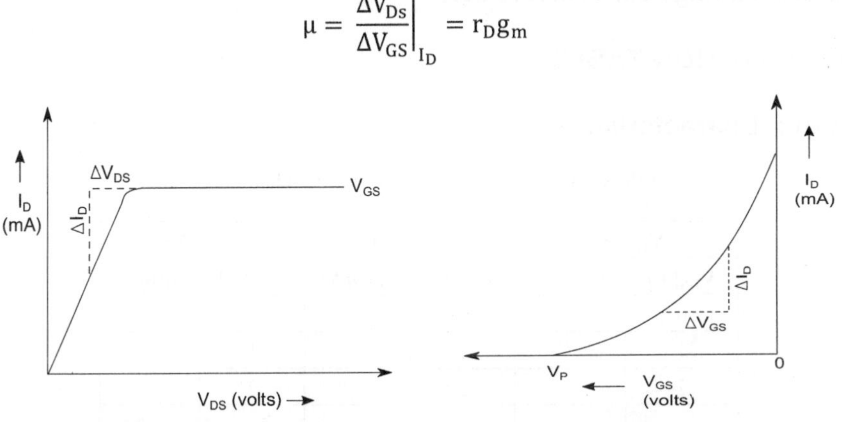

Fig. 2.5.9 Drain resistance of JFET   **Fig. 2.5.10** Transconductance of JFET

PROCEDURE

Output Characteristics

1. Connect the circuit as shown in Fig. 2.5.6 on breadboard.
2. Initially, fix the gate voltage V_{GS} to 0V.
3. Increase the drain to source voltage V_{DS} slowly from 0V by varying V_{DD} and measure the corresponding drain current I_D.
4. Take the readings and note them in tabular form.
5. Repeat steps 3 and 4 for different values of V_{GS}.
6. Plot the graph between V_{DS} and I_D for different values of V_{GS} and calculate the drain resistance r_D.

Transfer Characteristics

1. For the same circuit, fix drain to source voltage V_{DS} to 1V.
2. Increase the gate to source voltage V_{GS} slowly from 0V by varying V_{GG} and measure the corresponding drain current I_D.
3. Take the readings and note them in tabular form.
4. Repeat steps 2 and 3 for different values of V_{DS}.
5. Plot the graph between V_{GS} and I_D for different values of V_{DS} and calculate the trans-conductance g_m.
6. Calculate the amplification factor $\mu = g_m * r_D$.

Note

Gate to source junction is always reverse biased and gate to source voltage V_{GS} should always be kept lower than drain to source voltage V_{DS}, otherwise, carriers will start moving from drain to source.

OBSERVATION TABLE

Output Characteristics

Table 2.5.1 Observation table for output characteristics

V_{GS1}(V)		V_{GS2}(V)		V_{GS3}(V)	
V_{DS}(V)	I_D(mA)	V_{DS}(V)	I_D(mA)	V_{DS}(V)	I_D(mA)
0		0		0	
1		1		1	
2		2		2	
...		...		...	

Transfer Characteristics

Table 2.5.2 Observation table for transfer characteristics

$V_{DS1}(V)$		$V_{DS2}(V)$		$V_{DS3}(V)$	
$V_{GS}(V)$	$I_D(mA)$	$V_{GS}(V)$	$I_D(mA)$	$V_{GS}(V)$	$I_D(mA)$
0		0		0	
1		1		1	
2		2		2	
...		...		...	

RESULT

The output characteristics and transfer characteristics of FET have been studied and graph has been plotted successfully. The value of the following parameters was calculated.

Drain resistance, r_D = ...

Transconductance, g_m = ...

Amplification factor, μ = ...

DISCUSSION

FET is an improvement over BJT and has many advantages over it. Differences between BJT and FET are

Table 2.5.3 Comparison between BJT and FET

S.No.	BJT	FET
1.	It is a Bipolar device where conduction takes place by both electrons and holes.	It is a Unipolar device where conduction takes place either by holes or by electrons.
2.	Input impedance of the order of few kΩ.	High input impedance of the order of hundreds of MΩ. Hence more suitable for voltage amplifiers.
3.	Temperature stability is less.	Temperature stability is more.
4.	Occupies more area on silicon chip.	Occupies less area on silicon chip.
5.	Current controlled device.	Voltage controlled device.
6.	High intrinsic noise	Less intrinsic noise and hence used as input stage of low-level amplifiers.

The disadvantage of FET over BJT is that the gain bandwidth product is smaller as compared to BJT and it can be easily damaged by static electricity.

FETs are widely used as: Input amplifiers in oscilloscopes and voltmeters; Low noise amplifiers; Buffer amplifiers; Chopper amplifiers; RF amplifiers; Switch in mixers and digital circuits.

<table><tr><td>

3

</td><td>

Experiments on Analog Electronics - I

</td></tr></table>

LIST OF EXPERIMENTS

1. To Study the Half Wave Rectifier and Full Wave Rectifier.

2. To Study Power Supply using C Filter and Zener Diode.

3. To Study Clipping Circuits.

4. To Study Clamping Circuits.

5. To Study Fixed Bias, Voltage Divider and Collector-to-Base Bias Feedback Configuration for Transistors.

6. To Design a Single Stage CE Amplifier for a Specific Gain.

7. To Study the Colpitt's Oscillator.

8. To Study the Phase Shift Oscillator.

9. To Study the Frequency Response of Common Source FET Amplifier.

Experiment 1

FULL WAVE and HALF WAVE RECTIFIERS

AIM

To Study the Half Wave Rectifier and Full Wave Rectifier.

APPARATUS REQUIRED

Breadboard, transformer 6-0-6, diode IN 4007, resistor - 1kΩ, connecting wires, cathode ray oscilloscope, multimeter, probe.

THEORY

Power Supply

Most of the electronic devices need a dc source of power to operate. If we use portable batteries as dc power source, we need to replace them again and again because of their limited operating period which becomes costly. Hence, a better and economical way is the use of power supply which is used to convert ac to dc.

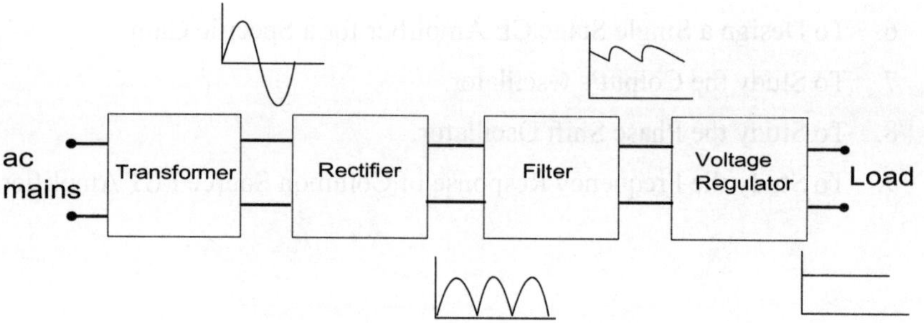

Fig. 3.1.1 Block diagram of power supply

Different blocks in power supply, as shown in Fig. 3.1.1 are as follows:

Transformer

The transformer steps up or steps down the input voltage as well as isolates the power supply from the power line, thereby, reduces the electrical shock risk.

Rectifier

Rectifier consists of diodes which are used to convert the sinusoidal ac voltage to non-sinusoidal pulsating dc.

Filter

The rectified output called as pulsating dc contains some unwanted ripples. Role of filter is to filter out the unwanted ripples and make the output ripple free. Combination of capacitors and inductors are used which passes dc component to the load and blocks ac components.

Regulator

The power supply output can vary due to changes in load current or input line voltage. To avoid this, zener diode is used as a voltage regulator to provide a constant dc output voltage across load terminals.

Half Wave Rectifier (HWR)

In HWR, current flows through the load only for the positive half cycle of the input waveform. It consists of a diode and resistor as shown in Fig. 3.1.2.

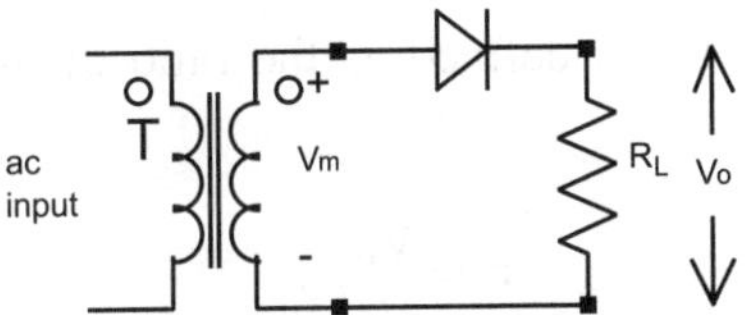

Fig. 3.1.2 Half wave rectifier

Working

As the input is applied across the primary of the transformer, an ac voltage is induced across its secondary. The transformer can be step up or step down depending on the required output voltage.

During the positive half cycle of input voltage, diode is forward biased and conducts. It acts as a short with very low resistance (ideally zero) and hence voltage drop across the diode is also very small (ideally zero but practically 0.3V for Ge diode and 0.7V for Si diode) and therefore, output voltage across load resistance R_L is same as the input voltage.

During the negative half cycle, diode is reversed biased and does not conduct. It acts as an open switch and hence no current flows through it. Therefore, voltage across the load resistance is zero.

The output dc voltage (across load) is given as

$$V_{dc} = {V_m}/{\Pi}$$

Where, V_m is the peak ac voltage at the input of the rectifier.

Input and output waveforms

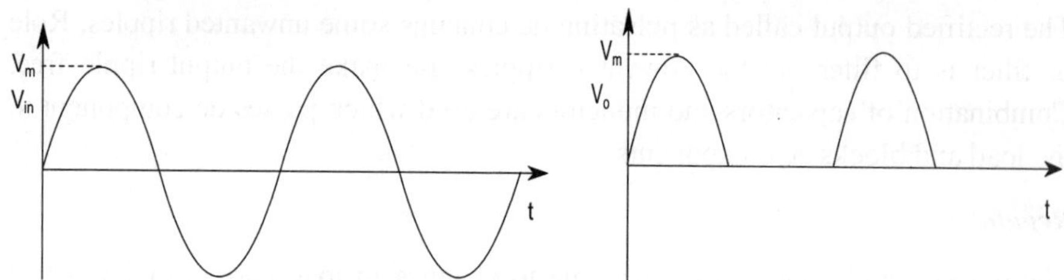

Fig. 3.1.3 Input and output voltage for HWR

Ripple factor

The half wave rectified output voltage across load resistance R_L is unidirectional as shown in Fig. 3.1.3. But due to the fluctuating magnitude, output voltage has dc as well as ac component. The presence of ac component in the rectified output is called as the ripple and the value of ripple factor indicates the amount of ac component present in the output voltage.

Therefore, ripple factor is defined as the ratio of the ac voltage to the dc voltage and is given as

$$ r = {V_{ac}}/{V_{dc}} $$

For HWR, r = 1.21

Full Wave Rectifier (FWR)

In FWR, the current flows through the load for whole of the input cycle *i.e.* for positive as well as negative half cycles of the input waveform. It is of two types - center tap rectifier and bridge rectifier.

Center tap rectifier

It consists of two diodes D_1 and D_2 and a center tapped transformer as shown in Fig. 3.1.4.

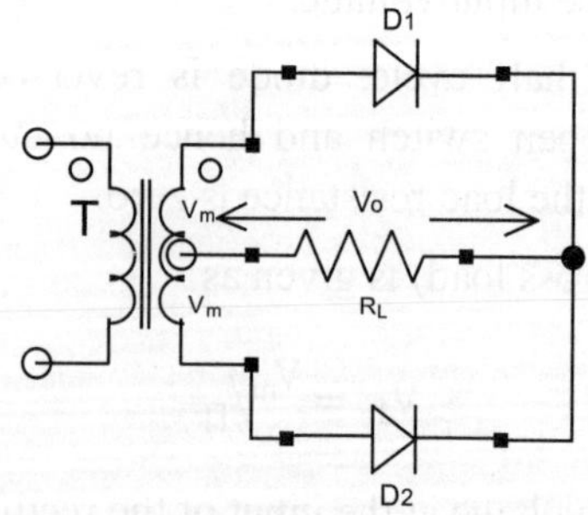

Fig. 3.1.4 Center tap full wave rectifier

Working

As the input ac signal is applied across the primary of transformer, an ac voltage is induced across the secondary of a center tapped transformer. The center tapped connection provides out of phase voltages of same magnitude to two diodes.

During the positive half cycle of input voltage, diode D_1 is forward biased and D_2 is reverse biased. Hence, current flows from D_1 to the load resistance and then to the upper half of winding as shown in Fig. 3.1.5a.

During the negative half cycle of input voltage, diode D_2 is forward biased and D_1 is reverse biased. Now, current flows from D_2 to the load resistance and then to the lower half of winding as shown in Fig. 3.1.5b.

Hence, current flows through the load in the same direction and voltage appears across it during both half cycles and therefore, a full wave rectified output voltage is produced across the load resistance as shown in Fig. 3.1.7b.

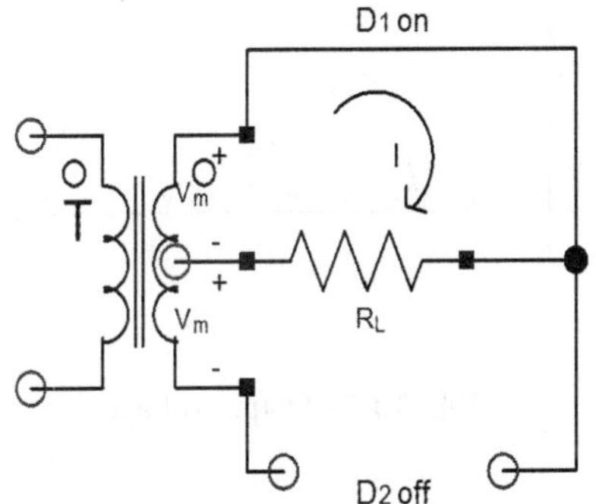

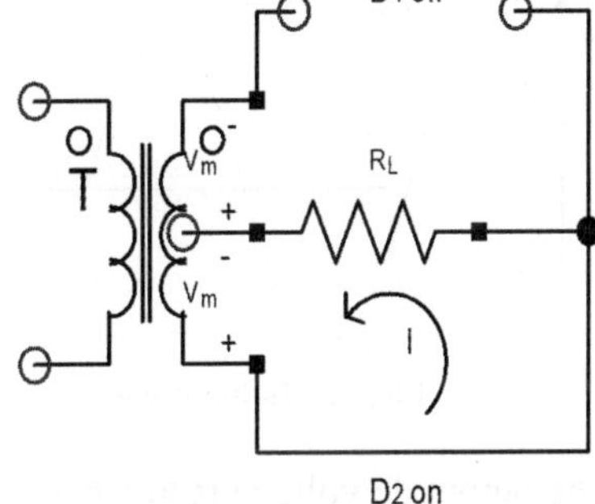

Fig. 3.1.5a Positive half cycle   **Fig. 3.1.5b** Negative half cycle

Bridge rectifier

It is the most widely used full wave rectifier and consists of four diodes D_1, D_2, D_3 and D_4 as shown in Fig. 3.1.6.

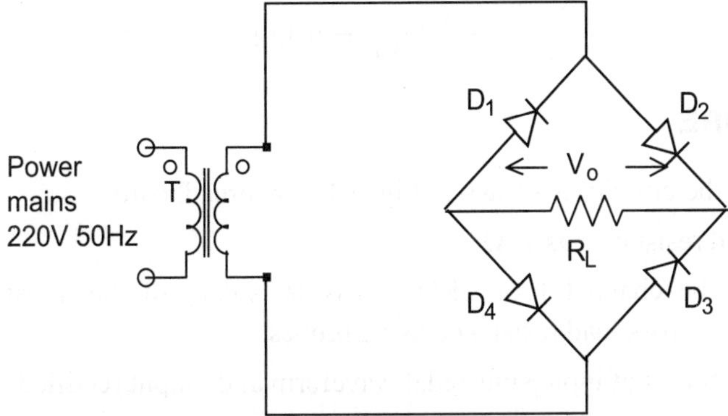

Fig. 3.1.6 Bridge rectifier

Working

As the input is applied across the primary of transformer, an ac voltage is induced across the secondary of transformer. During the positive half cycle of input voltage, diodes D_2 and D_4 are forward biased and D_1 and D_3 are reverse biased. Hence, current flows from secondary winding to diode D_2, then to load resistance and finally to diode D_4.

During the negative half cycle of input voltage, diodes D_1 and D_3 are forward biased and D_2 and D_4 are reverse biased. Now, current flows from secondary winding to diode D_3, then to load resistance and finally to diode D_1.

Hence, current flows through the load in the same direction and voltage appears across it during both half cycles and therefore, a full wave rectified output voltage is produced across the load resistance as shown in Fig. 3.1.7b.

Input and output waveforms

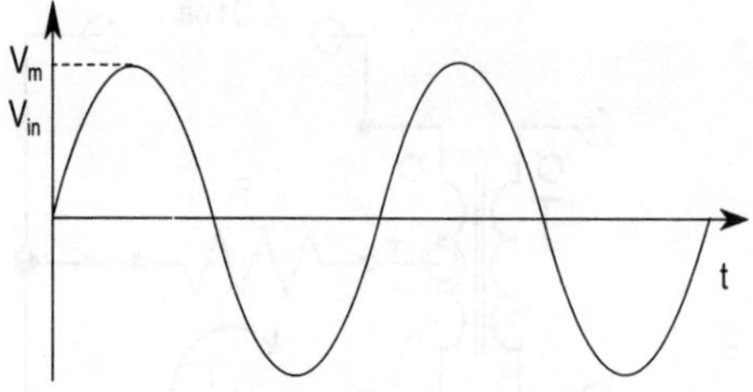
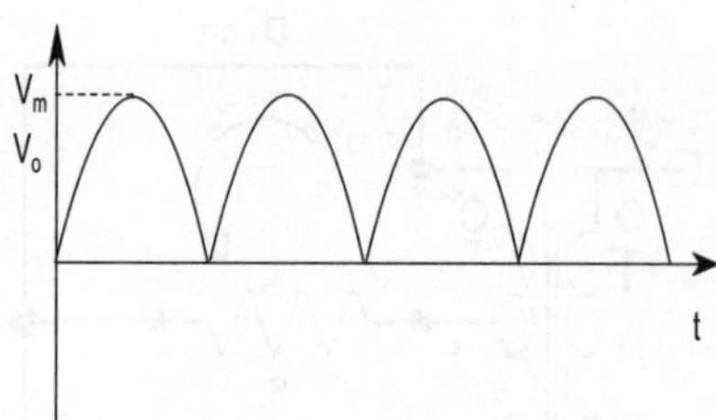

<table>
<tr><td align="center">Fig. 3.1.7a Input voltage</td><td align="center">Fig. 3.1.7b Output voltage</td></tr>
</table>

The output dc voltage is given as

$$V_{dc} = \frac{2V_m}{\Pi}$$

Where, Vm is the peak ac voltage at the input of the rectifier.

Ripple factor

Ripple factor for FWR is given as

$$r = \frac{V_{ac}}{V_{dc}} = 0.482$$

PROCEDURE

1.   Connect the circuits as shown in Fig. 3.1.2 on breadboard.
2.   Take load resistance as 1kΩ.
3.   Connect the channel 1 of CRO across secondary of the transformer and channel 2 across load resistance using probes.
4.   Take the traces of input sinusoidal waveform and output rectified waveform.
5.   Measure the peak voltage V_m on CRO.
6.   Measure ac and dc voltage across load resistance using CRO.

7. Calculate the value of ripple factor practically.

8. Change the value of load resistance and repeat the above steps.

9. Now, connect the circuit as shown in Fig. 3.1.4 and Fig. 3.1.6 and repeat the above steps.

Note:

AC and DC voltage can be measured by using multimeter as well.

OBSERVATION TABLE

Half Wave Rectifier

Table 3.1.1 Observation table for HWR

S.No.	Load resistance R_L (kΩ)	Peak value of input voltage V_m	Vdc		Vac	Ripple factor	
			Th. V_m/Π	Pr.		Th.	Pr. V_{ac}/V_{dc}
1.						1.21	
2.						1.21	

Full Wave Rectifier

Table 3.1.2 Observation table for FWR

S.No.	Load resistance R_L (kΩ)	Peak value of input voltage V_m	Vdc		Vac	Ripple factor	
			Th. $2V_m/\Pi$	Pr.		Th.	Pr. V_{ac}/V_{dc}
1.						0.482	
2.						0.482	

OBSERVATIONS

Attach the traces of input and output waveforms for HWR and FWR.

RESULT

HWR and FWR have been studied and traces have been taken successfully. Theoretical and practical value of ripple factor comes out to be almost same.

DISCUSSION

Rectifiers are used in power supplies to convert ac to dc. Ripple factor which shows the presence of ac component in rectified output is 1.21 in case of HWR and 0.482 in case of FWR. Therefore, FWR is better than HWR.

Bridge rectifier is better than center tap rectifier because it does not need center tapped secondary winding which makes the circuit costly and bulky. Moreover, PIV of diode in case of bridge rectifier is V_m whereas it is $2V_m$ in case of center tap rectifier.

The disadvantage of using bridge rectifier over center tap rectifier is the use of four diodes. While working with low voltages *i.e.* if voltage across secondary winding is low, the drop across two diodes in one half of the cycle is 1.4V (0.7V + 0.7V) which becomes significant. For *e.g.* if voltage across the secondary winding is 2V, then the voltage drop across two diodes is 1.4V. Hence, remaining voltage that appears across the load resistance is equal to 0.6V only. Therefore, while working with low voltages center tap FWR is useful.

Experiment 2

POWER SUPPLY with C FILTER and ZENER DIODE

AIM

To Study Power Supply using C Filter and Zener Diode.

APPARATUS REQUIRED

Breadboard, transformer 6-0-6, ordinary diode IN4007, zener diode, resistors - 1kΩ, 10kΩ, 50kΩ, 100kΩ, capacitors - 0.1μF, 0.01μF, 0.001μ, 0.047μF, connecting wires, cathode ray oscilloscope, multimeter, probe, 0-20V variable dc power supply, resistance box.

THEORY

Filters

The output produced by rectifier is unidirectional and continuous but fluctuates with time. This fluctuating output consists of dc component and number of ac components of different frequencies. The presence of ac components, known as ripples, in the rectified output is highly undesirable. Therefore, in order to make the output steady, it is important to reduce the ac component to a minimum value by using filters.

Filter is a combination of inductor and capacitor which allows dc current to pass through the load and blocks ac current, hence makes the output ripple free. There are different types of filters that are used to make the rectifier output ripple free.

- Shunt capacitor filter
- Series inductor filter
- Choke input LC filter
- Π filter

Shunt Capacitor Filter

This is the simplest filter available to remove the ripples present in pulsating dc. In this, a capacitor is connected in parallel to the load.

Half Wave Rectifier with C Filter

HWR with C filter consists of a diode and a capacitor connected in parallel to the load resistance as shown in Fig. 3.2.1.

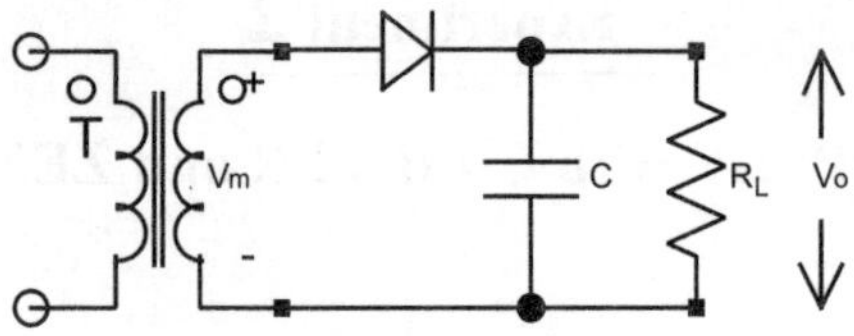

Fig. 3.2.1 HWR with C filter

Role of capacitor

Reactance of Capacitor is given by the formula

$$X_C = \frac{1}{j\omega C}$$

For dc, $\omega = 0$. Therefore, X_C is infinite and thus capacitor acts as an open circuit. Therefore, all the dc current passes through the load. For ac, it offers a low reactance path due to which most of the ac current passes through it and a very small ac current passes through the load. Hence, the output voltage across the load is a dc voltage with a very small ripple in it.

Working

During the positive half cycle of input voltage, diode conducts and acts as a closed switch. Hence, capacitor connected across the load gets connected to the input voltage as well and charges itself to the peak value of the input voltage Vm and reaches point 'b' as shown in Fig. 3.2.2. As the input voltage decreases and falls below Vm, the capacitor is still at voltage Vm. Therefore, the voltage at cathode of diode (connected to capacitor) becomes more than voltage at the anode (connected to input voltage). This reverse biases the diode and thus diode acts as an open circuit. Due to this, the load gets disconnected from the input voltage and capacitor starts discharging itself through the load. Therefore, voltage across the load is not zero as in the case of the rectified output but it is equal to the capacitor voltage.

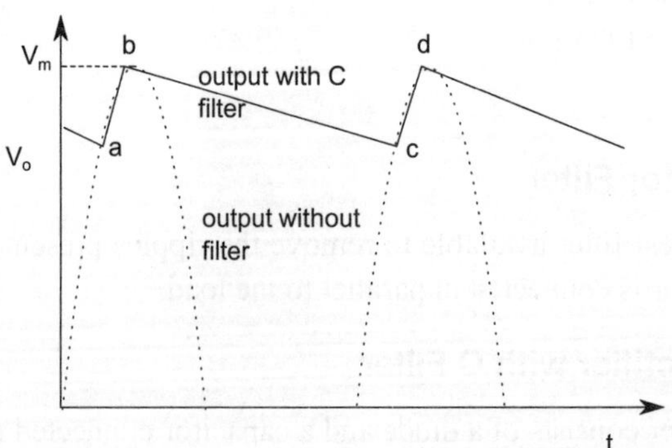

Fig. 3.2.2 Output waveform for HWR with C filter

During the negative half cycle of input voltage, diode is reversed biased and hence capacitor continues to discharge itself through the load. For the next positive half cycle of input voltage, as the input voltage increases and becomes more than capacitor voltage (point c), the diode becomes forward biased and it starts conducting. Capacitor is again connected to the input voltage and charges itself to the peak voltage Vm and reaches point d. Again the diode becomes reverse biased and capacitor starts discharging and the same cycle continues.

The time constant R_LC decides the rate at which capacitor will discharge. It should be such that capacitor is not discharged quickly. As the value of R_L or C increases, the output becomes steadier. Hence, ripples will be less. For HWR with shunt capacitor filter, ripple factor is given as

$$r = 1/(2\sqrt{3}\,f\,R_LC)$$

Where,

f: frequency of the rectified signal, R_L: load resistance, C: capacitance.

Full Wave Rectifier with C Filter

Center tap FWR with C filter consists of two diodes and a capacitor connected in parallel to the load resistance as shown in Fig. 3.2.3.

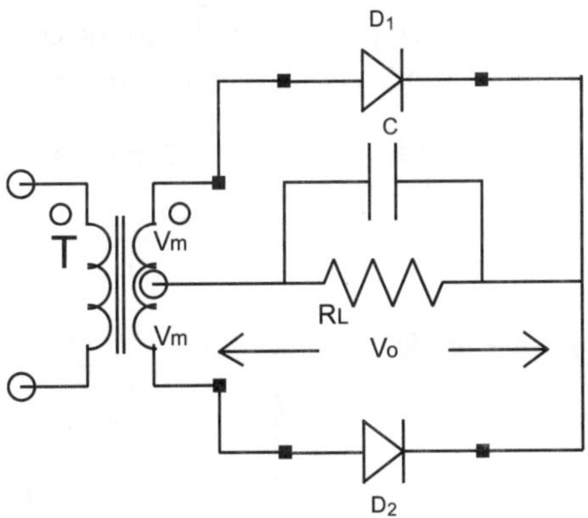

Fig. 3.2.3 FWR with C filter

Working

The working is similar to the HWR with C filter. In this, the presence of diode D_2 allows capacitor to charge itself to peak voltage Vm during the negative half cycle of input voltage as well. The output voltage will be as shown in Fig. 3.2.4.

For FWR with shunt capacitor filter, ripple factor is given as

$$r = 1/(4\sqrt{3}\,f\,R_LC)$$

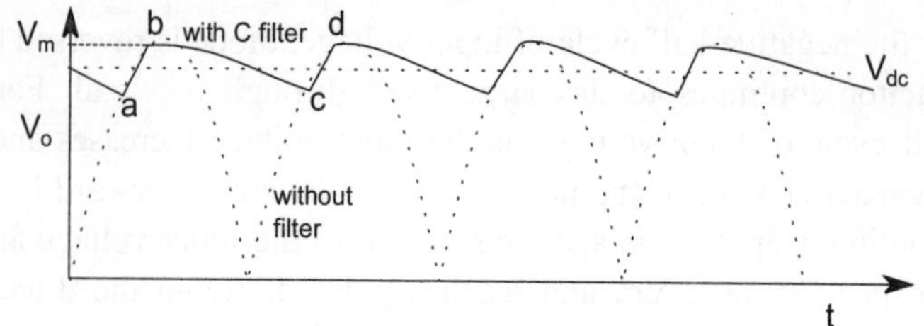

Fig. 3.2.4 Output waveform

Zener Diode

A zener diode is a highly doped p-n junction device with a very narrow junction that allows flow of current in both the directions *i.e.* forward as well as reverse. They behave same as ordinary p-n junction diode when forward biased.

When reverse biased, initially small leakage current flows through the device but as the reverse voltage is increased beyond the breakdown voltage known as zener voltage, zener breakdown occurs. This is due to the fact that the zener diode is heavily doped. As a result, the potential barrier and depletion width are reduced and high electric field is set up. This results in tunneling of carriers which give rise to sudden increase in reverse current.

The zener diode may be of silicon or germanium but silicon is preferred because of higher operating temperature and current capability. Moreover, the knee point in case of silicon diodes is sharper as compared to germanium diodes.

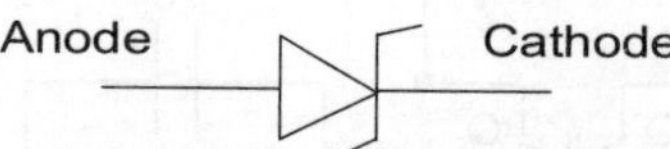

Fig. 3.2.5 Symbol of zener diode

I-V characteristics of zener diode

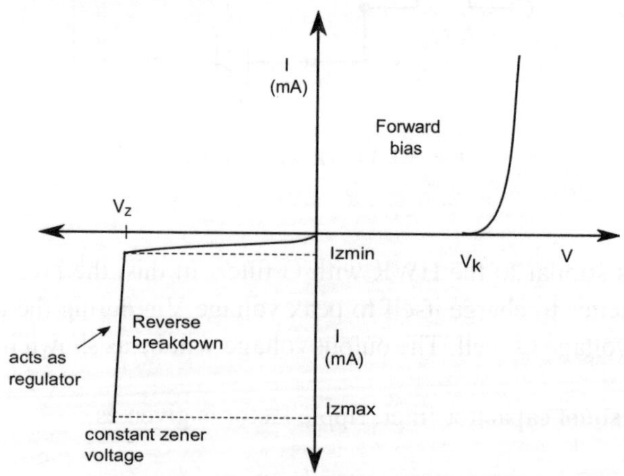

Fig. 3.2.6 I-V characteristics of zener diode

Zener Diode as Voltage Regulator

After rectification and filtration, a ripple free dc output is produced. But as the load resistance or input ac voltage changes, the output also changes which is undesirable. To make the dc output constant irrespective of changes in line voltage or load voltage, zener diode is used in parallel with the load. This is called as regulated power supply.

I-V characteristics of zener diode as shown in Fig. 3.2.6 show that in the reverse breakdown region, the voltage Vz across diode remains almost constant even if current flowing through it changes from Izmin to Izmax. This region can be used to maintain a constant voltage across the load even if there are any variations in the current. The minimum current Izmin should be maintained through the diode so that diode operates in reverse breakdown region. If current through the device is more than Izmax, the device may get damage. Hence, to avoid this situation an external resistance Rs is used in series with zener diode to limit the maximum current flowing through it.

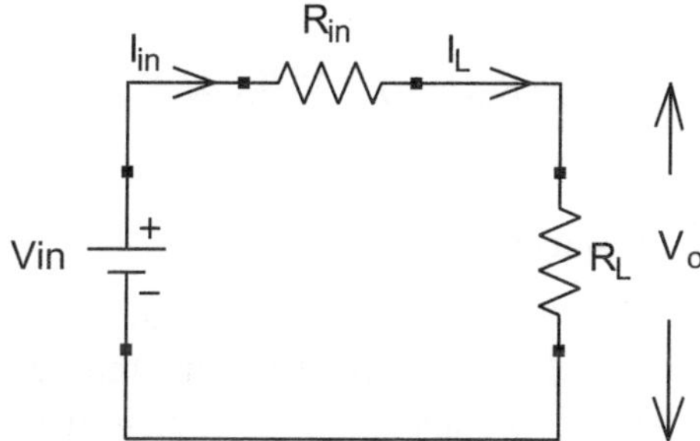

Fig. 3.2.7 Unregulated power supply

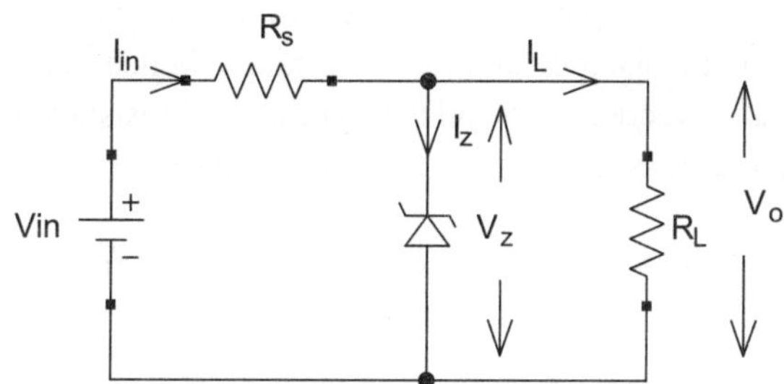

Fig. 3.2.8 Zener diode as voltage regulator

Where,

V_{in}: unregulated input voltage.

R_S: series resistance to limit maximum current flowing through the device.

I_L: current flowing through the load resistance.

V_Z: voltage across zener diode.

I_Z: current flowing through zener diode.

R_{in}: internal resistance of unregulated power supply

Izmax: maximum allowable current through zener diode

Zener diode is connected in parallel with the load as shown in Fig. 3.2.8. It acts as a voltage regulator *i.e.* it maintains constant voltage across the load irrespective of the variations in the load resistance or line voltage.

Series resistance

It is used to limit the maximum current flowing through the zener diode and its value should be greater than Rs (min) which is given as

$$R_s(min) = (V_{in} - V_z)/I_{zmax}$$

Load resistance

R_L should be more than R_L (min) so that zener diode operates in breakdown region and is given as

$$R_L(min) = V_z/I_{zmax}$$

Working

1. Variation in load resistance

    a. Without zener diode: Consider the circuit of unregulated power supply as shown in Fig. 3.2.7. The current flowing through the load is equal to the current flowing through the internal resistance *i.e.* $I_L = I_{in}$. As the load resistance increases, the current flowing through the circuit decreases and hence voltage drop across internal resistance also decreases. Since, input voltage is equal to the sum of voltage drop across internal resistance and load resistance; hence voltage drop across load resistance increases. Similarly, as the load resistance decreases, the voltage drop across load resistance also decreases.

    $$R_L \uparrow \rightarrow I_L = I_{in} \downarrow \rightarrow V_{Rin} \downarrow \rightarrow \text{Since } V_{in} = V_{Rin} + V_L \rightarrow \therefore V_L \uparrow$$

    Similarly,

    $$R_L \downarrow \rightarrow I_L = I_{in} \uparrow \rightarrow V_{Rin} \uparrow \rightarrow \text{Since } V_{in} = V_{Rin} + V_L \rightarrow \therefore V_L \downarrow$$

    Hence, output voltage across the load is not constant and it varies as the load changes.

    b. With zener diode: Consider the circuit as shown in Fig. 3.2.8. As load resistance decreases, load current I_L increases. Since, $I_{in} = I_z + I_L$, to maintain the constant input current I_{in} and constant voltage drop across source resistance R_S, current through zener diode I_Z reduces by the same amount. Similarly, as the load resistance increases, load current decreases. To maintain the constant input current I_{in} and constant voltage drop across source resistance R_S, current through zener diode I_Z increases by the same amount. Hence, zener diode absorbs the change in current and provides constant voltage across the load.

2. Variation in input voltage or line voltage

   a. Without zener diode: Consider the circuit of unregulated power supply as shown in Fig. 3.2.7. As the input voltage increases, the current flowing through the circuit also increases. This results in increase in the voltage drop across the load resistance. Similarly as the input voltage decreases, output voltage also decreases. Hence, with the fluctuations in the input voltage, output voltage also changes.

$$V_{in} \uparrow \rightarrow I_L = I_{in} \uparrow \rightarrow V_L \uparrow$$

   Similarly,

$$V_{in} \downarrow \rightarrow I_L = I_{in} \downarrow \rightarrow V_L \downarrow$$

   b. With zener diode: Consider the circuit as shown in Fig. 3.2.8. As the input voltage V_{in} increases, current flowing through the circuit I_{in} also increases. Since, $I_{in} = I_Z + I_L$, zener diode passes the extra current through it maintaining the same voltage V_Z across it. Similarly, as the input voltage V_{in} decreases, I_{in} also decreases. Zener diode passes less current through it maintaining the same voltage V_Z across it. Hence, the load current is not affected by the change in the input voltage and this provides the constant voltage across the load.

CALCULATIONS

Voltage Regulator

1. Set input voltage $V_{in} = 5V$.

2. Note the following values for zener diode from the data sheet.

$$Vz = \dots.$$

$$Izmax = \dots.$$

3. Find Rsmin and $R_L$min

$$R_S \text{ (min)} = (V_{in} - V_z)/I_{zmax} = \dots.$$

$$R_L(\text{min}) = V_z/I_{zmax} = \dots.$$

PROCEDURE

Shunt Capacitor Filter

1.   Connect the circuit as shown in Fig. 3.2.1 on breadboard.

2.   Connect the channel 1 of CRO across the secondary of transformer.

3.   Connect the channel 2 of CRO across the load resistance without connecting the capacitor.

4.   Take the traces of the input and output waveforms.

5.   Now, connect capacitor in parallel with the load resistance.

6.   Keeping load resistance constant, trace the output waveform for different values of capacitors.

7.   Measure ac and dc voltage across the load by using CRO.

8.   Keeping capacitor constant, trace the output waveforms for different values of load resistances.

9.   Measure ac and dc voltage across the load by using CRO.

10. Calculate the ripple factor using the formula.

11. Now, connect the circuit shown in Fig. 3.2.3 on breadboard and repeat the above steps.

Voltage Regulator

1.   Connect the circuit as shown in Fig. 3.2.8 on breadboard.

2.   Connect the variable dc power supply to give the input voltage V_{in}.

3.   Keeping the load resistance constant (= R_{Lmin}), increase the input voltage and note down the corresponding output voltage across the load using multimeter.

4.   Increase the input voltage till the time output voltage becomes constant.

5.   Plot the graph between input and output voltage.

6.   Now, set the input voltage greater than V_Z and connect resistance box for load resistance.

7.   Keeping the input voltage constant, increase the load resistance R_L starting from value less than R_L min and note down the corresponding output voltage across the load using multimeter.

8.   Increase the value of load resistance till the time output voltage becomes constant.

9.   Plot the graph between load resistance and output voltage.

OBSERVATION TABLE

HWR with C Filter

Load Resistance R_L= ...., Frequency f = ...

Table 3.2.1 Observation table for HWR with C filter for different values of capacitors

S.No.	C (μF)	Vac (V)	Vdc (V)	Ripple factor (theoretical) $r = 1/(2\sqrt{3}\, f\, R_L C)$	Ripple factor (practical) Vac / Vdc
1.	0.1				
2.	0.01				
3.	0.001				
4.	0.047				

Capacitance C = ...

Table 3.2.2 Observation table for HWR with C filter for different values of load resistances

S.No.	R_L (kΩ)	Vac (V)	Vdc (V)	Ripple factor (theoretical) $r = 1/(2\sqrt{3}\, f\, R_L C)$	Ripple factor (practical) Vac/Vdc
1.	1				
2.	10				
3.	50				
4.	100				

FWR with C Filter

Load Resistance R_L= ....

Frequency f = ...

Table 3.2.3 Observation table for FWR with C filter for different values of capacitors

S.No.	C (μF)	Vac (V)	Vdc (V)	Ripple factor (theoretical) $r = 1/(4\sqrt{3}\, f\, R_L C)$	Ripple factor (practical) Vac / Vdc
1.	0.1				
2.	0.01				
3.	0.001				
4.	0.047				

Capacitance C = ...

Table 3.2.4 Observation table for FWR with C filter for different values of load resistances

S.No.	R_L (kΩ)	Vac (V)	Vdc (V)	Ripple factor (theoretical) $r = 1/(4\sqrt{3}\, f\, R_L C)$	Ripple factor (practical) Vac/Vdc
1.	1				
2.	10				
3.	50				
4.	100				

Voltage Regulator

Series resistance Rs = ...

Load resistance R_L ($= R_{Lmin}$) = ...

Table 3.2.5 Observation table for voltage regulator for constant load resistance

S. No.	Input voltage Vin (V)	Output voltage Vo (V)
1.		
2.		
3.		
4.		

Series resistance Rs = ...

Input voltage Vin ($> V_Z$) = ...

Table 3.2.6 Observation table for voltage regulator for constant input voltage

S. No.	Load resistance R_L (Ω)	Output voltage Vo (V)
1.		
2.		
3.		
4.		

OBSERVATIONS

Attach the traces of input and output waveforms for HWR and FWR with C filter.

GRAPH

HWR and FWR with C Filter

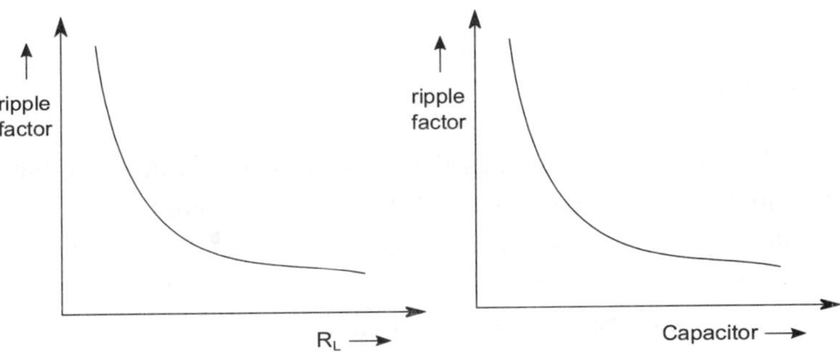

Fig. 3.2.9 Variation of ripple factor with load resistance and capacitor

Voltage Regulator

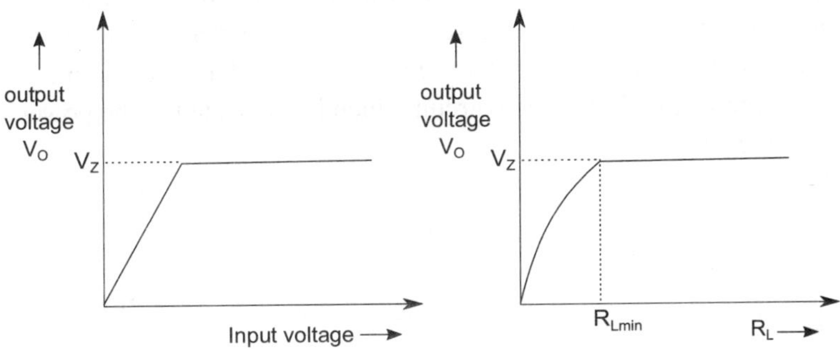

Fig. 3.2.10 Variation of output voltage with input voltage and load resistance

RESULT

Filter

HWR and FWR with shunt capacitor filter have been studied and traces have been taken successfully and it has been seen that as we increase the value of load resistance or capacitor, the ripple factor decreases. Ripple factor in case of FWR with C filter comes out to be small as compared to HWR with C filter.

Voltage Regulator

Voltage regulator using zener diode has been studied successfully. It has been seen that initially as the value of input voltage increases, output voltage varies linearly (since zener diode is not operating in its breakdown region). But as the input voltage reaches the value which is enough to make the zener diode into breakdown region, output voltage becomes constant irrespective of the input voltage. Similarly, as the value of load resistance increases, output voltage also varies. But for R_L greater than $R_L$min, zener diode starts operating in its breakdown region and the output voltage becomes constant irrespective of the load resistance.

DISCUSSION

In case of shunt capacitor filters, as the value of load resistance or capacitance increases, ripple factor decreases *i.e.* filter action becomes better. In case of inductor filter, ripple factor is directly proportional to the load resistance. Hence, as load resistance increases, ripple factor also increases. But by using LC filters, ripple factor becomes independent of the load resistance.

Zener diodes are widely used as voltage regulators. They are available in zener voltage range from 3V to 200V. They provide good voltage regulation over a wide range of currents. But the disadvantage of using zener diode is that the output load voltage depends on the zener breakdown voltage and it can be used only with the low value of current. For higher current applications, power transistors are used along with the zener diode. Voltage regulators are also available in the form of integrated circuit which is connected in the power supply at the output of the filter.

Experiment 3

CLIPPERS

AIM

To Study Clipping Circuits.

APPARATUS REQUIRED

Breadboard, diode IN4007, resistor - 1kΩ, cathode ray oscilloscope, function generator, probe, connecting wires, 2V dc power supply.

THEORY

Clipper Circuit

Clipping circuits are the wave shaping circuits which are used to remove some portion of the signal. They are also called as voltage/current limiters, amplitude selectors or slicers. It consists of diode and a resistor. Role of resistor is used to limit the current flowing through the diode when it is forward biased.

Positive Clippers

When the positive cycle of the waveform is clipped, then it is called as positive clipper.

Unbiased clipper i.e. without external supply

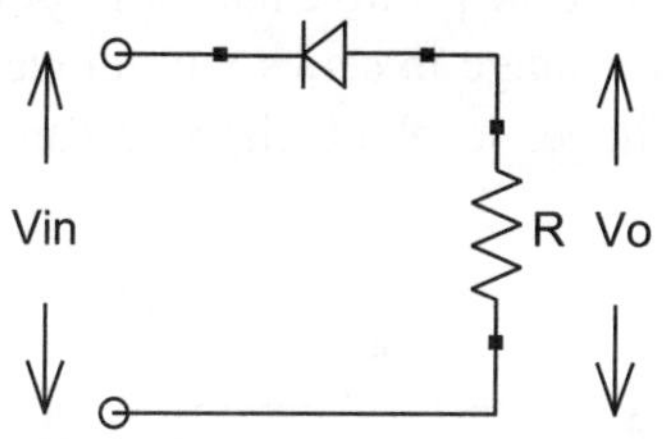

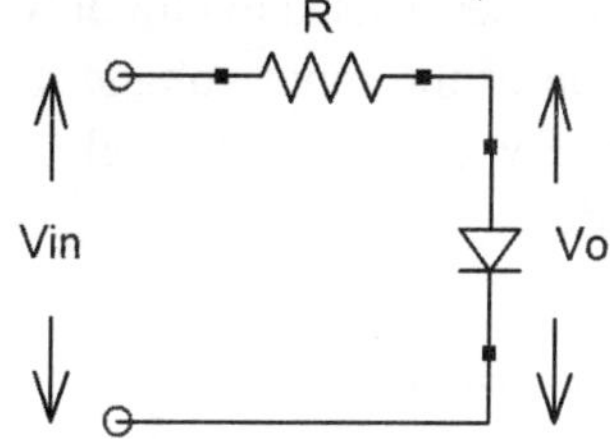

Fig. 3.3.1 Series unbiased positive clipper   **Fig. 3.3.2** Shunt unbiased positive clipper

Series unbiased positive clipper

During the positive half cycle of input voltage, diode is reversed biased and does not conduct. Hence, it acts as an open switch with infinite resistance. Therefore, no current flows through the circuit and thus output voltage across the resistance R is zero.

During the negative half cycle of input voltage, diode is forward biased and conducts. Hence, it acts as a short with a very low resistance (ideally zero) and thus a very small (negligible) voltage drop appears across it (0.3V for Ge diode and 0.7V for Si diode). Therefore, output voltage across the resistance R is same as the input voltage.

Shunt unbiased positive clipper

During the positive half cycle of input voltage, diode is forward biased and conducts. Hence, it acts as a short with a very low resistance and thus a very small (negligible) voltage drop appears across it. Therefore, output voltage across the diode is zero (ideally).

During the negative half cycle of input voltage, diode is reversed biased and does not conduct. Hence, it acts as an open switch with infinite resistance. Therefore, no current flows through the circuit and voltage across resistance R is zero and hence whole of the input voltage appears across the reversed biased diode. Therefore, output voltage is same as the input voltage.

Input and output waveform

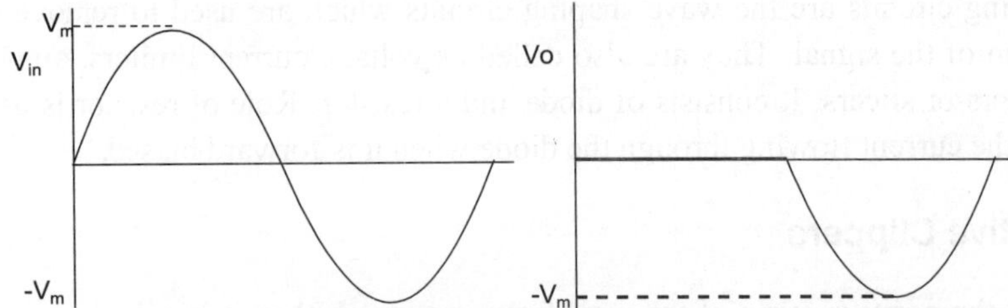

Fig. 3.3.3 Input and output voltage

Biased clipper i.e. with external power supply

Biased clipper is used to clip only a small portion of the positive half or negative half cycle of the input voltage. It consists of dc voltage in series with diode or resistor. By changing the value of biasing voltage, level of clipping can be adjusted.

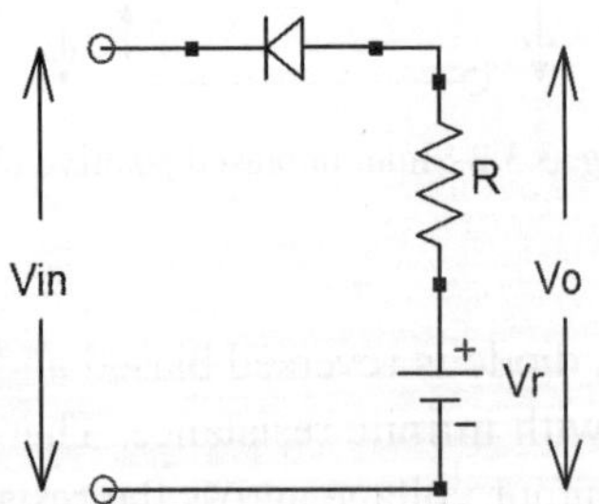

Fig. 3.3.4 Series biased positive clipper

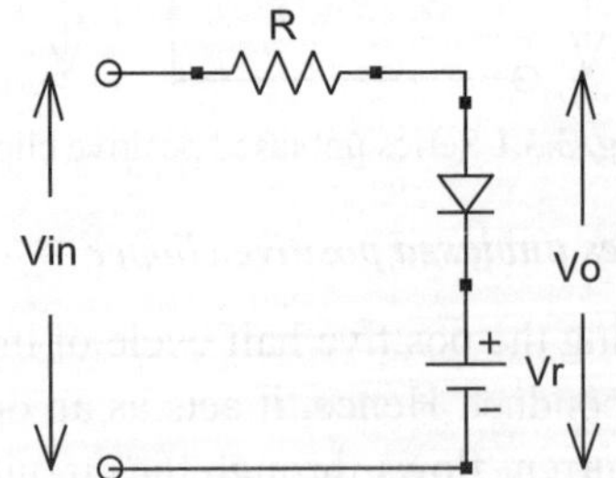

Fig. 3.3.5 Shunt biased positive clipper

Series biased positive clipper

When input voltage Vin is greater than Vr, diode is reversed biased and does not conduct. Hence, it acts as an open switch with infinite resistance. Therefore, no current flows through the circuit and thus voltage across resistance R is zero. Hence, output voltage which is equal to the sum of voltage Vr and voltage across resistance R is Vr only.

When input voltage Vin is less than Vr, diode is forward biased and conducts. Hence, it acts as a short with a very low resistance (ideally zero) and thus a very small (negligible) voltage drop appears across it. Therefore, the complete voltage (Vin-Vr) appears across resistance R and hence output voltage which is equal to sum of voltage across resistance (Vin-Vr) and voltage Vr is the input voltage only.

Shunt biased positive clipper

When input voltage Vin is greater than Vr, diode is forward biased and conducts. Hence, it acts as a short with a very low resistance and thus a very small (negligible) voltage drop appears across it. Hence, output voltage which is equal to the sum of voltage Vr and voltage across diode is Vr only.

When input voltage Vin is less than Vr, diode is reversed biased and does not conduct. Hence, it acts as an open switch with infinite resistance. Therefore, no current flows through the circuit and voltage across resistance R is zero and hence, whole of the input voltage appears across the reverse biased diode. Therefore, output voltage is same as the input voltage.

Input and output waveform

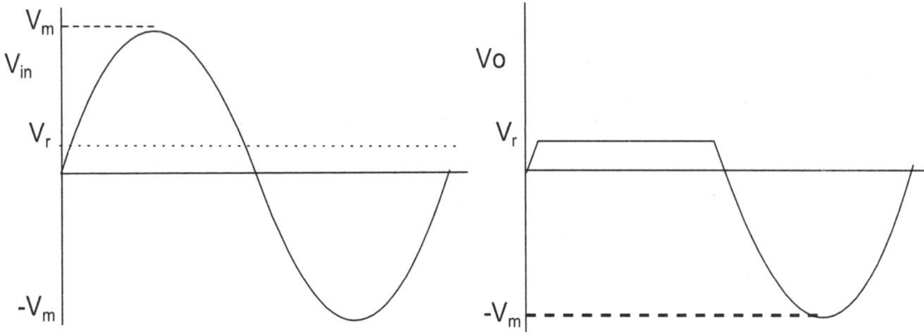

Fig. 3.3.6 Input and output voltage

Negative Clippers

When the negative cycle of the waveform is clipped, then it is called as negative clipper.

Unbiased clipper i.e. without external supply

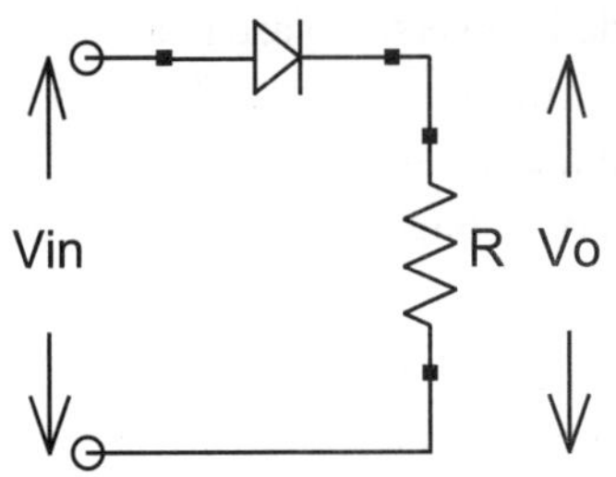

Fig. 3.3.7 Series unbiased negative clipper

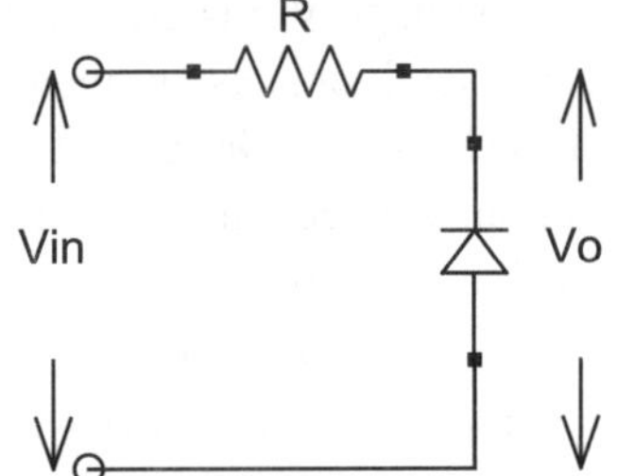

Fig. 3.3.8 Shunt unbiased negative clipper

Series unbiased negative clipper

During the positive half cycle of input voltage, diode is forward biased and conducts. Hence, it acts as a short with a very low resistance (ideally zero) and thus a very small (negligible) voltage drop appears across it (ideally zero but practically 0.3V for Ge diode and 0.7V for Si diode). Therefore, output voltage across the resistance R is same as the input voltage.

During the negative half cycle of input voltage, diode is reversed biased and does not conduct. Hence, it acts as an open switch with infinite resistance. Therefore, no current flows through the circuit and thus output voltage across the resistance R is zero.

Shunt unbiased negative clipper

During the positive half cycle of input voltage, diode is reversed biased and does not conduct. Hence, it acts as an open switch with infinite resistance. Therefore, no current flows through the circuit and voltage across resistance R is zero and hence, whole of the input voltage appears across the reversed biased diode. Therefore, output voltage is same as the input voltage.

During the negative half cycle of input voltage, diode is forward biased and conducts. Hence, it acts as a short with a very low resistance and thus a very small (negligible) voltage drop appears across it. Therefore, output voltage across the diode is zero.

Input and output waveform

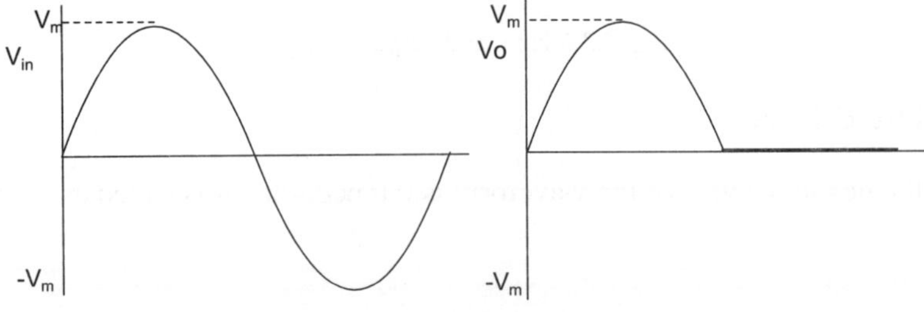

Fig. 3.3.9 Input and output voltage

Biased clipper i.e. with external power supply

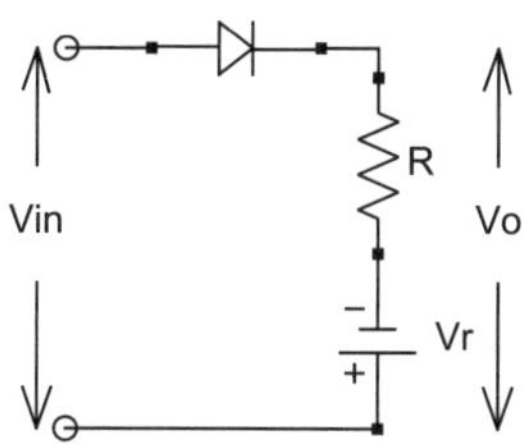

Fig. 3.3.10 Series biased negative clipper

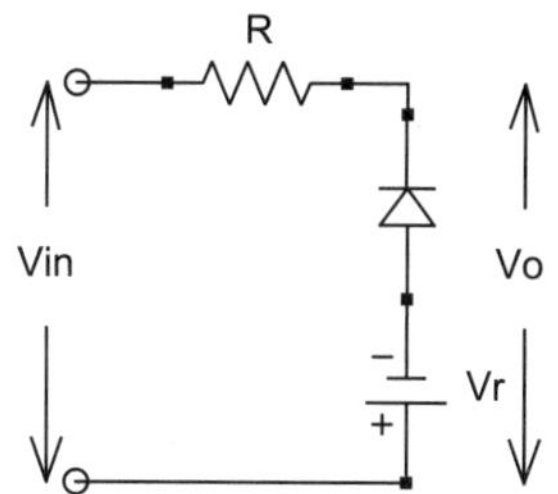

Fig. 3.3.11 Shunt biased negative clipper

Series biased negative clipper

When input voltage Vin is greater than -Vr, diode is forward biased and conducts. Hence, it acts as a short with a very low resistance (ideally zero) and thus a very small (negligible) voltage drop appears across it. Therefore, the complete voltage (Vin+Vr) appears across resistance R and hence output voltage which is equal to sum of voltage across resistance (Vin+Vr) and voltage −Vr is the input voltage only.

When input voltage Vin is less than -Vr, diode is reversed biased and does not conduct. Hence, it acts as an open switch with infinite resistance. Therefore, no current flows through the circuit and thus voltage across the resistance R is zero. Hence, output voltage which is equal to the sum of voltage -Vr and voltage across resistance R is -Vr only.

Shunt biased negative clipper

When input voltage Vin is greater than -Vr, diode is reversed biased and does not conduct. Hence, it acts as an open switch with infinite resistance. Therefore, no current flows through the circuit and voltage across resistance R is zero and hence, whole of the input voltage appears across the reversed biased diode. Therefore, output voltage is same as the input voltage.

When input voltage Vin is less than -Vr, diode is forward biased and conducts. Hence, it acts as a short with a very low resistance and thus a very small (negligible) voltage drop appears across it. Hence, output voltage which is equal to the sum of voltage -Vr and voltage across diode is -Vr only.

Input and output waveform

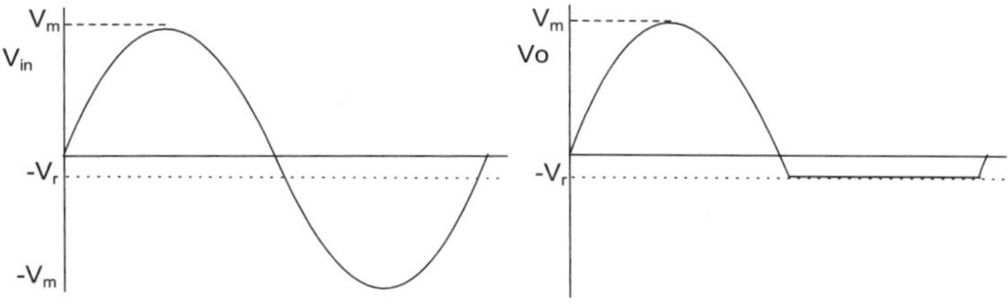

Fig. 3.3.12 Input and output voltage

PROCEDURE

1. Connect the clipper circuit as shown in Fig. 3.3.1 on breadboard.
2. Take R = 1kΩ.
3. Connect function generator to give sinusoidal input using a probe.
4. Connect channel 1 of CRO to the input using a probe.
5. Connect channel 2 of CRO to the output.
6. Take the traces of input and output waveforms on CRO.
7. Change the amplitude of input signal and observe the output.
8. Take the traces for triangular as well as square wave input.
9. Now, connect the circuits shown in Fig. 3.3.2, 3.3.4, 3.3.5, 3.3.7, 3.3.8, 3.3.10, 3.3.11 on breadboard and repeat the above steps.

OBSERVATIONS

Attach the traces of input and output waveforms.

RESULT

Positive and negative clippers have been studied successfully.

DISCUSSION

While transmitting a message signal from transmitter to receiver, the message signal is greatly affected by noise present in the communication channel. Therefore, clipper circuits are used in communication receivers to remove the noise present in the incoming message signal.

Clipper circuits are also used in radars, digital computers and radio and television receivers. They are also used to convert sine wave into a square wave by clipping the sine wave from both halves. Half wave rectifiers also act as clipping circuit since it removes negative half of the cycle.

Experiment 4

CLAMPERS

AIM

To Study Clamping Circuits.

APPARATUS REQUIRED

Breadboard, diode IN4007, resistor - 10kΩ, capacitor – 0.1uF, cathode ray oscilloscope, probe, function generator, connecting wires.

THEORY

Clamper

Clampers are used to add dc component in the waveform without changing its shape and amplitude. They are also called as dc restorers or level shifters. A clamper circuit consists of capacitor, diode and a resistor.

Unbiased negative clamper

Negative unbiased clamper as shown in Fig. 3.4.1 clamps the input waveform towards the negative level.

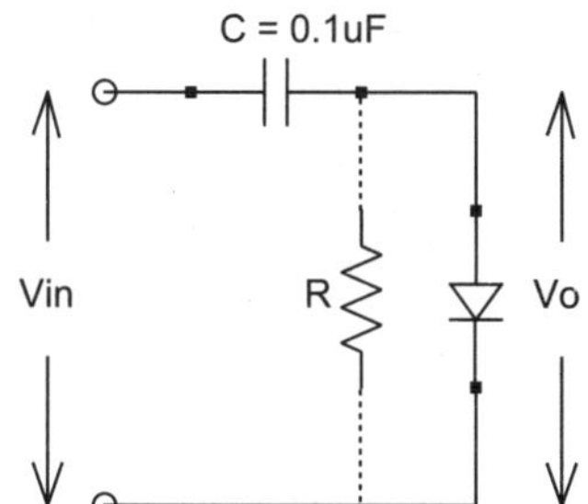

Fig. 3.4.1 Unbiased negative clamper

Working

During the positive half cycle of input voltage *i.e.* for the time interval between 0 and T/2, diode is forward biased and conducts. Hence, it acts as a short with a very low resistance (ideally zero). Hence, resistance R is also effectively shorted out by the conducting diode. Due to this, the time constant RC becomes very small and hence, capacitor C charges quickly to the peak value of the input voltage Vm with negative polarity on the right side as shown in Fig. 3.4.2. Since, diode acts as a short thus output voltage across diode is zero.

$$V_c = V_m \qquad \text{– eqn 1}$$

And,

$$V_o = 0$$

Where,

Vc: voltage across capacitor, Vm: peak input voltage.

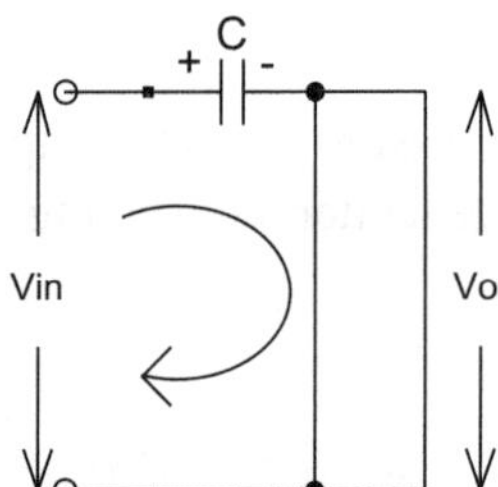

Fig. 3.4.2 Negative clamper for positive half cycle of input voltage

During the negative half cycle of input voltage *i.e.* for the time interval between T/2 and T, diode is reverse biased and does not conduct. Hence, it acts as an open switch with infinite resistance as shown in Fig. 3.4.3. Due to this, R and C get connected in series and there is no path for capacitor to discharge other than resistance R. But the time constant RC is large enough which ensures that voltage across capacitor does not discharge quickly. Hence, capacitor tries to retain the charge during the negative half cycle and the output voltage is sum of input voltage and voltage stored in capacitor.

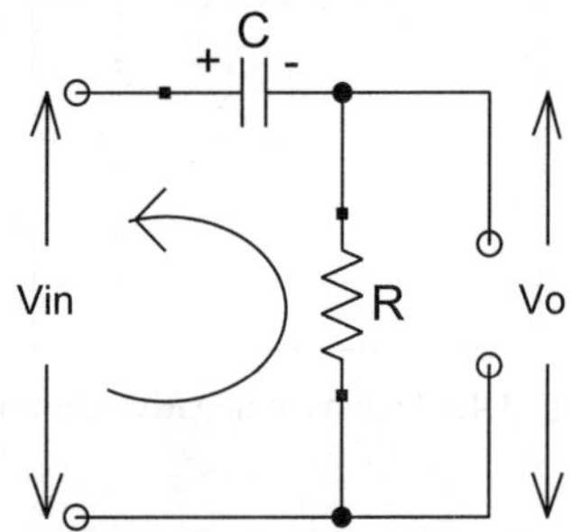

Fig. 3.4.3 Negative clamper for negative half cycle of input voltage

Applying the Kirchhoff's voltage law in the loop, we get

$$V_{in} = V_c + V_o$$

$$\Rightarrow \qquad V_o = V_{in} - V_c$$

Since, peak value of input voltage for negative half cycle is $-V_m$ *i.e.* $V_{in} = -V_m$.

$$\Rightarrow \qquad V_o = -V_m - V_c$$

Using equation 1, we get

$$\Rightarrow \qquad V_o = -V_m - V_m$$

$$\Rightarrow \qquad V_o = -2V_m$$

Input and output waveforms

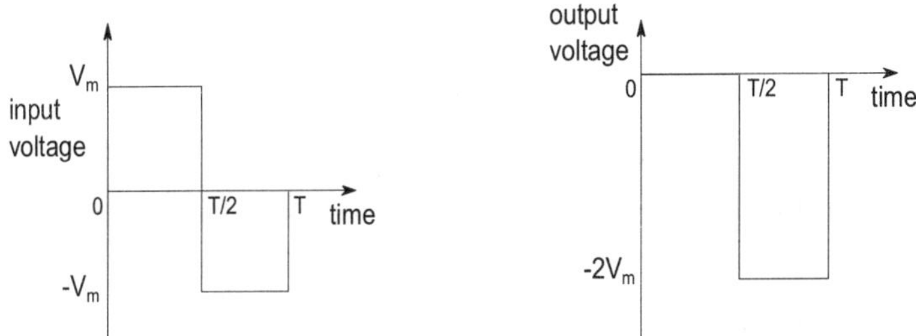

Fig. 3.4.4 Input and output voltage

Unbiased positive clamper

Positive unbiased clamper as shown in Fig. 3.4.5 clamps the input waveform towards the positive level.

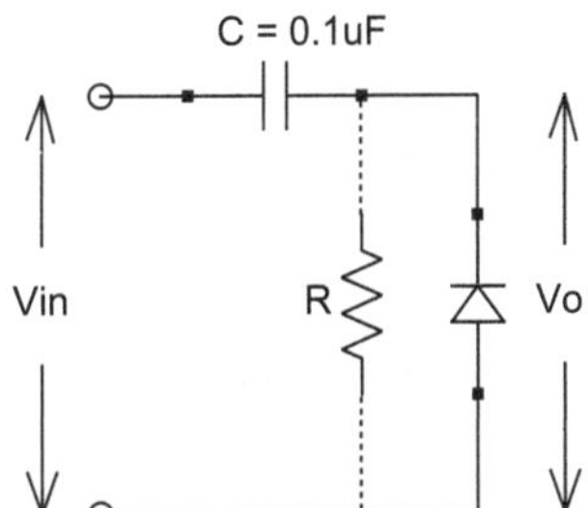

Fig. 3.4.5 Unbiased positive clamper

Working

During the negative half cycle of input voltage *i.e.* for the time interval between T/2 and T, diode is forward biased and conducts. Hence, it acts as a short with a very low resistance (ideally zero). Hence, resistance R is also effectively shorted out by the conducting diode. Due to this, the time constant RC becomes very small and hence, capacitor C charges quickly to the peak value of the input voltage Vm with negative polarity on the left side as shown in Fig. 3.4.6. Since, diode acts as a short thus output voltage across diode is zero.

$$V_c = -V_m \qquad\qquad -\text{eqn 2}$$

And

$$V_o = 0$$

Where,

Vc: voltage across capacitor, Vm: peak input voltage.

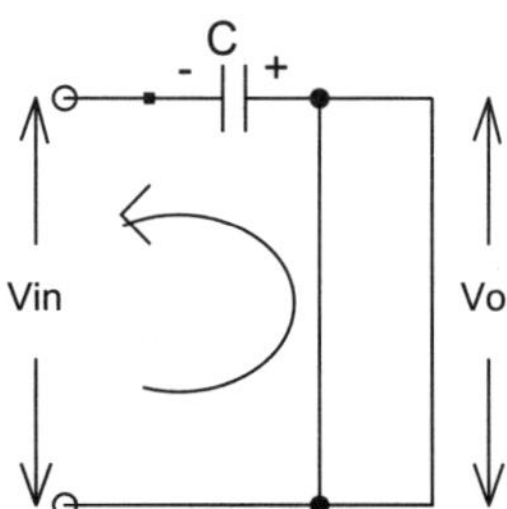

Fig. 3.4.6 Positive clamper for negative half cycle of input voltage

During the next positive half cycle of input voltage *i.e.* for the time interval between T and 3T/2, diode is reversed biased and does not conduct. Hence, it acts as open switch with infinite resistance as shown in Fig. 3.4.7. Due to this, R and C get connected in series and there is no path for capacitor to discharge other than resistance R. But the time constant RC is large enough which ensures that voltage across capacitor does not discharge quickly. Hence, capacitor tries to retain the charge during the positive half cycle and the output voltage is sum of input voltage and voltage stored in capacitor.

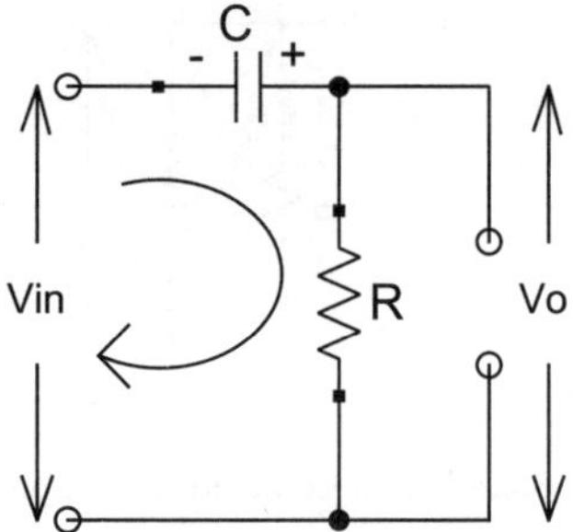

Fig. 3.4.7 Positive clamper for positive half cycle of input voltage

Applying the Kirchhoff's voltage law in the loop, we get

$$V_{in} = V_c + V_o$$

$$\Rightarrow \qquad V_o = V_{in} - V_c$$

Since, peak value of input voltage for positive half cycle if Vm *i.e.* $V_{in} = V_m$.

$$\Rightarrow \qquad V_o = V_m - V_c$$

Using equation 2, we get

$$\Rightarrow \qquad V_o = V_m + V_m$$

$$\Rightarrow \qquad V_o = 2V_m$$

Input and output waveform

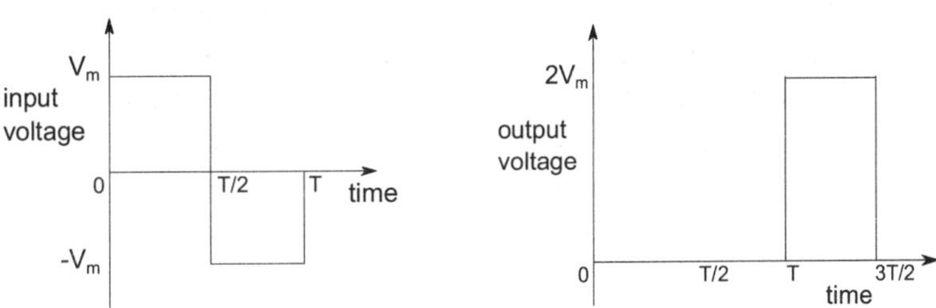

Fig. 3.4.8 Input and output voltage

CALCULATIONS

Let $R = 10k\Omega$ and $C = 0.1\mu F$

Time Constant $RC = 10k\Omega * 0.1\mu F = 1ms$

Therefore, capacitor takes $5RC = 5ms$ time to discharge completely.

Let frequency of input signal $= 10kHz$

Since, time period of input signal is $0.1ms$ which is much less than discharge time of capacitor; therefore, capacitor will not be able to discharge itself and will retain its charge with the same polarity during the time when diode is non-conducting.

PROCEDURE

1.   Connect the circuit as shown in Fig. 3.4.1 on breadboard.
2.   Connect function generator to give square wave input using a probe.
3.   Connect channel 1 of CRO across the input terminals.
4.   Connect channel 2 of CRO across the diode to view the output.
5.   Take the traces of input and output waveforms.
6.   Now, connect the circuit shown in Fig. 3.4.5 on breadboard and repeat the above steps.

OBSERVATIONS

Attach the traces of input and output waveforms.

RESULT

Positive and negative clampers have been studied successfully.

DISCUSSION

They are widely used as dc restorers in TV receivers. To process the incoming video signals, TV receivers have the capacitive coupled amplifier. Due of the presence of capacitor, dc components are lost which leads to the loss of black and white reference levels and the blanking levels as well. Clampers circuits are used to restore back these reference levels by adding dc component before the video signal is applied to the picture.

Experiment 5

BIASING CIRCUITS

AIM

To Study Fixed Bias, Voltage Divider and Collector-to-Base Bias Feedback Configuration for Transistors.

APPARATUS REQUIRED

Breadboard, BJT - BC108/BC547, resistors, connecting wires, 12V dc power supply, multimeter.

THEORY

Bipolar Junction Transistor

It is a three terminal device with terminals named as emitter (E), base (B) and collector (C). The term 'bipolar' signifies that two types of carriers *i.e.* electrons and holes are involved in the flow of current through the transistor, the term 'junction' indicates that it has two junctions known as emitter base junction and base collector junction and the term 'transistor' is a combination of two words *i.e.* 'transfer + resistor' which means current is transferred from region of 'low resistance' to 'high resistance'.

It has three regions of operation: cut off, saturation and active region. When the transistor is made to operate between cut off and saturation, it acts as a switch which is the basis of digital electronics and when the transistor is made to operate in active region, it works as an amplifier which is the basis of analog electronics.

Based on construction, there are two basic types of bipolar transistor: npn and pnp transistor.

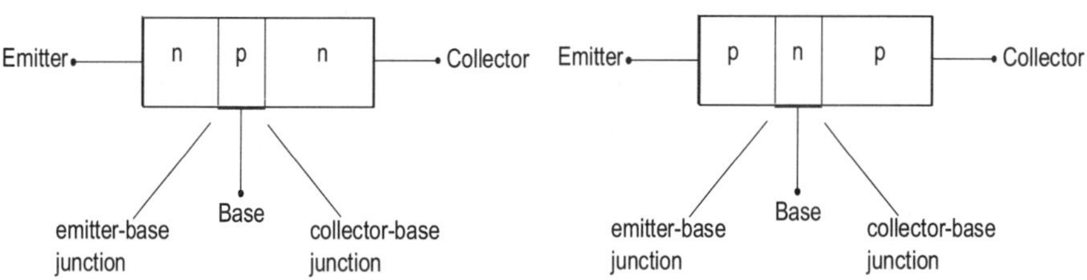

Fig. 3.5.1 Structure for npn and pnp transistor

In pnp transistor, two p-type semiconductor regions are separated by a thin n-type region and the conduction is by holes whereas in npn transistor, two n-type semiconductor regions are separated by a thin p-type region and the conduction is by electrons (Fig. 3.5.1).

In the structure of bipolar transistor, the thin lightly doped central region is called as base and on the one side, there is highly doped region called as emitter which is a source of majority carriers and on the other side, there is a collector region which is comparatively lightly doped and collects the carriers.

The input circuit *i.e.* emitter-base junction is forward biased and offers low resistance whereas the output circuit *i.e.* collector-base junction is reverse biased and offers high resistance. Therefore, a transistor transfers the input signal current from a low resistance circuit to high resistance circuit which is responsible for the amplifying action of a transistor.

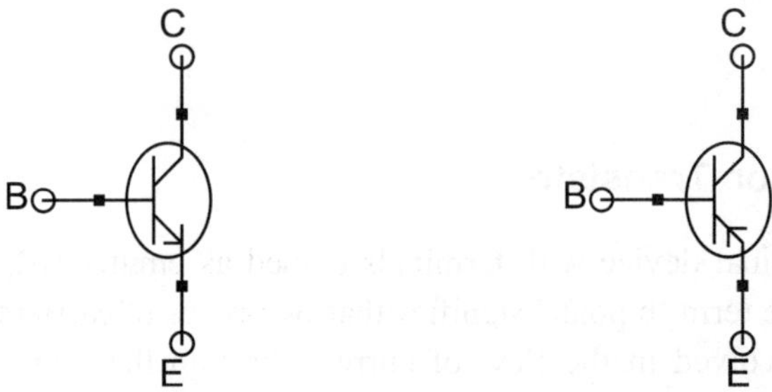

Fig. 3.5.2 Symbol for npn and pnp transistor

In the symbol shown in Fig. 3.5.2, the arrow indicates the direction of flow of conventional current between the base and emitter terminals.

Biasing of Transistor

The word biasing means application of external dc voltages to the transistor. In order to operate transistor in different region of operation, different biasing voltages are applied to the two junctions *i.e.* emitter base and collector base junction. Different operating regions are listed below.

Table 3.5.1 Different modes of operation of transistor

S.No.	Mode of operation	Emitter-base junction	Collector-base junction	Use
1.	Active	Forward Biased	Reverse Biased	Amplifier
2.	Saturation	Forward Biased	Forward Biased	Closed switch
3.	Cut off	Reverse Biased	Reverse Biased	Open switch
4.	Inverted	Reverse Biased	Forward Biased	Not used

Operating point or Quiescent point or Q point

As the external dc voltage is applied to the transistor, a certain dc collector current starts flowing in the circuit at a certain dc collector to emitter voltage. These values of current and voltages without any input signal are expressed by the term 'operating point' or 'quiescent point' or 'Q point'. In other words, quiescent point can be defined as a fixed point that represents the value of I_C and V_{CE} in transistor circuit when no input signal is applied.

Selection of operating point

Consider the output characteristics of a common emitter configuration transistor as shown in Fig. 3.5.3. Let V_{CC} be the collector voltage supply and R_C be the load resistance.

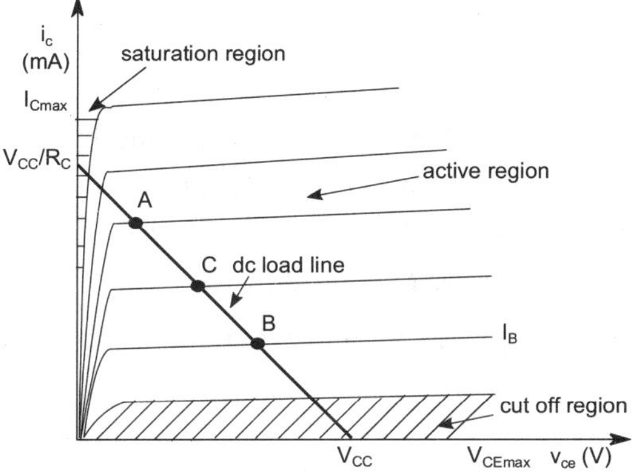

Fig. 3.5.3 Output characteristics of transistor in common emitter configuration

As shown in Fig. 3.5.3, Q point can lie anywhere on the dc load line depending on the input base current i_B. For transistor to work as an amplifier, Q point should lie in the active region only.

As the ac input signal is applied to the emitter base junction, the base current changes. Due to this, collector current as well as collector to emitter voltage also changes. Hence, Q point varies with the varying input signal. Let us consider three cases and see which point is the best operating point for transistor to work as an amplifier.

a. Point A: Biasing is done in such a way that point A becomes the operating point. As the input signal is applied, the base current changes. This results in shifting of the Q point. Since, point A lies near to the saturation region, Q point can shift to the saturation region on application of the input signal. Therefore, output signal will get clipped at the positive peaks that result in distortion of the signal. Hence, point A is not a suitable operating point.

b. Point B: Similarly, if point B becomes the operating point, then as the input signal is applied, the Q point can shift to the cut-off region. Hence, output signal will get clipped at the negative peaks. Hence, point B is not a suitable operating point.

c.   Point C: If point C becomes the operating point, it lies in the middle of the load line. As the input signal is applied, Q point also shifts. If the amplitude of input signal is very large, the Q point lying at point C can shift to cut-off and saturation region both. Hence, output signal will get clipped at the positive peaks as well as negative peaks. But if the amplitude of input signal is kept small, variation in Q point will also be small. Therefore, operating point will remain in active region only and it will produce an amplified output signal without any distortion. Hence, this is the best operating point for transistor to work as an amplifier. Since, only a small amplitude signal can be applied as the input hence, such an amplifier is called as small signal amplifier.

Need of biasing

Once the Q point is fixed at the center of load line by applying appropriate voltages, it is not necessary that the Q point will remain fixed at the same point. Operating point can shift due to the following reasons.

* As the temperature changes, Q point shifts.

* As the transistor is replaced by another transistor in the circuit, β changes and hence Q point shifts.

* Thermal runaway

   As the current flows in the collector circuit, it produces heat at the collector junction. This increases temperature which as a result generates more number of minority carriers in base-collector region. This increases the leakage current I_{CBO}. Since,

$$I_{CEO} = (1 + \beta)I_{CBO}$$

and

$$I_C = \beta I_B + I_{CEO}$$

   Therefore, as I_{CBO} increases, I_{CEO} also increases which further increases the collector current I_C. This collector current again produces more heat at the collector junction that results in further increase of temperature. As a result of this, more number of minority carriers is generated. As minority carrier increases, leakage current again increases which further increases the collector current and the whole cycle repeats again.

$I_C\uparrow \rightarrow$ temperature$\uparrow \rightarrow$ minority carriers$\uparrow \rightarrow I_{CBO}\uparrow \rightarrow I_{CEO}\uparrow \rightarrow I_C\uparrow \rightarrow$ temperature$\uparrow \rightarrow$ minority carriers$\uparrow \rightarrow I_{CEO}\uparrow \rightarrow I_C\uparrow$.

This cycle increases the collector current to a large value because of which Q point shifts to the saturation region. This excessive heat can even damage the transistor. This is known as thermal runaway.

Therefore, biasing is done for the following reasons.

- To fix Q point at the center of the load line.
- To stabilize collector current against any temperature variations.
- To make Q point independent of transistor parameters.

Biasing Circuits

These are the circuits to obtain the operating point of transistor. Different biasing circuits are:

1. Fixed bias circuit
2. Collector to base bias circuit
3. Voltage divider circuit

Fixed bias circuit

It is the simplest biasing configuration as shown in Fig. 3.5.4.

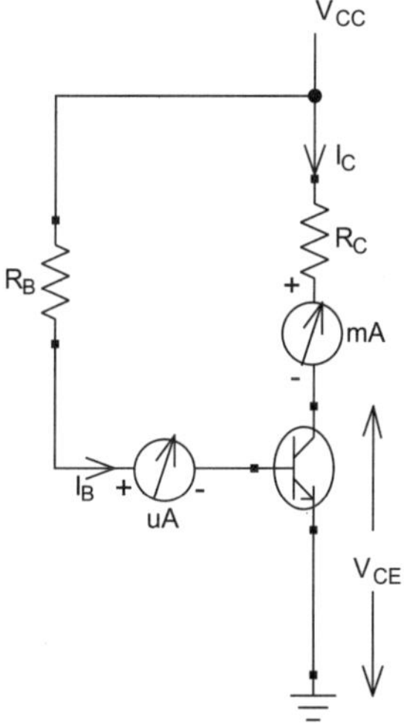

Fig. 3.5.4 Circuit for fixed bias

Base-Emitter loop

Applying Kirchhoff's voltage law to the base-emitter loop, we get

$$V_{CC} = I_B R_B + V_{BE} \qquad\qquad - \text{eqn 1}$$

Therefore, base current is

$$\Rightarrow \qquad I_B = {}^{(V_{CC} - V_{BE})}\!/_{R_B}$$

Since, V_{BE} (= 0.7V for Si transistor) is very small as compared to V_{CC},

$$\Rightarrow \qquad I_B \approx V_{CC}/R_B$$

Since, V_{CC} is fixed in the circuit and R_B is also fixed, therefore base current I_B is also fixed and hence this circuit is called as fixed bias circuit.

Collector-Emitter loop

Applying Kirchhoff's voltage law to the collector-emitter loop, we get collector-to-emitter voltage as

$$V_{CC} = I_C R_C + V_{CE} \qquad\qquad - \text{eqn 2}$$

Since, $I_C \approx I_E$

$$\Rightarrow \qquad\qquad V_{CE} = V_{CC} - I_E R_C$$

Collector current is given as

$$I_C = \beta I_B + I_{CEO}$$

Where,

I_{CEO}: leakage current, β: current gain.

Ignoring the leakage current I_{CEO}, re-writing the above equation we get

$$\Rightarrow \qquad\qquad I_C = \beta I_B$$

When transistor is in saturation,

$$I_{C(sat)} = {V_{CC}}/{R_C}$$

and

$$V_{CE(sat)} = 0$$

When transistor is in cut-off,

$$I_{C(cut-off)} = 0$$

and

$$V_{CE(cut-off)} = V_{CC}$$

Note:

Calculate the collector current I_C and this value should be less than $I_{C(sat)}$. If $I_C > I_{C(sat)}$, that means transistor is in saturation region.

Designing

1. Let $V_{CC} = 12V$ and $I_C = 2mA$.

2. Find h_{fe} of transistor using multimeter.

3. Calculate base current by using the relation

$$h_{fe} = {I_C}/{I_B}$$

4.   Calculate R_B by using equation 1.

5.   Take $V_{CE} = V_{CC}/2 = 6V$.

6.   Calculate R_C by using equation 2.

Collector-to-Base bias configuration

It is a modification over fixed bias circuit. In this, base resistance R_B is connected to the collector instead of connecting it to the dc voltage V_{CC} directly as shown in Fig. 3.5.5.

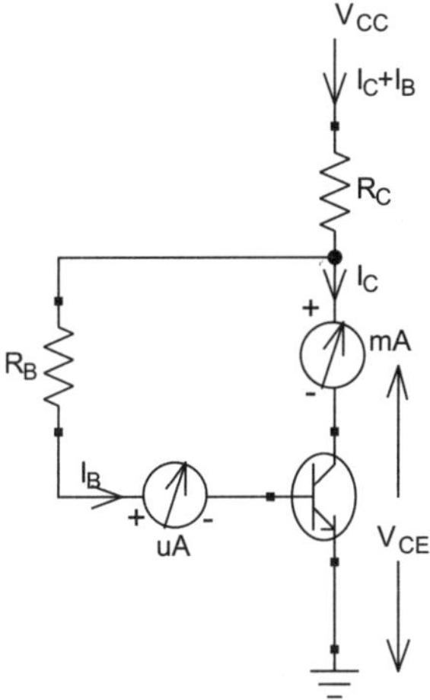

Fig. 3.5.5 Circuit for collector to base bias

Base-Emitter loop

Applying Kirchhoff's voltage law to the input circuit, we get

$$V_{CC} = (I_B + I_C)R_C + I_B R_B + V_{BE} \qquad - \text{eqn 3}$$

Since, $I_C = \beta I_B$

$\Rightarrow$
$$V_{CC} = (I_B + \beta I_B)R_C + I_B R_B + V_{BE}$$

$\Rightarrow$
$$I_B = \frac{V_{CC} - V_{BE}}{(1 + \beta)R_C + R_B}$$

Since, $\beta \gg 1$

$\Rightarrow$
$$I_B = \frac{V_{CC} - V_{BE}}{\beta R_C + R_B} \qquad - \text{eqn 4}$$

Since, V_{BE} is very small ($= 0.7V$ for Si transistor)

$\Rightarrow$
$$I_B \approx V_{CC}/(\beta R_C + R_B)$$

Collector current is given as

$$I_C = \beta I_B$$

Using equation 4,

$$\Rightarrow \qquad I_C = \frac{(V_{CC} - V_{BE})}{(R_C + R_B/\beta)}$$

$$\Rightarrow \qquad I_C \approx \frac{V_{CC}}{(R_C + R_B/\beta)}$$

Collector-Emitter loop

Applying Kirchhoff's voltage law to the output circuit, we get

$$V_{CC} = (I_C + I_B)R_C + V_{CE} \qquad\qquad - \text{eqn 5}$$

$$\Rightarrow \qquad V_{CE} = V_{CC} - (I_C + I_B)R_C$$

Since, $I_B \ll I_C$

$$\Rightarrow \qquad V_{CE} \approx V_{CC} - I_C R_C$$

Designing

1.   Let $V_{CC} = 12V$ and $I_C = 2mA$.

2.   Find h_{fe} of transistor using multimeter.

3.   Calculate base current by using the relation

$$h_{fe} = \frac{I_C}{I_B}$$

4.   Take $V_{CE} = V_{CC}/2 = 6V$.

5.   Calculate R_C by using equation 5.

6.   Calculate R_B by using equation 3.

Voltage divider bias configuration

It is the most widely used biasing circuit as shown in Fig. 3.5.6. Resistance R_1 and R_2 forms the voltage divider network, R_E is the self-biasing resistor. In this configuration, operating point is independent of β hence it is also called as 'biasing circuit independent of β'.

Collector to emitter voltage is given as

$$V_{CE} = V_{CC} - I_C(R_C + R_E)$$

Since, $I_C \approx I_E$

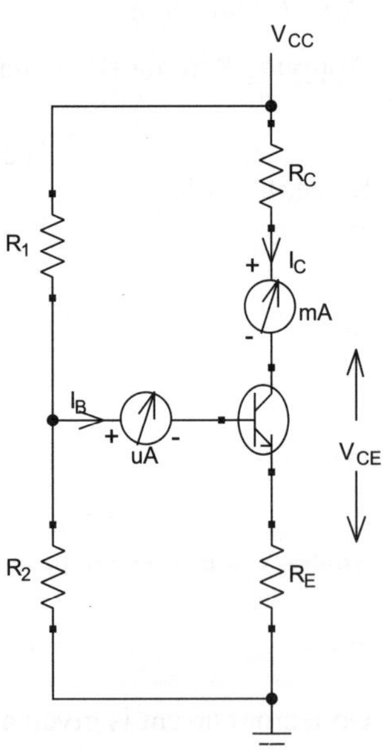

Fig. 3.5.6 Circuit for self-bias

$$\Rightarrow \qquad V_{CE} = V_{CC} - I_E(R_C + R_E) \qquad\qquad - \text{eqn 6}$$

Collector current which is equal to emitter current is given as

$$I_C = \left(\frac{V_{CC}R_2}{R_1 + R_2} - V_{BE}\right)/R_E$$

Designing

1. Let $V_{CC} = 12V$ and $I_C = 2mA$.

2. Find h_{fe} of transistor using multimeter.

3. Calculate base current I_B by using the relation

$$h_{fe} = I_C/I_B$$

4. Take $I_E \approx I_C$.

5. Take emitter voltage $V_E = 2V$.

6. Calculate emitter resistance R_E by using equation

$$V_E = I_E.R_E$$

7. Calculate base voltage as

$$V_B = V_{BE} + V_E$$

Where, $V_{BE} = 0.7V$ (Si transistor) or $0.3V$ (Ge transistor).

8. Choose resistors R_1 and R_2 such that the current through them is about 10 times the required base current.

$$R_1 = (V_{CC} - V_B)/10I_B$$
$$R_2 = V_B/9I_B$$

9. Take $V_{CE} = V_{CC}/2 = 6V$.

10. Calculate R_C by using equation 6 *i.e.*

$$V_{CC} = I_E R_C + V_{CE} + V_E$$
$$\Rightarrow \qquad R_C = (V_{CC} - V_{CE} - V_E)/I_E$$

PROCEDURE

1. Design the circuits according to the method as mentioned above.
2. Connect the circuit shown in Fig. 3.5.4 on breadboard.
3. Note the value of hfe (β) of BJT using multimeter.
4. Note the theoretical value of I_C, I_B and V_{CE}.
5. Switch on the dc power supply and measure the practical values of I_B, I_C and V_{CE} in the circuit using voltmeter and ammeter.

6. Increase the temperature of transistor by using hot water or by putting a lamp near to it and then reconnect it in the circuit.

7. Measure I_C, V_{CE} and I_B again using voltmeter and ammeter.

8. Observe the change in I_C by change in temperature.

9. Replace the transistor with different β value and repeat steps 3 to 5. Observe the change in I_C due to change in current gain.

10. Take different value of R_B and repeat steps 4 and 5. Observe the change in I_C due to change in I_B because of change in R_B.

11. Repeat the above steps for circuits shown in Fig. 3.5.5 and Fig. 3.5.6.

OBSERVATION TABLE

Table 3.5.2 Observation table for transistor at different temperature

Parameter	I_C		V_{CE}	
	Theoretical	Practical	Theoretical	Practical
temperature T_1				
temperature T_2				

Table 3.5.3 Observation table for transistors with different gain

Parameter	I_C		V_{CE}	
	Theoretical	Practical	Theoretical	Practical
$\beta_1 = ...$				
$\beta_2 = ...$				

Table 3.54 Observation table for transistor with different values of R_B

Parameter	I_C		V_{CE}	
	Theoretical	Practical	Theoretical	Practical
$R_B = ...$ kΩ				
$R_B = ...$ kΩ				

RESULT

Different biasing configurations have been studied successfully. In case of fixed bias configuration, Q point changes with change in temperature or change in transistor parameters. In collector-to-base bias configuration, Q point shifts slowly as compared to fixed bias circuit. In voltage divider circuit, Q point does not shift with change in any of the parameters.

DISCUSSION

Npn transistors are preferred over pnp transistors because of the following reasons: Electrons are the majority carriers in case of npn and they have better mobility as compared to holes; High frequency response is better in case of npn transistors.

Fixed bias is a very simple but seldom used circuit because of poor stabilization of Q point. If the transistor is replaced by another transistor with different value of β, the collector current I_C changes which results in shifting of the Q point. As temperature increases, reverse leakage current I_{CEO} increases and hence I_C also increases. Hence, with change in temperature or transistor parameters, Q point shifts. Therefore, this configuration is rarely used.

Collector-to-Base bias circuit is also called as voltage feedback circuit because of the presence of resistance R_B. It provides feedback path by connecting collector to the base. This configuration is better than the fixed bias because it provides better stabilization of Q point against changes in temperature and β. Still this circuit is not used because R_B provides dc as well as ac feedback. Dc feedback stabilizes the Q point but ac feedback results in reduction of voltage gain of the amplifier. Hence, signals will not get amplified properly.

Voltage divider circuit, also known as self-bias circuit, is the most widely used biasing circuit because of its best stabilization. The equations for I_C and V_{CE} show that they are not dependent on β. Hence, even on replacing the transistor with different value of β, the operating point will remain same. It is called as voltage divider because of the network formed by R_1 and R_2.

Experiment 6

COMMON EMITTER AMPLIFIER

AIM

To Design a Single Stage CE Amplifier for a Specific Gain.

APPARATUS

Breadboard, connecting wires, BJT - BC547, cathode ray oscilloscope, function generator, probes, resistors, capacitors, dc power supply.

THEORY

Amplifier

It is an electronic device which is capable of increasing the strength of a weak signal while preserving its shape.

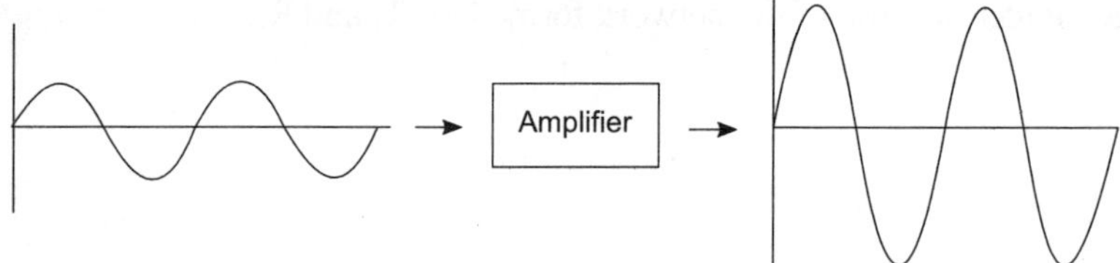

Fig. 3.6.1 Role of amplifier

BJT as Amplifier

Bipolar junction transistor is a three terminal device with three terminals named as emitter, base and collector. It has three regions of operation: cut off, saturation and active region. When the transistor is made to operate between cut off and saturation, it acts as a switch which is the basis of digital electronics and when the transistor is made to operate in active region, it works as an amplifier which is the basis of analog electronics.

Small signal amplifier

Transistors are widely used as an amplifier to amplify weak signals and in order to use transistor as an amplifier, it should be biased in active region only. Therefore, proper biasing is required to ensure that the Q point remains in the active region.

As the input signal is applied, Q point shifts. If the applied signal is large, Q point can shift to saturation and cut-off region which distorts the output signal. Therefore, the amplitude of the input signal should be small only so that Q point remains in the active region and hence such an amplifier is called as small signal amplifier.

Amplifying action

Consider an npn transistor connected in common base configuration as shown in Fig. 3.6.2. Emitter base junction is forward biased and hence offers a very low impedance ($R_i \sim 20\Omega$ to 100Ω) to input voltage and collector base junction is reverse biased and hence offers a very high impedance ($R_O \sim 100k\Omega$ to $1M\Omega$). The biasing is done to keep the transistor in active region.

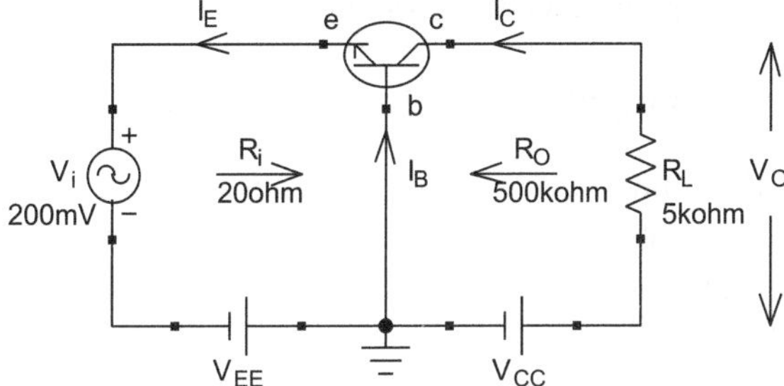

Fig. 3.6.2 Npn transistor in common base configuration

Consider an input voltage V_i connected at the input terminal. As the input voltage varies, the emitter to base voltage V_{EB} also varies. This results in variation in the emitter current I_E and is given as

$$I_E = \frac{V_i}{R_i} = \frac{200mV}{20} = 10mA$$

Since, collector current is equal to emitter current *i.e.*

$$I_C \approx I_E = 10mA$$

Therefore, collector current also varies with variation is emitter current. The output resistance of transistor is very high as compared to the load resistance (*i.e.* $R_O \gg R_L$), therefore the complete current I_C passes through the load resistance only and develops a varying voltage V_O across it. The output voltage is given as

$$V_O = I_C R_L$$

$$V_O = 10mA * 5k\Omega = 50V$$

Therefore, voltage gain A_V which is defined as the ratio of output voltage to input voltage is given as

$$A_V = {V_0}/{V_i} = {50V}/{200mV} = 250$$

The above value shows that the output voltage is amplified to a large extent by using a transistor. The reason behind the voltage amplification is the capability of transistor to transfer signals from low resistance circuit to high resistance circuit and that is why transistor name has come from the combination of two words *i.e.* transfer + resistor.

Different configurations

There are three basic configurations named as common emitter, common base and common collector in which transistor can be used as an amplifier. The most commonly used configuration is common emitter configuration because it provides both voltage gain as well as current gain.

Common Emitter Amplifier

In CE configuration, input terminals in transistor are base and emitter and the output terminals are collector and emitter. Since, emitter is made common to both input and output section, hence transistor is said to be in common emitter configuration.

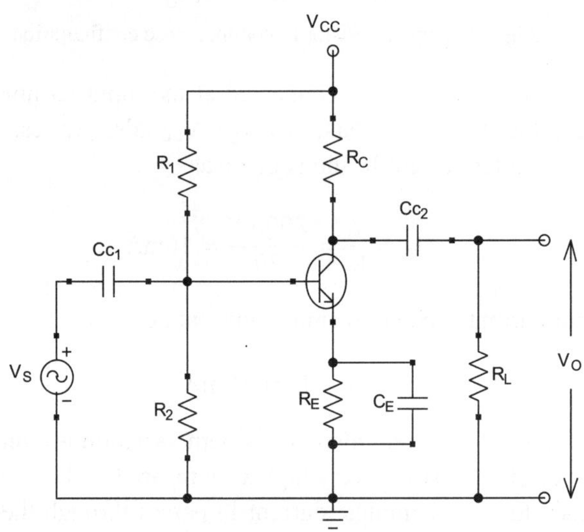

Fig. 3.6.3 Circuit for CE amplifier

CE amplifier as shown in Fig. 3.6.3 uses voltage divider biasing circuit because it provides the best stabilization of Q point against any changes in temperature or transistor's parameters. As the input signal is applied at the base of the transistor, CE amplifier amplifies the input signal and produces an amplified output 180° out of phase with respect to the input. Coupling capacitors Cc_1 and Cc_2, also known as blocking capacitors, passes the ac signal from one stage to another and blocks the dc signal. Role of resistor R_E is to provide stabilization of Q point by providing dc feedback. Since, it provides ac feedback as well which reduces the ac gain of the circuit hence, in order to eliminate this effect, bypass capacitor C_E is used in parallel with R_E. It acts as a short for ac signals and hence provides a low reactance path to the ac signal. Therefore, R_E also gets shorted due to which there is no ac feedback and hence ac gain remains same. R_L is the resistance of the load connected to the amplifier.

The voltage gain of the amplifier is given as

$$A_V = \frac{V_o(\text{ output voltage})}{V_{in}(\text{input voltage})}$$

Frequency response curve

The graph between the voltage gain and signal frequency is called as the frequency response.

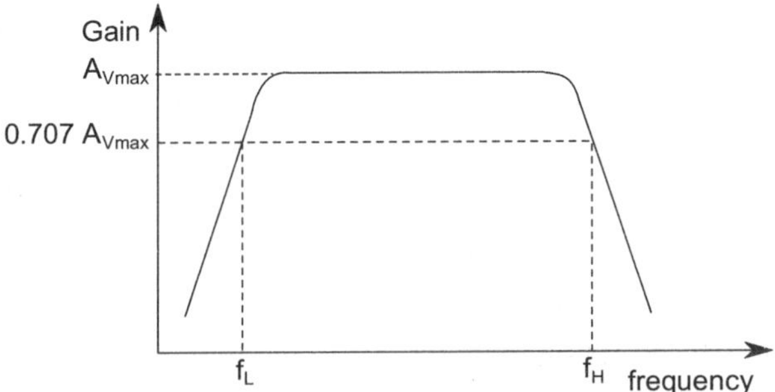

Fig. 3.6.4 Frequency response for CE amplifier

The graph shown in Fig. 3.6.4 shows the variation of voltage gain of an amplifier with signal frequency. The reason of this variation is due to the presence of different capacitors in the circuit whose reactance changes with the change in frequency. Voltage gain decreases for frequencies below f_L and above f_H and remains constant for the region between f_L and f_H.

Low frequencies

For the frequencies below lower cut off frequency *i.e.* f_L, there is a fall in the gain of an amplifier. It is because of the presence of coupling capacitors at the input and output section. At low frequency, reactance offered by the coupling capacitors is high and hence, a large amount of input voltage appears across it. Moreover, the reactance offered by bypass capacitor C_E also becomes comparable with emitter resistance R_E which reduces the bypass capacitor action. C_E will not act as short and as a result R_E will provide ac feedback as well which reduces the gain of an amplifier.

High frequencies

For the frequencies above higher cut off frequency *i.e.* f_H, there is a fall in the gain of the amplifier. It is because of the presence of intrinsic capacitance which provides the low impedance path to the high frequency ac input signals. Hence, the whole input signal does not appear across the transistor and therefore, output voltage also reduces. The effect due to coupling capacitance and bypass capacitance is almost negligible in this case because reactance offered by these capacitors is very less at high frequencies and hence negligible voltage appears across it.

Medium frequencies

For the frequencies between f_L and f_H, the effect of coupling capacitance, bypass capacitance and inter-electrode capacitance is very small. Hence, we get the maximum gain in this region. At f_L and f_H, gain is 3dB down the maximum value or 0.707 times the maximum gain.

Bandwidth

It is defined as the difference between the upper cut off and the lower cut off frequency at which gain is 3 dB down the maximum value.

DESIGNING

The foremost requirement of the transistor to act as an amplifier is that it should be biased such that the operating point lies in the middle of the active region of operation. The steps for designing of an amplifier for a specific gain A are as follows.

1.   Assume the value of load resistance (R_L), say $R_L = 10k\Omega$.

2.   Note down the value of h_{ie} from datasheet of BC 547.

3.   Let $V_{CC} = 12V$ and $I_{CQ} = 2mA$.

4.   For a given value of gain A, find R_C using the relation.

$$A = {R_C}/{r_e}$$

$\Rightarrow \qquad\qquad R_C = A.\dfrac{25mV}{I_{CQ}(mA)}$

5.   Take $V_{CEQ} = V_{CC}/2 = 6V$ and take $I_{EQ} \approx I_{CQ}$.

6.   Find the value of R_E by using the given equation.

$$V_{CC} = I_{CQ}R_C + V_{CEQ} + I_{EQ}R_E$$

$\Rightarrow \qquad\qquad V_{CC} = V_{CEQ} + I_{CQ}(R_C + R_E)$

$\Rightarrow$
$$R_E = \left(V_{CC} - V_{CEQ} - I_{CQ}R_C\right)/I_{CQ}$$

7. Calculate voltage across emitter resistance by using equation

$$V_E = I_{EQ}.R_E$$

8. Calculate base voltage or voltage across resistance R_2 as

$$V_B = V_{R2} = V_{BE} + V_E$$

Where, $V_{BE} = 0.7V$ (Si transistor) or $0.3V$ (Ge transistor).

9. Assume any value of resistance R_2 such that it is much greater than R_E. ($\sim 10R_E$).

10. Find the value of resistance R_1 by using the given equation.

$$V_{R2} = \frac{R_2}{R_1 + R_2} V_{CC}$$

$\Rightarrow$
$$R_1 = R_2 \left(\frac{V_{CC}}{V_{R2}} - 1\right)$$

11. Bypass capacitor C_E is chosen according to the lowest cut-off frequency f such that

$$R_E \gg X_{CE}$$

or
$$10X_{CE} = R_E$$

$\Rightarrow$
$$C_E = \frac{10}{2\Pi f R_E}$$

Let $f = 100Hz$, find the value of C_E by using the above relation and this gives the minimum value of bypass capacitor.

12. Coupling capacitors C_{C1} and C_{C2} are chosen such that they have little effect on amplifier frequency response. The value of these capacitors should be such that the reactance of each coupling capacitor is approximately equal to one tenth of the impedance in series of it at the lowest operating frequency for the circuit *i.e.*

$$X_{C1} = \frac{Z_i}{10}$$

$\Rightarrow$
$$\frac{1}{2\Pi f C_{C1}} = \frac{R_1 || R_2 || h_{ie}}{10}$$

$\Rightarrow$
$$C_{C1} = \dots$$

and
$$X_{C2} = \frac{R_L}{10}$$

$$\Rightarrow \qquad \frac{1}{2\Pi f C_{C2}} = \frac{R_L}{10}$$

$$\Rightarrow \qquad C_{C2} = \ldots$$

PROCEDURE

1. Design the CE amplifier according to the method as mentioned above.
2. Connect the circuit as shown in Fig. 3.6.3 on breadboard.
3. Connect function generator for applying the input signal. Set it for sine wave with frequency around 100Hz and amplitude in mV range.
4. Connect the channel 1 of CRO to the input signal.
5. Connect the channel 2 of CRO across load resistance R_L.
6. Measure the input voltage and the output voltage using CRO.
7. Keeping the amplitude of input signal constant, increase the frequency and measure the output voltage.
8. Tabulate the readings and plot the graph between gain and frequency.
9. Find the bandwidth of CE amplifier using the frequency response curve.

OBSERVATION TABLE

Table 3.6.1 Observation table for common emitter amplifier

Frequency (Hz)	Vin (V)	Vo (V)	Gain Vo / Vin
100			
200			
300			
…			
…			
10MHz			

RESULT

CE amplifier has been designed successfully and the frequency response graph has been plotted. Bandwidth of CE amplifier comes out to be

$$BW = f_H - f_L = \cdots$$

Frequency response graph shows that the gain remains constant for mid-range of frequencies and at frequency f_L and f_H, gain becomes 0.707 times of its maximum value.

DISCUSSION

Amplifiers play a major role in electronic systems. No system can be imagined without the use of an amplifier. For *e.g.* amplifiers are widely used as audio amplifiers. The audio signal is first converted into electrical signal by the use of microphone which is then amplified by the amplifier. The amplified signal is connected to a loud speaker which gives final output as an audio signal. This system is widely used in public gathering, large auditoriums or conference rooms. Amplifiers are also used in radio and TV receivers, tape recorders, home theaters, stereos, instruments, telephones, cell phones, video conference systems, computer communications and satellite communications *etc.*

Common base, common emitter and common collector configurations have different characteristics which are listed in the table shown below.

Table 3.6.2 Comparison of CB, CE and CC configurations

Characteristics	CB	CE	CC
Input impedance	Very low	Medium	Very high
Output impedance	Very high	Medium	Very low
Current gain	Less than 1	High	Very high
Voltage gain	High	High	Less than 1
Power gain	Medium	Highest	Medium
Phase relationship between input and output	In phase	Out of phase by 180°	In phase
Applications	Constant current source	Power amplifier	Buffer For impedance matching applications

Experiment 7

COLPITT'S OSCILLATOR

AIM

To Study the Colpitt's Oscillator.

APPARATUS

Breadboard, connecting wires, BJT - BC547, cathode ray oscilloscope, probes, resistors, capacitors, inductor, dc power supply.

THEORY

Oscillator

An oscillator is a circuit that is used to generate a repetitive current or voltage waveform of specific amplitude and frequency without any external input signal.

Types of oscillators

1. Depending on the type of waveform generated

    - Sinusoidal oscillator which generates pure sinusoidal waveform.

    - Non sinusoidal or relaxation oscillator which generates square, pulses, triangular and saw tooth waveforms.

2. Depending on the type of components used

    - RC oscillator

    - LC oscillator

    - Crystal oscillator

3. Depending on the frequency of oscillations

    - Audio frequency oscillator

    - Radio frequency oscillator

Principle

Oscillator is a positive feedback amplifier that is capable of producing an ac output signal without any ac input signal. The word 'feedback' means that the output is fed back to its input via a feedback circuit and the word 'positive' means that the output that is fed back is in phase with the input.

An oscillator needs a starting voltage to produce oscillations. This starting voltage is provided by the noise voltage produced by the random motion of electrons in resistors or active devices. Noise, which is random in nature having all the frequencies, is amplified by an amplifier and is fed to the feedback network. The feedback network provides maximum gain and 180° phase shift only for a single frequency. Hence, only a single frequency present in the noise is fed back with proper phase to the amplifier. Therefore, a positive feedback amplifier *i.e.* an oscillator generates oscillations only at a single frequency as decided by the feedback network. For positive feedback, the gain of an amplifier is

$$A_F = {}^A\!/(1 - A\beta)$$

Where,

A_F: gain with positive feedback, A: open loop gain (voltage gain) *i.e.* without feedback, β: feedback gain or gain of the feedback circuit.

Two conditions necessary to be fulfilled for the oscillations

- The magnitude of loop gain $A\beta$ must be ≥ 1. The value of $A\beta$ must be equal to 1 for oscillations to start and for the sustained oscillations $A\beta$ must be greater than 1, otherwise, the oscillations would die out. $A\beta = 1$ is called as the Barkhausen criterion of oscillation.

- Total phase shift of the loop gain must be 0° or 360°.

Colpitt's Oscillator

Colpitt's oscillator is a LC type of oscillator that consists of a transistor with the feedback network as shown in Fig. 3.7.1.

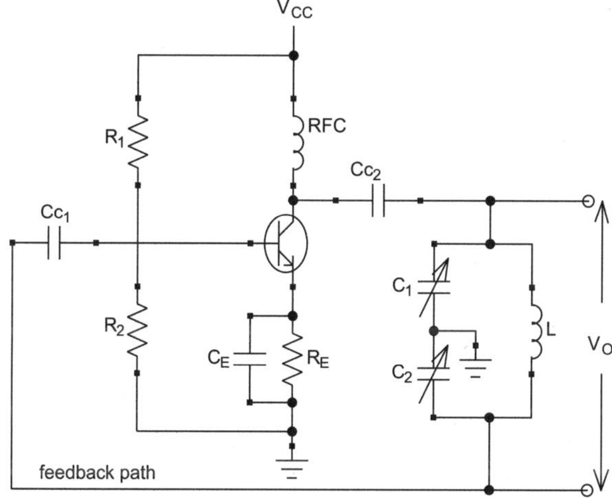

Fig. 3.7.1 Circuit for colpitt's oscillator

Transistor is used as an amplifier in common emitter configuration. It amplifies the input signal as well as provides a phase shift of 180°. Another 180° phase shift is provided by the feedback network which consists of two series capacitors C_1 and C_2 connected in parallel with inductor L. The two capacitors form a voltage divider network and voltage across capacitor C_2 provides the feedback voltage. The total phase shift around the loop is 360°.

The feedback factor is given as

$$\beta = \frac{C_1}{C_2}$$

The minimum value of amplifier gain for maintaining oscillations is given as

$$A_{min} = \frac{1}{\frac{C_1}{C_2}} = \frac{C_2}{C_1} \qquad\qquad - \text{eqn 1}$$

CE amplifier

It uses voltage divider biasing circuit because it provides the best stabilization of Q point against any changes in temperature or transistor's parameters. Role of resistor R_E is used to provide stabilization of Q point by providing dc feedback. C_E acts as a bypass capacitor and is used in parallel with R_E to provide a low reactance path to the ac signal. Radio frequency choke (RFC) is used to allow easy flow of dc current. It offers high impedance to ac signals and act as short for dc current. Hence, it prevents ac signal from entering the dc power supply V_{CC}. Moreover, it also provides the necessary dc load resistance R_C. C_{C1} and C_{C2} are known as coupling capacitors or blocking capacitors that passes the ac signal from one stage to another and blocks the dc signal.

Feedback network

The two series capacitors C_1 and C_2 connected in parallel to the inductor L form the tank circuit. Both inductor and capacitor are capable of storing energy. The capacitor stores the energy in the form of electric field whenever there is a potential difference across it and inductor stores the energy in the form of magnetic field whenever there is a flow of current through it. As the power is switched on, the capacitor gets charged. The electrons start moving from first plate to second through coil L and as a result the electric field strength reduces. As the electric current flows in the circuit, the self induced emf in the coil starts opposing the current flow and moreover, magnetic field is generated around the coil. The capacitor keeps on discharging and energy gets stored in the form of magnetic field. As the capacitor gets fully discharged, there is no electric field but the magnetic field around the coil is the maximum. However, due to the self induction of the coil, more electrons gets transferred to the second plate of capacitor *i.e.* second plate now has more electrons as compared to first plate. As a result of this, capacitor once again starts charging in the opposite direction. The magnetic field starts reducing and the energy is now stored in the form of electric field. The capacitor gets fully charged with opposite polarity. Now, capacitor starts discharging in opposite direction *i.e.* electrons move from plate second to first through the coil which generates magnetic field around the coil in opposite direction. This process of charging and discharging continues and if there are no losses in the circuit, this process keeps on repeating indefinitely. But practically, with each oscillation, the circuit loses some of its voltage which results in damped oscillations. Hence, an active device capable of amplifying the signal is used to sustain the oscillations.

The frequency of oscillation depends on value of L, C_1 and C_2 and is given as

$$f = \frac{1}{2\pi \sqrt{(LC_{eq})}} \qquad - \text{eqn 2}$$

Where,

$$C_{eq} = \frac{C_1 . C_2}{C_1 + C_2} \qquad - \text{eqn 3}$$

Designing

1. For designing the feedback network of the colpitt's oscillator, follow the given steps.

   a. Fix the frequency f for which colpitt's oscillator is to be designed.

   b. Find the h_{fe} of the transistor by using multimeter.

   c. Fix the value of C_2 and using the given relation, find the value of C_1.

   $$C_1 = h_{fe} \times C_2$$

   d. Compute the value of C_{eq} by using equation 3.

   e. Calculate the value of inductor L by using equation 2 *i.e.*

   $$L = \frac{1}{4C_{eq}\Pi^2 f^2}$$

2. For designing the common emitter amplifier, choose the gain according to the equation 1 and refer to experiment 6 of chapter 3.

PROCEDURE

1. Design the colpitt's oscillator according to the method as mentioned above.
2. Connect the circuit as shown in Fig. 3.7.1 on breadboard.
3. Connect the channel 1 of CRO to output terminal of the colpitt's oscillator using a probe.
4. Measure the output frequency and take the trace.

5. Compare the theoretical and practical value of the frequency.

6. Change the value of L and C for different frequency and take the readings again.

OBSERVATION TABLE

Table 3.7.1 Observation table for colpitt's oscillator

S.No.	L(mH)	$C_1(\mu F)$	$C_2(\mu F)$	Ceq	Theoretical frequency	Practical frequency
1.						
2.						
3.						

OBSERVATIONS

Attach the traces for output waveforms.

RESULT

Colpitt's oscillator has been studied and traces of sinusoidal output waveforms for different combinations of L and C have been taken successfully. Difference between practical and theoretical value of frequency comes out to be very small.

DISCUSSION

Colpitt's oscillators are widely used to generate sinusoidal signals of very high frequencies of the order of 1MHz. It is used in RF generators, radio and TV receivers *etc*. It is also used for the development of mobile and radio communications.

For generating audio frequencies, very large value of inductance is required which is practically difficult to achieve. Hence, LC oscillators are not preferred in audio frequency range.

Experiment **8**

PHASE SHIFT OSCILLATOR

AIM

To Study the Phase Shift Oscillator.

APPARATUS

Breadboard, connecting wires, BJT - BC547, cathode ray oscilloscope, probes, resistors, capacitors, dc power supply.

THEORY

Oscillator

An oscillator is a circuit that is used to generate a repetitive current or voltage waveform of specific amplitude and frequency without any external input signal.

Types of oscillators

1. Depending on the type of waveform generated

    *   Sinusoidal oscillator which generates pure sinusoidal waveform.

    *   Non sinusoidal or relaxation oscillator which generates square, pulses, triangular and saw tooth waveforms.

2. Depending on the type of components used

    *   RC oscillator

    *   LC oscillator

    *   Crystal oscillator

3. Depending on the frequency of oscillations

    *   Audio frequency oscillator

    *   Radio frequency oscillator

Principle

Oscillator is a positive feedback amplifier that is capable of producing an ac output signal without any ac input signal. The word 'feedback' means that the output is fed back to its input via a feedback circuit and the word 'positive' means that the output that is fed back is in phase with the input.

An oscillator needs a starting voltage to produce oscillations. This starting voltage is provided by the noise voltage produced by the random motion of electrons in resistors or active device. Noise, which is random in nature having all the frequencies, is amplified by an amplifier and is fed back to the feedback network. The feedback network provides maximum gain and 180° phase shift only for a single frequency. Hence, only a single frequency present in the noise is fed back with proper phase to the amplifier. Therefore, a positive feedback amplifier *i.e.* an oscillator generates oscillations only at a single frequency as decided by the feedback network. For positive feedback, the gain of an amplifier is

$$A_F = {}^A\!/_{(1 - A\beta)}$$

Where,

A_F: gain with positive feedback, A: open loop gain (voltage gain) *i.e.* without feedback, β: feedback gain or gain of the feedback circuit.

Two conditions necessary to be fulfilled for the oscillations

- The magnitude of loop gain $A\beta$ must be ≥ 1. The value of $A\beta$ must be 1 for oscillations to start and for the sustained oscillations $A\beta$ must be greater than 1, otherwise, the oscillations would die out. $A\beta = 1$ is called as the Barkhausen criterion of oscillation.

- Total phase shift of the loop gain must be 0° or 360°.

Phase Shift Oscillator

Phase shift oscillator is a RC type of an oscillator that consists of transistor and three RC networks cascaded together as shown in Fig. 3.8.1

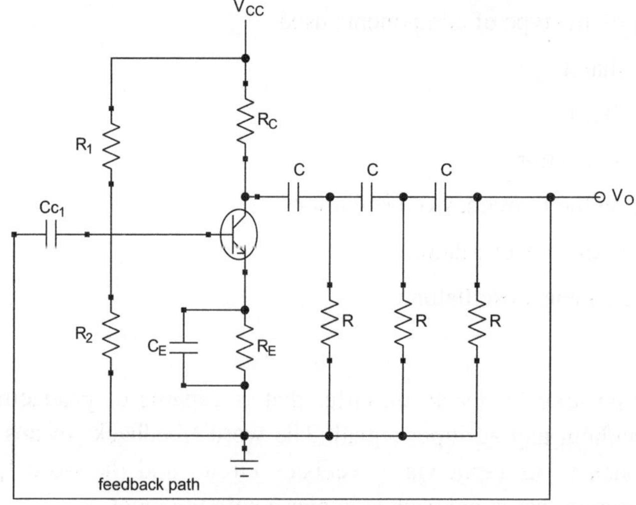

Fig. 3.8.1 Phase shift oscillator

Transistor is used as an amplifier in common emitter configuration. It amplifies the input signal as well as provides a phase shift of 180°. Another 180° phase shift is provided by the RC network. One RC section is capable of producing a maximum phase shift of 60° practically (90° ideally). Hence, 3 RC sections are cascaded together to provide a total of 180° phase shift. Thus, the total phase shift around the loop is 360°.

Therefore, the circuit oscillates at some specific frequency only where the phase shift provided by the cascaded RC networks is exactly 180° and gain of the amplifier is sufficiently large. This frequency of oscillation is given as

$$f = \frac{1}{2\pi RC\sqrt{6 + 4\,{}^{R_C}/_R}} \qquad - \text{eqn 1}$$

CE amplifier

It uses voltage divider biasing circuit because it provides the best stabilization of Q point against any changes in temperature or transistor's parameters. Role of resistor R_E is used to provide stabilization of Q point by providing dc feedback. C_E acts as a bypass capacitor and is used in parallel with R_E to provide a low reactance path to the ac signal. C_{C1} and C_{C2} are known as coupling capacitors or blocking capacitors that passes the ac signal from one stage to another and blocks the dc signal.

RC phase shift network

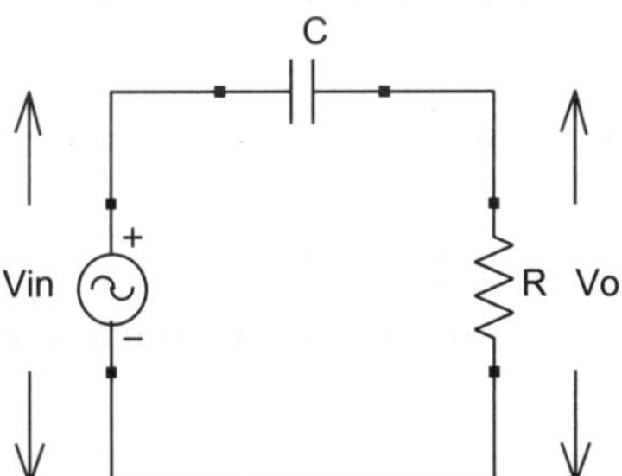

Fig. 3.8.2 RC phase shift network

The output voltage across R is given as

$$V_0 = \frac{V_{in}R}{R + {}^1/_{j\omega C}}$$

$$\Rightarrow \qquad V_0 = \frac{V_{in}R}{\left(R + {}^1/_{j\omega C}\right)} \cdot \frac{\left(R - {}^1/_{j\omega C}\right)}{\left(R - {}^1/_{j\omega C}\right)}$$

$$\Rightarrow \qquad V_0 = \frac{V_{in} R \left(R - \frac{1}{j\omega C} \right)}{R^2 + \frac{1}{\omega^2 C^2}}$$

Magnitude of output voltage is given as

$$V_0 = \frac{R V_{in}}{R^2 + \frac{1}{\omega^2 C^2}} \sqrt{R^2 + \frac{1}{\omega^2 C^2}}$$

Phase of the output voltage with respect to input voltage is given as

$$\theta = \tan^{-1} \left(\frac{1}{\omega RC} \right)$$

The above equation shows that the RC network produces a phase shift between input and output voltage and the maximum phase shift that can be produced by a single RC network is 90°.

Designing

1. For designing the common emitter amplifier, choose the transistor with $h_{fe} >$ 44.5 and choose the gain A to be around 120 and refer to experiment 6 of chapter 3.

2. For designing the feedback network of phase shift oscillator, follow the given steps.

   a. Fix the frequency of oscillations f for which phase shift oscillator is to be designed.

   b. Choose the value of capacitor C and calculate the value of resistance R by using equation 1.

   c. Or choose the value of $R = 2R_C$ to avoid loading by RC network and calculate the value of capacitor C by using equation 1.

PROCEDURE

1. Design the phase shift oscillator according to the method as mentioned above.

2. Connect the circuit as shown in Fig. 3.8.1 on breadboard.

3. Connect the channel 1 of CRO to output terminal of the phase shift oscillator using a probe.

4. Measure the output frequency and take the trace.

5. Compare the theoretical and practical value of the frequency.

6. Change the value of R and C for different frequencies and take the readings again.

OBSERVATION TABLE

Table 3.8.1 Observation table for phase shift oscillator

S.No.	R(Ω)	C(μF)	Theoretical frequency	Practical frequency
1.				
2.				
3.				

OBSERVATIONS

Attach the traces for output waveforms.

RESULT

Phase shift oscillator has been studied and traces of the sinusoidal waveforms for different combinations of R and C have been taken successfully. Difference between practical and theoretical value of frequency comes out to be very small.

DISCUSSION

Phase shift oscillators are the RC oscillators which are used for generating audio frequencies. They have a wide frequency range from few Hz to several hundred kHz with good frequency stability. Circuitry is simple and cheap and they do not require inductors for their operation. However, phase shift oscillators cannot be used for high frequencies because at high frequency, capacitor offers very low impedance and acts as a short. Therefore, resistances will come in parallel to each other and hence RC combinations will lose its action of producing 180° phase shift.

Experiment 9

FET AMPLIFIER

AIM

To Study the Frequency Response of Common Source FET Amplifier.

APPARATUS

Breadboard, cathode ray oscilloscope, function generator, probes, JFET – 2N3819, resistors, capacitors, connecting wires, 15V dc power supply.

THEORY

Field Effect Transistor

A Field effect transistor abbreviated as FET is a semiconductor device with three terminals named as Drain, Source and Gate. It is a unipolar device because current conduction takes place by only one type of carriers *i.e.* majority carriers. It has very high input impedance (several hundred mega ohms) and high voltage gain which makes it very useful in designing the device as an amplifier.

FETs are classified into two types

- N-channel FETs with electrons as majority carriers.

- P-channel FETs with holes as majority carriers.

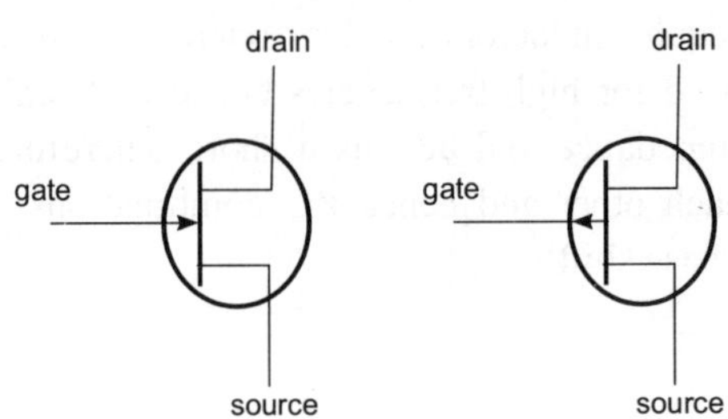

Fig. 3.9.1 Symbol for n-channel and p-channel JFET

Structure of JFET

JFET consists of uniformly doped semiconductor bar called as channel with two ohmic contacts at both the ends, known as Source S and Drain D. The bar has two junctions at its both sides which are connected internally to form a common terminal known as gate. In n-channel FET, the semiconductor bar is of n-type and junctions on both sides of the channel are highly doped p^+ region as shown in Fig. 3.9.2. In case of p-channel FET, the semiconductor bar is of p-type and junctions on both sides of the channel are highly doped n^+ region.

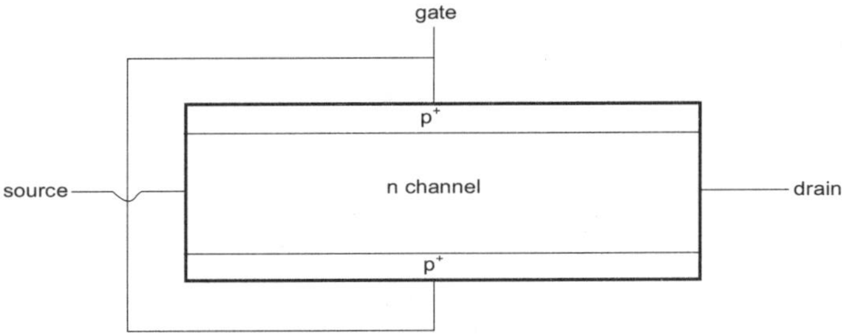

Fig. 3.9.2 Structure of n channel JFET

FET as an Amplifier

FETs are widely used as an amplifier to amplify weak signals. There are three basic configurations named as common source, common drain and common gate in which FET can be used as an amplifier. The most commonly used configuration is common source configuration because it provides good voltage gain. Common drain configuration, also known as source follower, is used as a buffer amplifier to provide unity gain and the common gate configuration is rarely used.

Common Source FET Amplifier

In CS configuration, input terminals in FET are gate and source and the output terminals are drain and source. Since, source is made common to both input and output section, hence FET is said to be in common source configuration.

CS FET amplifier as shown in Fig. 3.9.3 uses voltage divider biasing circuit because it provides the best stabilization of Q point against any changes in temperature or transistor's parameters. As the input signal is applied at the gate of transistor, CS amplifier amplifies the input signal and produces an amplified output signal which is 180° out of phase with respect to the input signal. Coupling capacitors Cc_1 and Cc_2, also known as blocking capacitors, passes the ac signal from one stage to another and blocks the dc signal. Role of resistor R_S is used to provide stabilization of Q point by providing dc feedback. Since, it provides ac feedback as well which reduces ac gain of the circuit hence, to eliminate this effect, bypass capacitor C_S is used in parallel with R_S which provides a low reactance path to the ac signal. It acts as a short for ac signals and hence, R_S also gets shorted by which there is no ac feedback. R_L is the resistance of the load connected to the amplifier.

The voltage gain of the amplifier is given as

$$A_V = \frac{V_o \,(\text{output voltage})}{V_{in}\,(\text{input voltage})}$$

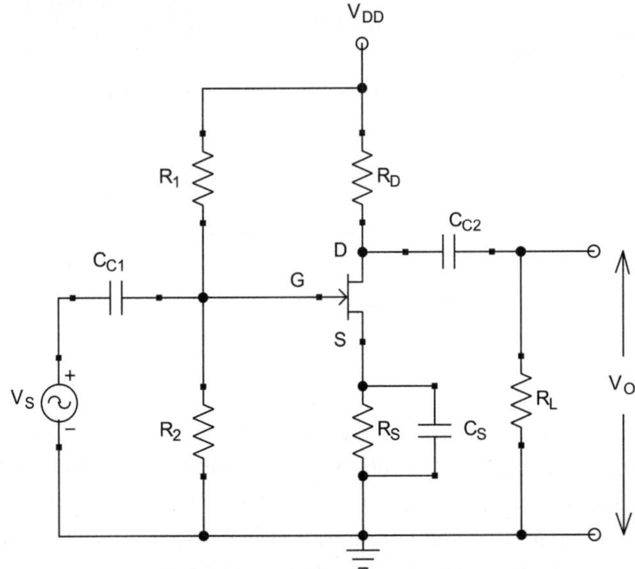

Fig. 3.9.3 Circuit for common source FET amplifier

Frequency response

The graph between the voltage gain and signal frequency is called as the frequency response.

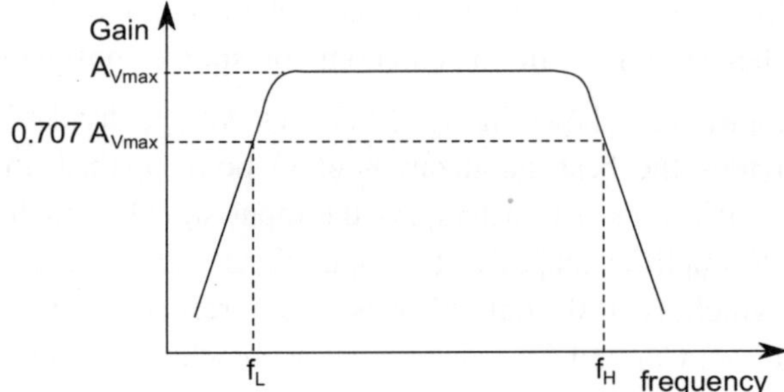

Fig. 3.9.4 Frequency response for CS amplifier

The graph shown in Fig. 3.9.4 shows the variation of voltage gain of an amplifier with signal frequency. The reason of this variation is due to the presence of different capacitors in the circuit whose reactance changes with the change in frequency. Voltage gain decreases for frequencies below f_L and above f_H and remains constant for the region between f_L and f_H.

Low frequencies

For the frequencies below lower cut off frequency *i.e.* f_L, there is a fall in the gain of an amplifier. It is because of the presence of coupling capacitors at the input and output section. At low frequency, reactance offered by the coupling capacitors is high and hence, a large amount of input voltage appears across it. Moreover, the reactance offered by bypass capacitor C_S also becomes comparable with source resistance R_S which reduces the bypass capacitor action. C_S will not act as short and as a result R_S will provide ac feedback as well which reduces the gain of an amplifier.

High frequencies

For the frequencies above higher cut off frequency *i.e.* f_H, there is a fall in the gain of the amplifier. It is because of the presence of intrinsic capacitance which provides the low impedance path to the high frequency ac input signals. Hence, the whole input signal does not appear across the transistor and therefore, output voltage also reduces. The effect due to coupling capacitance and bypass capacitance is almost negligible in this case because reactance offered by these capacitors is very less at high frequencies and hence negligible voltage appears across it.

Medium frequencies

For the frequencies between f_L and f_H, the effect of coupling capacitance, bypass capacitance and inter-electrode capacitance is very small. Hence, we get the maximum gain in this region. At f_L and f_H, gain is 3dB down the maximum value or 0.707 times the maximum gain.

Bandwidth

It is defined as the difference between the upper cut off frequency and the lower cut off frequency at which gain is 3 dB down the maximum value.

PROCEDURE

1. Take $V_{DD} = 15V$, $R_S = 2k\Omega$, $R_2 = 100k\Omega$, $R_1 = 200k\Omega$, $R_D = 1.33k\Omega$, $C_{C1} = C_{C2} = C_S = 0.01\mu F$, $R_L = 10k\Omega$.

2. Connect the circuit as shown in Fig. 3.9.3 on breadboard.

3. Connect function generator for giving the input signal. Set it for sine wave with frequency around 100Hz and amplitude in mV range.

4. Connect the channel 1 of CRO to the input signal.

5. Connect the channel 2 of CRO across load resistance R_L.

6. Measure the input voltage and the output voltage using CRO.

7. Keeping the amplitude of input signal constant, increase its frequency and measure the output voltage.

8. Tabulate the readings and plot the graph between gain and frequency.

9. Find the bandwidth of CS amplifier using the frequency response curve.

OBSERVATIONS

Table 3.9.1 Observation table for CS FET amplifier

Frequency (Hz)	Vin (V)	Vo (V)	Gain Vo / Vin
100			
200			
300			
…			
…			
10MHz			

RESULT

Common source FET amplifier has been designed successfully and the frequency response graph has been plotted. Bandwidth comes out to be

$$BW = f_H - f_L = \ldots.$$

Frequency response graph shows that the gain remains constant for mid-range of frequencies and at frequency f_L and f_H, gain becomes 0.707 times of its maximum value.

DISCUSSION

Amplifiers play a major role in electronic systems. No system can be imagined without the use of an amplifier. For *e.g.* amplifiers are widely used as audio amplifiers. The audio signal is first converted into electrical signal by the use of microphone which is then amplified by the amplifier. The amplified signal is connected to a loud speaker which gives final output as an audio signal. This system is widely used in public gathering, large auditoriums or conference rooms. Amplifiers are also used in radio and TV receivers, tape recorders, home theaters, stereos, instruments, telephones, cell phones, video conference systems, computer communications and satellite communications *etc.*

BJT is a current controlled device whereas FET is a voltage controlled device. In BJT, a small base current controls a large collector current whereas in FET a small gate voltage controls a large drain current. Therefore, in FET, voltage controls the conductivity of the device.

Main advantage of FET over BJT is that FET offers very high input impedance as compared to BJT which minimizes the effect of the loading. It also has a very high packing density which makes it possible to fabricate the device on a very small area. They are also more temperature stable as compared to BJT.

But the disadvantage of using FET is that their voltage gain is less as compared to BJT amplifier.

FETs/MOSFETs are widely used in calculators; computers; laptops; as input amplifiers in oscilloscopes, electronic voltmeters because of their high input impedance; in computer memories because of their very small size; as voltage variable resistors *etc*.

NOTE

Frequency response is the curve drawn between gain of an amplifier and input signal frequency. To plot the frequency response graph, semi log graph paper is used. Frequency is taken on x axis which is on log scale and gain is taken on y axis which is on linear scale.

4. CRO

<table><tr><td>

4

</td><td>

Experiments on Digital Electronics

</td></tr></table>

LIST OF EXPERIMENTS

1. Verify and Design AND, OR, NOT and XOR Gates using NAND and NOR Gates.

2. Convert a Boolean Expression into Logic Gate Circuit and Assemble it using Logic Gate IC's.

3. Design a Half and Full Adder.

4. Design a Half and Full Subtractor.

5. Design a Seven Segment Display Driver.

6. Design 4x1 Multiplexer using Gates.

7. To build a Flip-Flop Circuits using Elementary Gates (RS, Clocked RS, D-type).

8. Design a Counter using D/T/JK Flip-Flop.

9. Design a Shift Register and Study Serial and Parallel Shifting of Data.

10. Design a Digital to Analog Converter of Given Specification.

IC PIN DIAGRAMS

7400 (QUAD TWO INPUT NAND GATE)

It has four independent NAND gates each with two inputs and one output as shown in Fig. 4a.

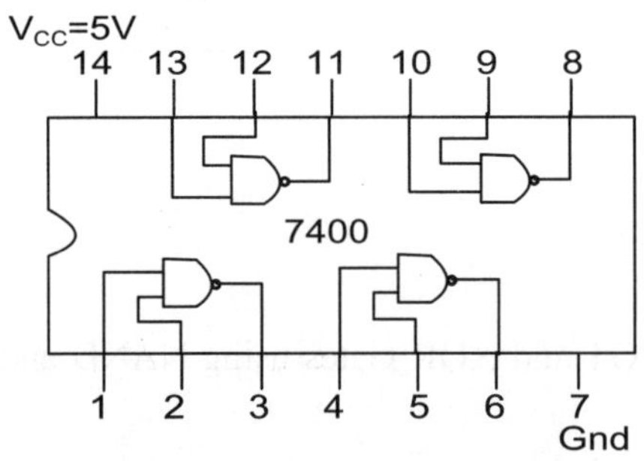

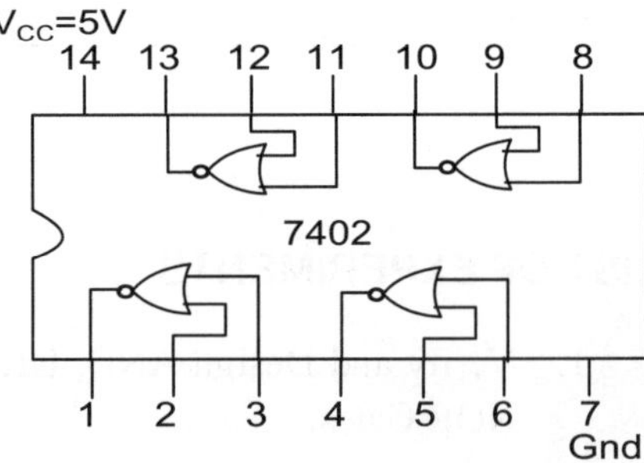

Fig. 4a 7400 quad two input NAND gate IC   **Fig. 4b** 7402 quad two input NOR gate IC

7402 (QUAD TWO INPUT NOR GATE)

It has four independent NOR gates each with two inputs and one output as shown in Fig. 4b.

7408 (QUAD TWO INPUT AND GATE)

It has four independent AND gates each with two inputs and one output as shown in Fig. 4c.

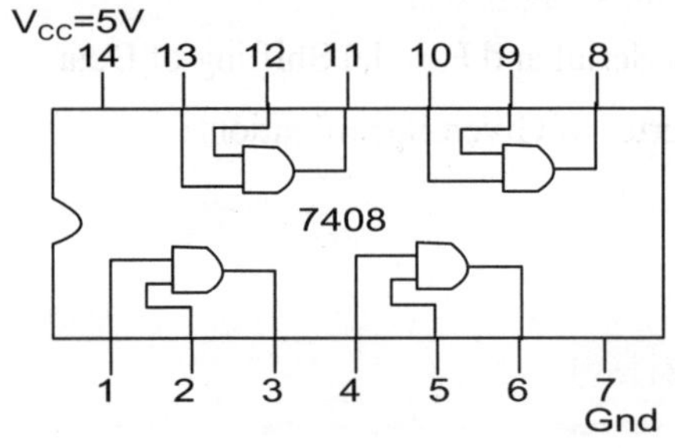

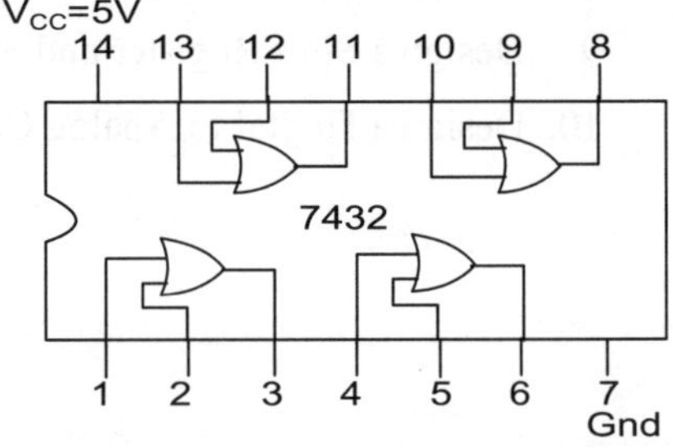

Fig. 4c 7408 quad two input AND gate IC   **Fig. 4d** 7432 quad two input OR gate IC

7432 (QUAD TWO INPUT OR GATE)

It has four independent OR gates each with two inputs and one output as shown in Fig. 4d.

7404 (NOT GATE)

It has six independent NOT gates each with one input and one output as shown in Fig. 4e.

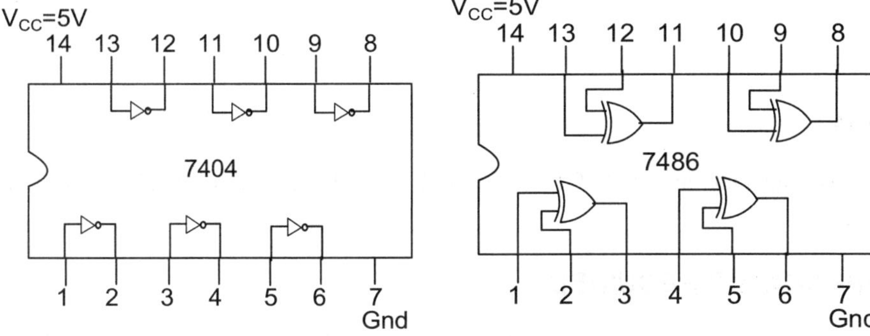

Fig. 4e 7404 NOT gate IC

Fig. 4f 7486 quad two input XOR gate IC

7486 (QUAD TWO INPUT XOR GATE)

It has four independent XOR gates with two inputs and one output as shown in Fig. 4f.

Note:

74 series indicates that logic gates are implemented by using TTL (transistor transistor logic) family. They require +5V for their operation.

Experiment 1

LOGIC GATES using NAND and NOR GATES

AIM

Verify and Design AND, OR, NOT and XOR Gates using NAND and NOR Gates.

APPARATUS REQUIRED

Breadboard, connecting wires, 5V dc power supply, resistor - 220Ω, IC – 7400 and 7402, LED, multimeter.

THEORY

Logic Gates

A logic gate is a basic building block of a digital circuit which performs logical operations on one or more logical inputs to give a single logical output. It has the ability to make decisions and hence, it is called as a logic gate. Three basic type of logic gates are AND, OR and NOT. These gates can be interconnected to make complex circuits.

Logic gates can be implemented by using diodes or transistors. Input and output of logic gates can have two levels: LOW or 0 and HIGH or 1.

AND gate

It can have two or more inputs but only single output. Output is high if all inputs are high otherwise output is low. Symbol for AND gate is shown in Fig. 4.1.1.

Truth table

A	B	A.B
0	0	0
0	1	0
1	0	0
1	1	1

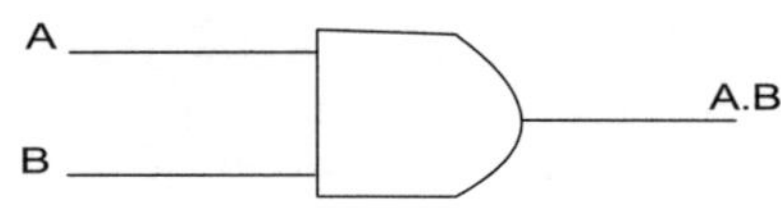

Fig. **4.1.1** Symbol for AND gate

OR gate

It can have two or more inputs but only single output. Output is high if any of the inputs is high. Symbol for OR gate is shown in Fig. 4.1.2.

Truth table

A	B	A+B
0	0	0
0	1	1
1	0	1
1	1	1

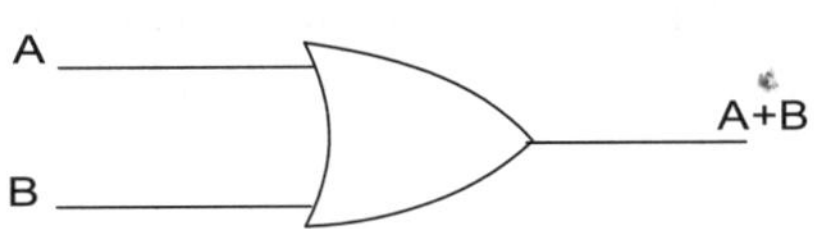
Fig. 4.1.2 Symbol for OR gate

NOT gate

It has single input and single output. Output is high if input is low and output is low if input is high. Hence, it is also called as inverter. Symbol for NOT gate is shown in Fig. 4.1.3.

Truth table

A	A'
0	1
1	0

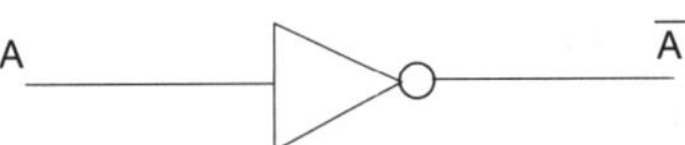
Fig. 4.1.3 Symbol for NOT gate

XOR gate

It is an Exclusive OR gate which gives high output only when one of the inputs is high and gives low output when both inputs are either low or high. Symbol for XOR gate is shown in Fig. 4.1.4.

Truth table

A	B	A⊕B
0	0	0
0	1	1
1	0	1
1	1	0

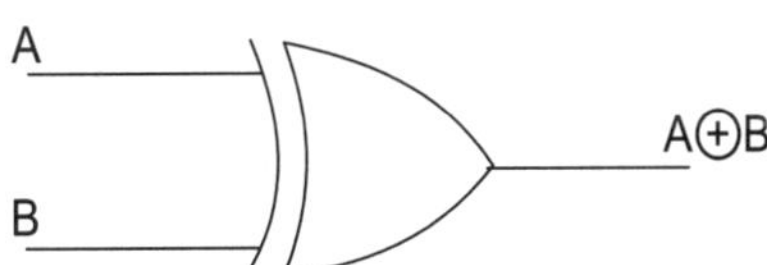
Fig. 4.1.4 Symbol for XOR gate

XNOR gate

It is an Exclusive NOR gate. It is a combination of XOR gate followed by a NOT gate. It gives high output when both the inputs are same *i.e.* either 1 or 0. Symbol for XNOR gate is shown in Fig. 4.1.5.

Truth table

A	B	A⊙B
0	0	1
0	1	0
1	0	0
1	1	1

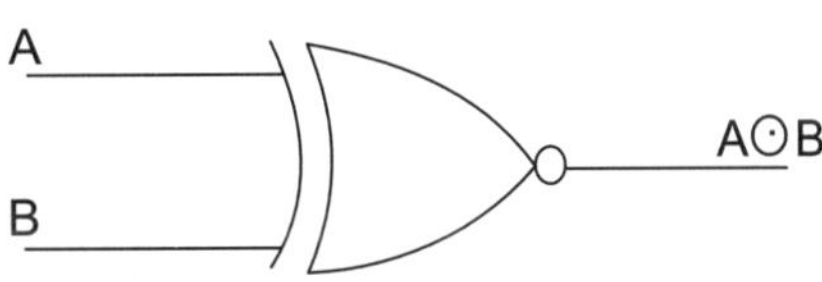
Fig. 4.1.5 Symbol for XNOR gate

NAND gate

NAND gate is a combination of AND gate followed by a NOT gate. Output of NAND gate is low when all inputs are high otherwise output will be high. Symbol for NAND gate is shown in Fig. 4.1.6.

Truth table

A	B	$\overline{A.B}$
0	0	1
0	1	1
1	0	1
1	1	0

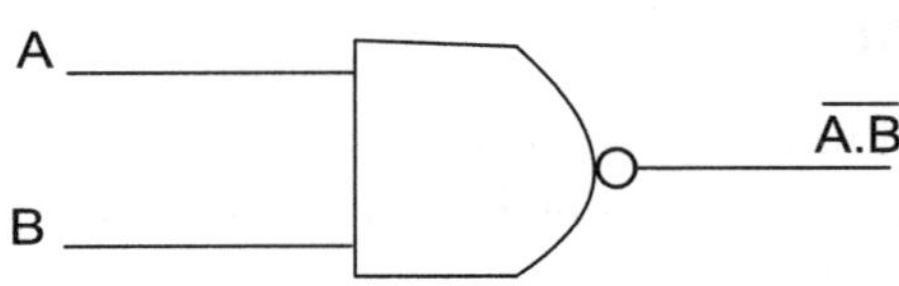

Fig. 4.1.6 Symbol for NAND gate

NOR gate

NOR gate is a combination of OR gate followed by a NOT gate. Output of NOR gate is low when any of the inputs is high. Symbol for NOR gate is shown in Fig. 4.1.7.

Truth table

A	B	$\overline{A+B}$
0	0	1
0	1	0
1	0	0
1	1	0

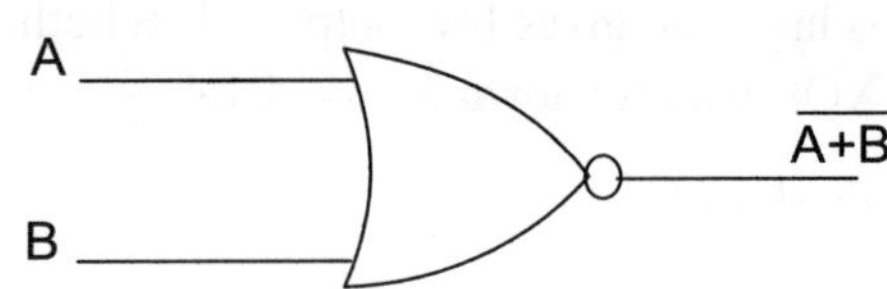

Fig. 4.1.7 Symbol for NOR gate

CIRCUIT DIAGRAM

1. Logic Gates using NAND Gate

NOT gate

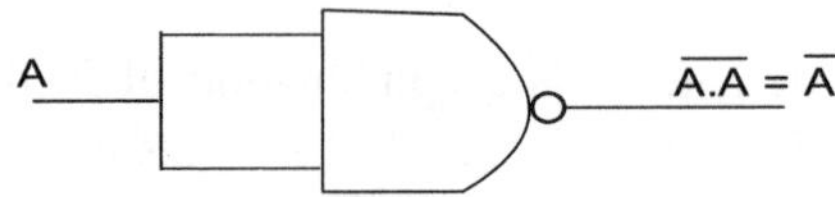

Fig. 4.1.8 NOT gate using NAND gate

AND gate

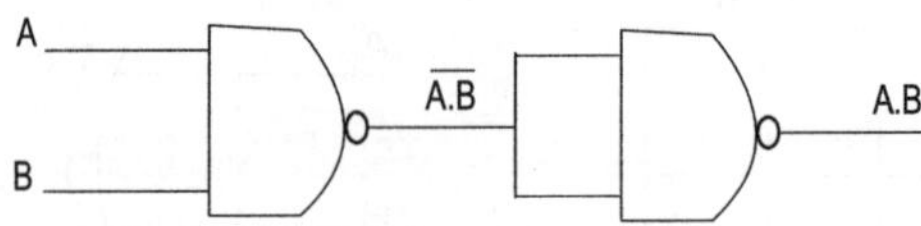

Fig. 4.1.9 AND gate using NAND gate

OR gate

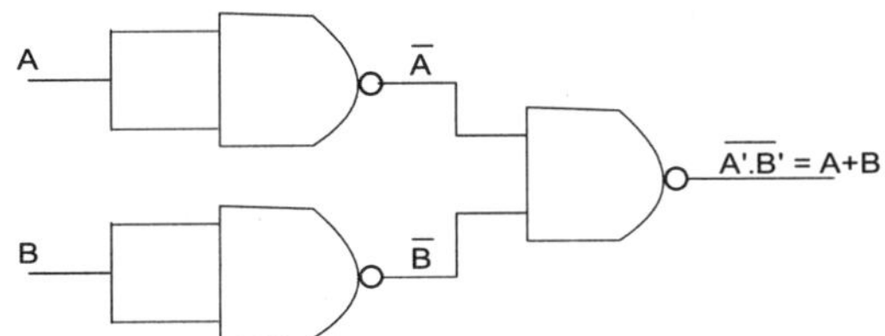

Fig. 4.1.10 OR gate using NAND gate

XOR gate

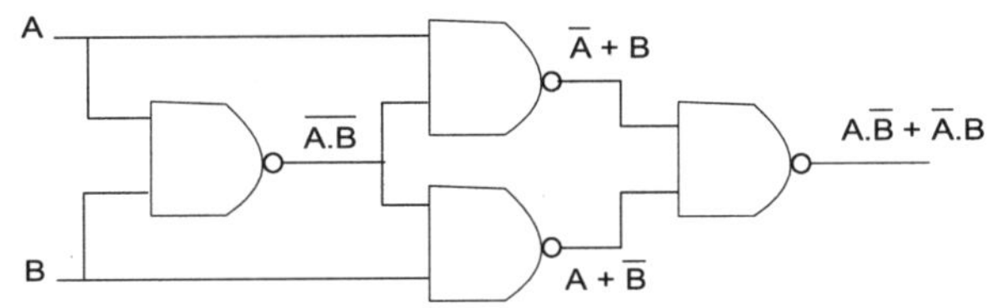

Fig. 4.1.11 XOR gate using NAND gate

XNOR gate

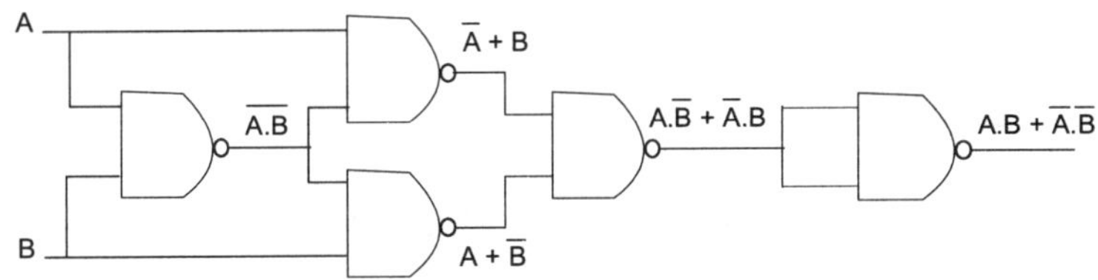

Fig. 4.1.12 XNOR gate using NAND gate

2. Logic Gates using NOR Gate

NOT gate

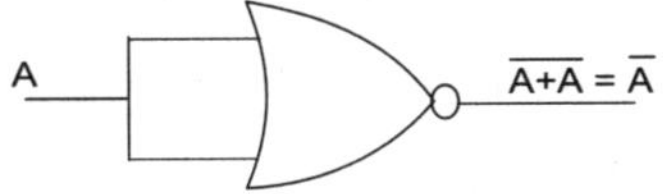

Fig. 4.1.13 NOT gate using NOR gate

AND gate

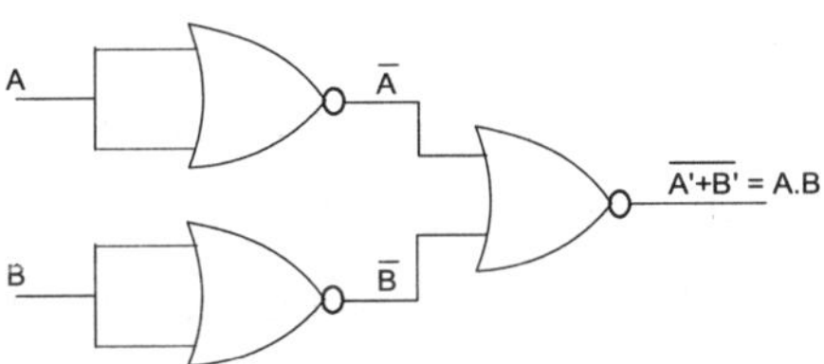

Fig. 4.1.14 AND gate using NOR gate

OR gate

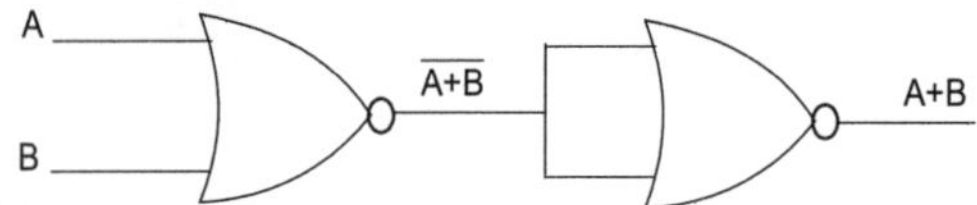

Fig. 4.1.15 OR gate using NOR gate

XOR gate

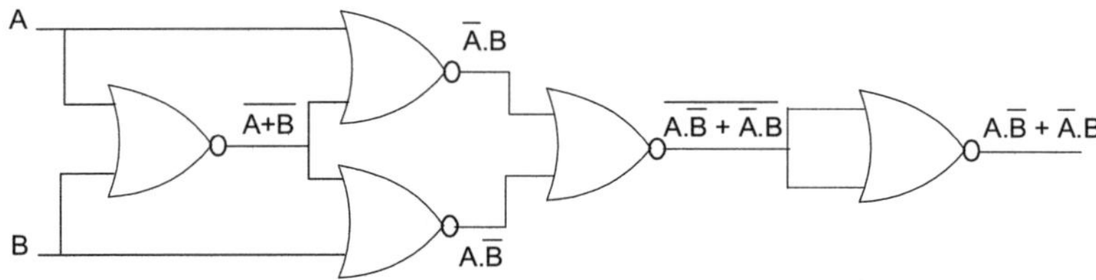

Fig. 4.1.16 XOR gate using NOR gate

XNOR gate

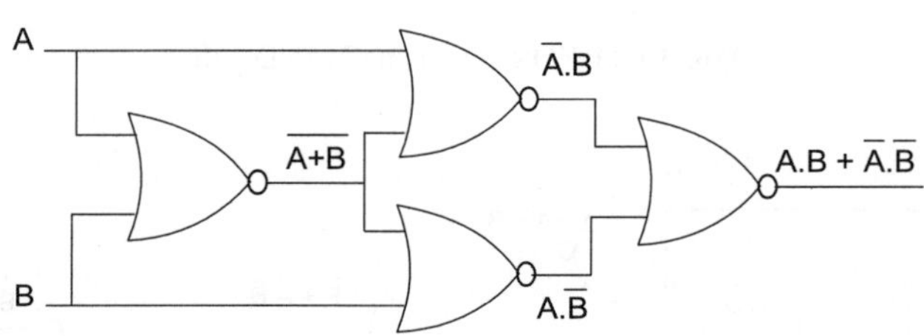

Fig. 4.1.17 XNOR gate using NOR gate

PROCEDURE

1. Mount IC 7400 on breadboard.

2. Connect 14th pin of IC to 5V and 7th pin to ground.

3. Check all four NAND gates before using them in the circuit.

4. For input, connect the input pin of gate to either ground (low) or 5V (high).

5. For output, connect the output pin of gate to the anode of LED and cathode to ground through 220Ω resistor. Role of resistor is to limit the current flowing through the LED.

6. Connect the circuits for AND, OR, NOT, XOR and XNOR gates using NAND gate IC as shown in Fig. 4.1.8 to 4.1.12.

7. Give different combinations of input and observe the output at the LED.

8. Verify the corresponding truth tables.

9. Now mount IC 7402 (NOR gate IC) on the breadboard and repeat steps 2 to 5.

10. Connect the circuits for AND, OR, NOT, XOR and XNOR gates using NOR gate IC as shown in Fig. 4.1.13 to 4.1.17.

11. Repeat steps 7 and 8.

OBSERVATIONS

Truth tables of all the gates using NAND and NOR gate ICs.

RESULT

Basic logic gates and their respective truth tables have been designed and verified successfully using NAND and NOR gates.

DISCUSSION

Logic gates are used to implement multiplexers, registers, counters, arithmetic logic units, computer memory, *etc.* NAND and NOR gates are called as universal gates because all the basic gates like AND, OR, NOT, XOR, XNOR and their combination as well can be implemented by using either NAND gate or NOR gate only.

XOR gate is used as an inequality detector because it gives high output only when the inputs are unequal. XNOR gate is used as an equality detector because it gives high output only when the inputs are equal.

Experiment 2

BOOLEAN EXPRESSION

AIM

Convert a Boolean Expression into Logic Gate Circuit and Assemble it using Logic Gate IC's.

APPARATUS REQUIRED

Breadboard, connecting wires, 5V dc power supply, resistor - 220Ω, LED, ICs – 7404, 7432, 7408, multimeter.

THEORY

Boolean Algebra

It describes the relationship between input and output and is used to express the logical functions algebraically.

In ordinary algebra, $A + A = 2A$ and $A.A = A^2$ where A has numerical value. For *e.g.* if A=1, A+A = 2. But in Boolean algebra, A+A = A and A.A = A where A has logical value. For *e.g.* if A = 1 (high), A+A = 1 (high).

In Boolean algebra, all the operations are logical. Logical multiplication is equivalent to AND operation and logical addition is equivalent to OR operation.

De Morgan's Theorem

De Morgan's theorem states that the complement of the function is obtained by interchanging AND and OR operators and complementing each literal. Two most important laws in Boolean algebra are

1. The complement of a sum of variables is equal to the product of their individual complements.

$$(A + B)' = A'.B'$$

2. The complement of product of variables is equal to the sum of their individual complements.

$$(A.B)' = A' + B'$$

Given Boolean Expression

Let the given Boolean expression be

$$f = A[B + C'(A.B + A.C')']$$

Minimization

Using De Morgan's theorem,

$$f = A[B + C'((A.B)'.(A.C')')]$$

$$\Rightarrow \qquad f = A[B + C'(A' + B').(A' + C'')]$$

$$\Rightarrow \qquad f = A[B + C'(A' + B').(A' + C)]$$

$$\Rightarrow \qquad f = A[B + C'(A'A' + A'C + B'A' + B'C)]$$

$$\Rightarrow \qquad f = A[B + C'(A' + A'C + B'A' + B'C)]$$

$$\Rightarrow \qquad f = A[B + C'A' + C'A'C + C'B'A' + C'B'C]$$

Since C.C' = 0.1 = 0

$$\Rightarrow \qquad f = A[B + C'A' + 0 + C'B'A' + 0]$$

$$\Rightarrow \qquad f = AB + AC'A' + AC'B'A'$$

$$\Rightarrow \qquad f = AB + 0 + 0$$

$$\Rightarrow \qquad f = A.B$$

CIRCUIT DIAGRAM

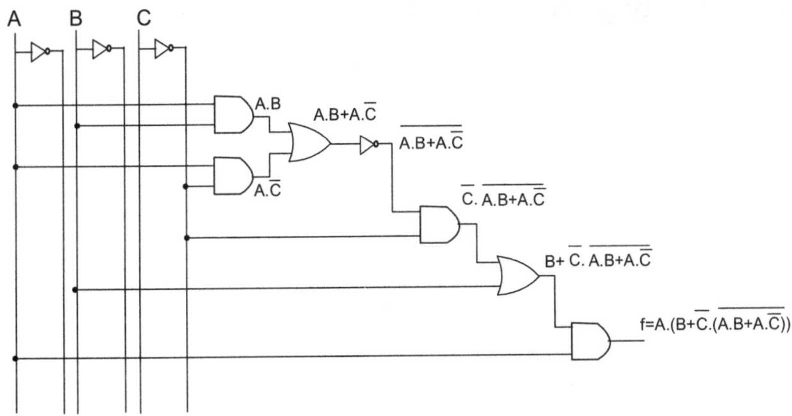

Fig. 4.2.1 Boolean expression using logic gates

After minimization, the given Boolean expression becomes

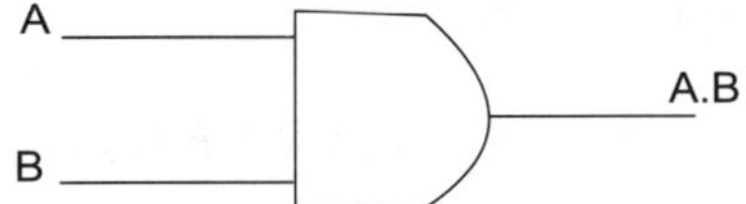

Fig. 4.2.2 Minimized Boolean expression

PROCEDURE

1. Mount IC 7404, 7432 and 7408 on breadboard.

2. Connect 14th pin of all ICs to 5V and 7th pin to ground.

3. Check gates in all the ICs before using them in the circuit.

4. For input, connect the input pin of gate to either ground (low) or 5V (high).

5. For output, connect the output pin of gate to the anode of LED and cathode to ground through 220Ω resistor. Role of resistor is to limit the current flowing through the LED.

6. Make the connections on breadboard as shown in Fig. 4.2.1.

7. Give different combinations of input and note the corresponding output in the form of truth table.

8. Now, connect the minimized circuit on the breadboard as shown in Fig. 4.2.2.

9. Give different combinations of input and note the corresponding output in the form of truth table.

10. Compare the two results.

OBSERVATIONS

Truth Table (for given expression)

Table 4.2.1 Truth table for given expression

A	B	C	F
0	0	0	0
0	0	1	0
0	1	0	0
0	1	1	0
1	0	0	0
1	0	1	0
1	1	0	1
1	1	1	1

Truth Table (after minimization)

Table 4.2.2 Truth table for minimized expression

A	B	Output
0	0	0
0	1	0
1	0	0
1	1	1

RESULT

The given Boolean expression has been minimized and realized successfully by using logic gate circuit. The truth tables for both the actual and minimized expressions are found to be same.

DISCUSSION

The observed truth table shows that output is independent of input C and will be high only when both A and B are high. Hence, the given Boolean expression can be realized by using AND gate with its inputs as A and B.

Experiment **3**

HALF and FULL ADDER

AIM

Design a Half and Full Adder.

APPARATUS REQUIRED

Breadboard, connecting wires, 5V dc power supply, resistor - 220Ω, LED, ICs – 7400, 7408, 7432, 7486.

THEORY

Combinational Circuits

Combinational circuits are those circuits in which output depends on present value of the input only. They do not have memory. For *e.g.* adders and subtractors.

Half Adder

It is a logical circuit that is used to add two binary digits A and B giving two outputs as Sum and Carry. Sum represents the sum of two digits and if there is an overflow of two digits, it is represented by carry. It can be implemented by using XOR gate and AND gate or just by using the universal gates *i.e.* either NAND gate or NOR gate.

Half adder using XOR and AND gate

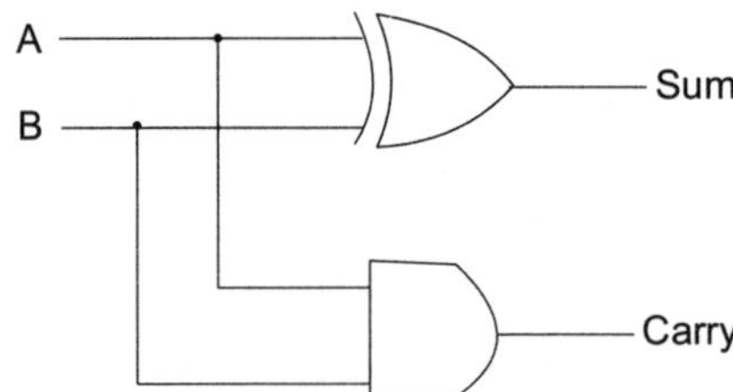

Fig. 4.3.1 Half adder using simple logic gates

Sum and Carry are given by the expressions

$$\text{Sum} = A \oplus B$$
$$\text{Carry} = A \cdot B$$

Half adder using NAND gates

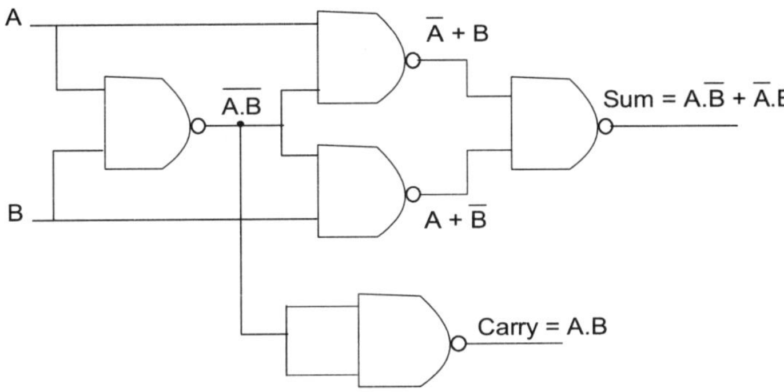

Fig. 4.3.2 Half adder using NAND gate

Truth table

Table 4.3.1 Truth table for half adder

Input		Output	
A	**B**	**Sum**	**Carry**
0	0	0	0
0	1	1	0
1	0	1	0
1	1	0	1

Full Adder

It is a logical circuit that is used to add three binary digits A, B and previous carry C_{in} and gives two outputs as Sum and Carry. Sum represents the sum of three digits and if there is an overflow of three digits, it is represented by carry C_{out}. It can be implemented by using XOR, AND and OR gates or just by using the universal gates *i.e.* either NAND gate or NOR gate.

Full adder using XOR, AND and OR gates

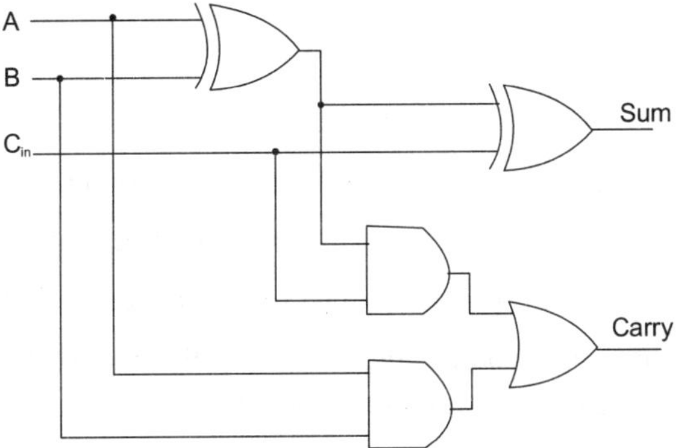

Fig. 4.3.3 Full adder using simple logic gates

Full adder using NAND gate

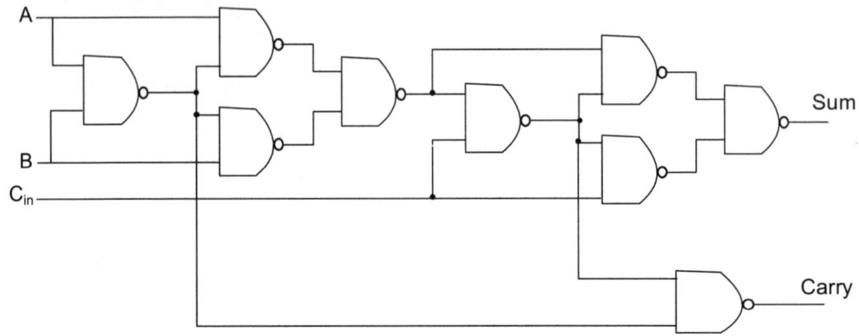

Fig. 4.3.4 Full adder using NAND gate

Sum and Carry out are given by the expressions

$$\text{Sum} = C_{in} \oplus (A \oplus B)$$

$$\text{Carry} = A.B + C_{in}.(A \oplus B)$$

Truth table

Table 4.3.2 Truth table for full adder

Input			Output	
A	**B**	**Cin**	**Sum**	**Carry**
0	0	0	0	0
0	0	1	1	0
0	1	0	1	0
0	1	1	0	1
1	0	0	1	0
1	0	1	0	1
1	1	0	0	1
1	1	1	1	1

PROCEDURE

1. Mount ICs 7408, 7432, 7486 on breadboard.
2. Connect 14th pin of all ICs to 5V and 7th pin to ground.
3. Check gates in all the ICs before using them in the circuit.
4. For input, connect the input pin of gate to either ground (low) or 5V (high).
5. For output, connect the output pin of gate to the anode of LED and cathode to ground through 220Ω resistor. Role of resistor is to limit the current flowing through the LED.

6. Connect the circuit for half adder as shown in Fig. 4.3.1 on breadboard.

7. Give different combinations of input and note the corresponding output in tabular form.

8. Verify the truth tables.

9. Now connect the circuit for full adder as shown in Fig. 4.3.3 and repeat the above steps.

10. Mount IC 7400 and repeat steps 2 to 5.

11. Connect the circuit for half adder and full adder using NAND gate as shown in Fig. 4.3.2 and Fig. 4.3.4.

12. Repeat step 7 and 8.

OBSERVATIONS

Truth table for half adder and full adder circuit.

RESULT

Half adder, full adder and their respective truth tables have been designed and verified successfully.

DISCUSSION

Adders are used in arithmetic logic circuitry of a computer to add binary numbers. They are also used in calculators. A full adder consists of two half adders. Suppose we want to add two numbers for *e.g.* $b_4\, b_3\, b_2\, b_1 + a_4\, a_3\, a_2\, a_1$, we need one half adder for adding LSB and three full adders for adding other higher bits.

Experiment 4

HALF and FULL SUBTRACTOR

AIM

Design Half and Full Subtractor.

APPARATUS REQUIRED

Breadboard, connecting wires, 5V dc power supply, resistor - 220Ω, LED, multimeter, ICs – 7400, 7404, 7408, 7432, 7486.

THEORY

Combinational Circuits

Combinational circuits are those circuits in which output depends on the present value of the input only. They do not have memory. For *e.g.* adders and subtractors.

Half Subtractor

It is a logical circuit that is used to subtract two binary digits A and B giving two outputs as Difference and Borrow. It can be implemented by using XOR gate, AND gate and NOT gate or just by using the universal gates *i.e.* either NAND or NOR gate.

Half subtractor using XOR, AND and NOT gate

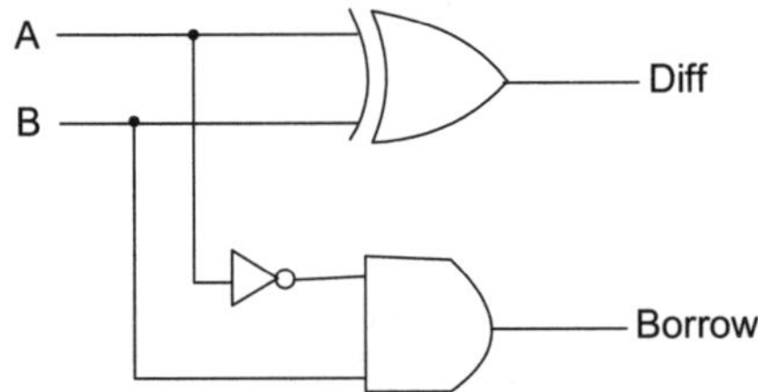

Fig. 4.4.1 Half subtractor using simple logic gates

Expressions for difference and borrow are given as:

$$\text{Diff} = A \oplus B$$

$$\text{Borrow} = \overline{A}.B$$

Half subtractor using NAND gate

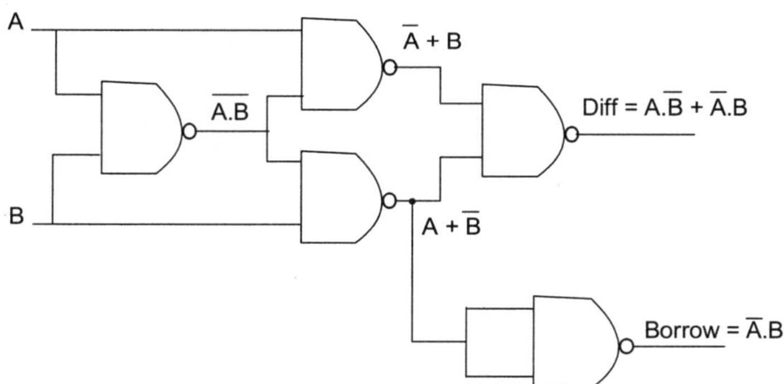

Fig. 4.4.2 Half subtractor using NAND gate

Truth table

Table 4.4.1 Truth table for half subtractor

Input		Output	
A	**B**	**Diff**	**Borrow**
0	0	0	0
0	1	1	1
1	0	1	0
1	1	0	0

Full Subtractor

It is a logical circuit that is used to subtract three binary digits A and B and previous borrow B_{in} and gives two outputs as Difference and Borrow. It can be implemented by using XOR, AND, OR and NOT gates or just by using the universal gates *i.e.* NAND or NOR gate.

Full subtractor using XOR, AND, OR and NOT gates

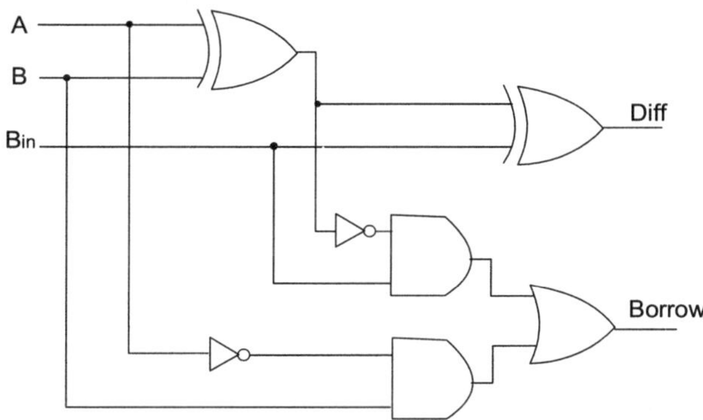

Fig. 4.4.3 Full subtractor using simple logic gates

Full subtractor using NAND gate

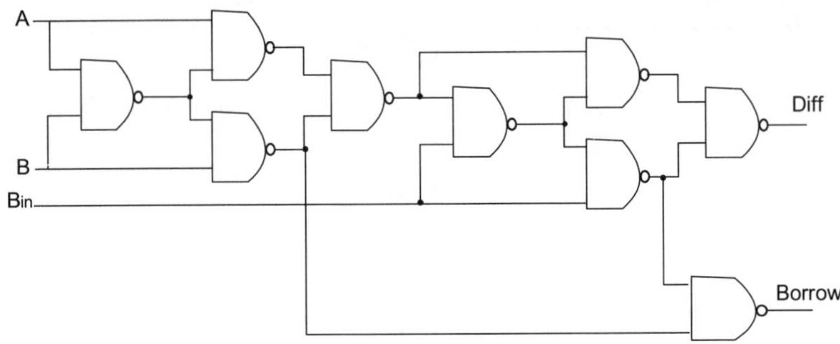

Fig. 4.4.4 Full subtractor using NAND gate

Expressions for difference and borrow are given as:

$$\text{Diff} = A \oplus B \oplus B_{in}$$

$$\text{Borrow} = \overline{A}.B + \overline{(A \oplus B)}.B_{in}$$

Truth table

Table 4.4.2 Truth table for full subtractor

Input			Output	
A	**B**	**Bin**	**Diff**	**Borrow**
0	0	0	0	0
0	0	1	1	1
0	1	0	1	1
0	1	1	0	1
1	0	0	1	0
1	0	1	0	0
1	1	0	0	0
1	1	1	1	1

PROCEDURE

1. Mount ICs 7408, 7432, 7486 on breadboard.

2. Connect 14th pin of all ICs to 5V and 7th pin to ground.

3. Check gates in all the ICs before using them in the circuit.

4. For input, connect the input pin of gate to either ground (low) or 5V (high).

5. For output, connect the output pin of gate to the anode of LED and cathode to ground through 220Ω resistor. Role of resistor is to limit the current flowing through the LED.

6. Connect the circuit for half subtractor as shown in Fig. 4.4.1 on breadboard.

7. Give different combinations of input and note the corresponding output in tabular form.

8. Verify the truth tables.

9. Now connect the circuit for full subtractor as shown in Fig. 4.4.3 and repeat the above steps.

10. Mount IC 7400 and repeat steps 2 to 5.

11. Connect the circuit for half and full subtractor using NAND gate as shown in Fig. 4.4.2 and Fig. 4.4.4.

12. Repeat step 7 and 8.

OBSERVATIONS

Truth table for half and full subtractor.

RESULT

Half subtractor, full subtractor and their respective truth tables have been designed and verified successfully.

DISCUSSION

Subtractors are used in arithmetic logic circuitry of computers to subtract binary numbers. They are also used in calculators. A full subtractor consists of two half subtractors. Suppose we want to subtract two numbers for *e.g.* b_4 b_3 b_2 b_1 - a_4 a_3 a_2 a_1, we need one half subtractor for LSB and three full subtractors for subtracting other higher bits.

Experiment 5

SEVEN SEGMENT DISPLAY DRIVER

AIM

Design a Seven Segment Display Driver.

APPARATUS REQUIRED

IC 7447, FND 507 seven segment LED display, breadboard, connecting wires, 5V dc power supply, multimeter, resistor - 220Ω.

THEORY

Seven Segment Display

It is an electronic display device used for displaying decimal numbers. There are two types of seven segment display - common cathode and common anode. In common cathode display, the cathode of all the LEDs are connected together to the ground and inputs are applied to the anode of LEDs. Input should be high to turn on the LED. In common anode display, anodes of all the LEDs are connected together to +5V and inputs are applied to the cathodes of LEDs. Input should be low to turn on the LED.

FND 507

It is a red GaAsP single digit seven segment LED display with a common anode configuration. It has seven LEDs named as a, b, c, d, e, f, g as shown in Fig. 4.5.1. There anodes are connected together to +5V and input is given to each of the cathode. The input should be low to turn on the LED. To protect these seven LEDs, resistors are used with each LED to limit the maximum current flowing through it.

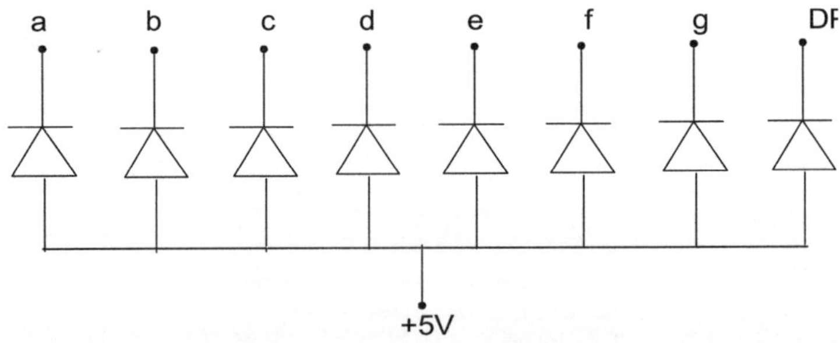

Fig. 4.5.1 Common anode seven segment

Where, DP: decimal point.

Pin diagram of FND 507 LED Display

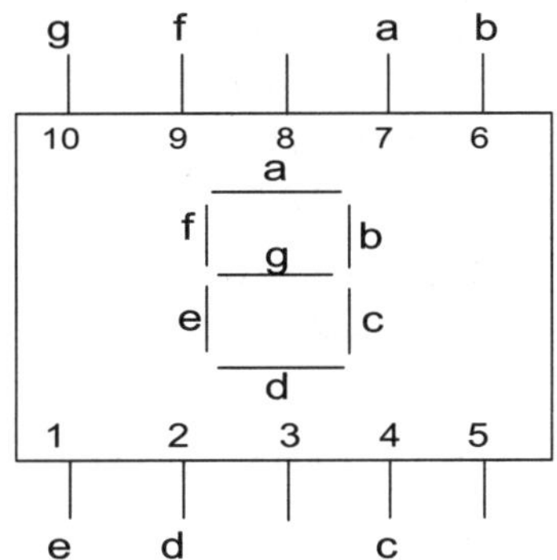

Fig. 4.5.2 FND 507 LED Display

Where,

1, 2, 4, 6, 7, 9, 10: input pins

3, 8: common anode

5: digit point

IC 7447

7447 IC as shown in Fig. 4.5.3 is used to drive common anode seven segment display.

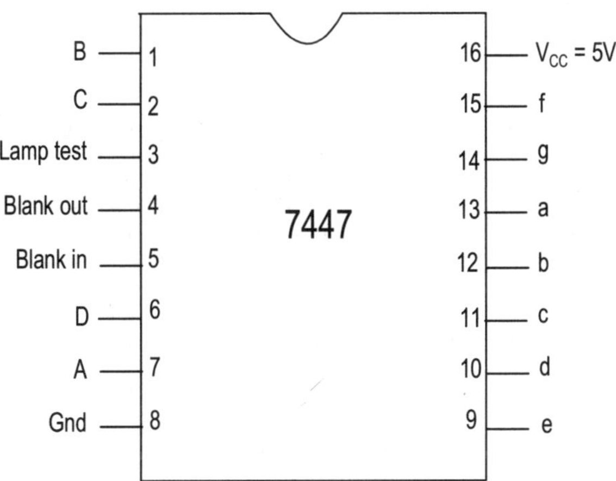

Fig. 4.5.3 7447 IC pin out diagram

Inputs are applied at pins A, B, C and D in BCD format. IC 7447 converts a 4 bit BCD number into a seven bit output at pins a, b, c, d, e, f, g which are active low that drives the seven segment display.

CIRCUIT DIAGRAM

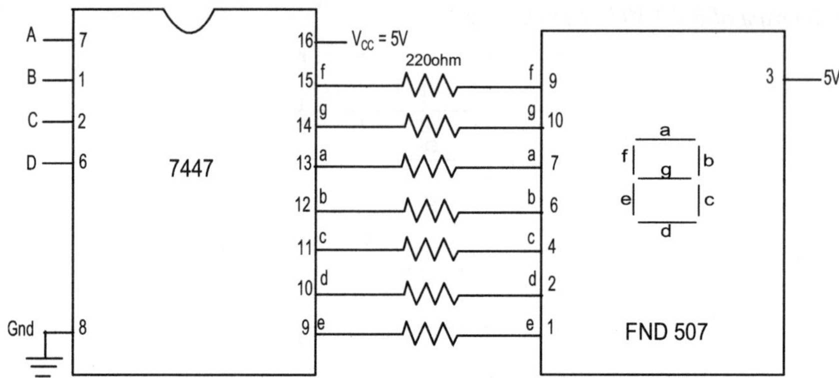

Fig. 4.5.4 Circuit diagram for seven segment display

PROCEDURE

1.   Connect the circuit as shown in Fig. 4.5.4 on breadboard.

2.   Connect 220Ω resistances between output pins of 7447 and cathode pins of FND 507 to limit the maximum current flowing through LEDs.

3.   Give different inputs at A, B, C and D and observe the number displayed on seven segment display.

OBSERVATIONS

Table 4.5.1 Observation table for seven segment display

BCD input				LED display							Displayed number
D	**C**	**B**	**A**	**a**	**b**	**c**	**d**	**e**	**F**	**g**	
0	0	0	0	0	0	0	0	0	0	1	0
0	0	0	1	1	0	0	1	1	1	1	1
0	0	1	0	0	0	1	0	0	1	0	2
0	0	1	1	0	0	0	0	1	1	0	3
0	1	0	0	1	0	0	1	1	0	0	4
0	1	0	1	0	1	0	0	1	0	0	5
0	1	1	0	0	1	0	0	0	0	0	6
0	1	1	1	0	0	0	1	1	1	1	7
1	0	0	0	0	0	0	0	0	0	0	8
1	0	0	1	0	0	0	0	1	0	0	9

RESULT

Seven segment display has been designed successfully to display digits 0 to 9.

DISCUSSION

Seven segment displays are used in digital clocks, electronic meters and other devices that display numeral information. They are also used to display various other letters and information like "no disc" on a CD player. Apart from 7 segment display, 14 or 16 segments are also available.

Experiment 6

MULTIPLEXER

AIM

Design 4 X 1 Multiplexer using Gates.

APPARATUS REQUIRED

Breadboard, connecting wires, 5 V dc power supply, resistor - 220Ω, IC – 7408, 7404, 7432, LED.

THEORY

Multiplexer

Multiplexer, also known as data selector is a combinational logic circuit having 2^n input lines, n select lines and one output line as shown in Fig. 4.6.1. It selects one of the inputs data according to the data on select lines, and sends it to the output. It can also be considered as many input and single output switch.

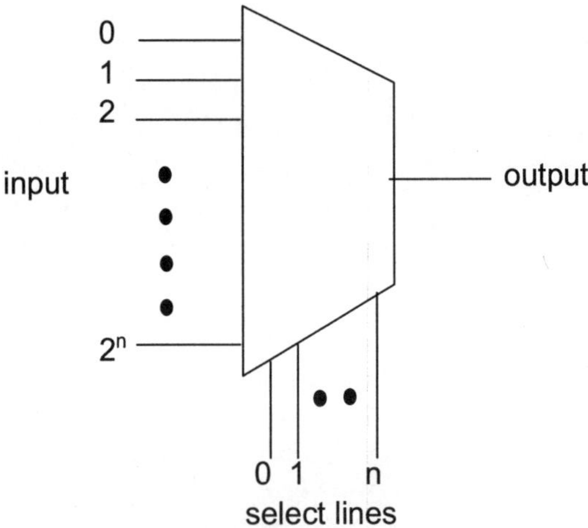

Fig. 4.6.1 Multiplexer

4 x 1 MUX

4x1 MUX has four inputs A, B, C and D, one output Z and two select lines S_0 and S_1 as shown in Fig. 4.6.2.

Truth table

Select lines		Output
S_1	S_0	Z
0	0	A
0	1	B
1	0	C
1	1	D

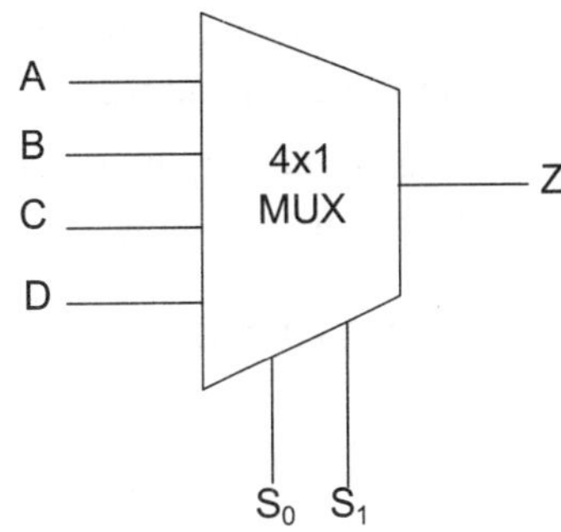

Fig. 4.6.2 4x1 MUX

Boolean expression

$$Z = A.\overline{S_1}.\overline{S_0} + B.\overline{S_1}.S_0 + C.S_1.\overline{S_0} + D.S_1.S_0$$

Implementation of 4x1 MUX using logic gates

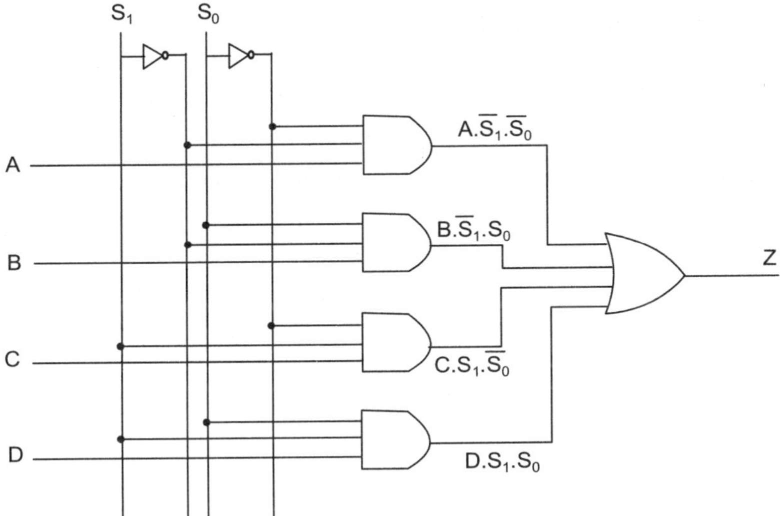

Fig. 4.6.3 4x1 MUX using gates

PROCEDURE

1. Mount ICs 7404, 7408, 7432 on breadboard.

2. Connect 14th pin of all ICs to 5V and 7th pin to ground.

3. Check gates in all the ICs before using them in the circuit.

4. For input, connect the input pin of gate to either ground (low) or 5V (high).

5. For output, connect the output pin of gate to the anode of LED and cathode to ground through 220Ω resistor. Role of resistor is to limit the current flowing through the LED.

6. Connect the circuit for 4x1 MUX as shown in Fig. 4.6.3.

7. Give different combinations of inputs at the input lines A, B, C, D and the select lines S_1 and S_0.

8. Observe the output according to the input given at select lines and record it in form of truth table.

OBSERVATIONS

Truth table for 4x1 MUX.

RESULT

4x1 MUX has been designed and verified successfully using simple logic gates.

DISCUSSION

Multiplexers are widely used to share the resources among multiple users. For *e.g.* (a) in communication, multiplexers are used to send more than one input message signal over the same communication channel and hence large data can be sent over a single line; (b) to connect more than one analog input to a single DAC.

Multiplexers have varied applications which include data selection, data routing, parallel-to-serial conversion, *etc.*

Experiment 7

FLIP FLOPS

AIM

To Build Flip-Flop Circuits using Elementary Gates (RS, Clocked RS, and D - type).

APPARATUS REQUIRED

Breadboard, connecting wires, 5V dc power supply, resistor - 220Ω, LED, IC – 7400, 7404, function generator, probe.

THEORY

Sequential Circuits

Sequential circuits are those circuits in which output depends on present as well as past value of the input. They are made of combinational circuits and memory elements to store the past values of input. For *e.g.* flip flops, registers, and counters.

Flip Flop

Flip flops, also known as bi-stable multivibrator, are the synchronous bi-stable devices. The term bi-stable means that the flip flop has two stable states *i.e.* 0 and 1 and the term synchronous means that the output of flip flop will change its state in accordance to the triggering input called as clock. It consists of logic gates that are connected in such a way so that it can store one bit of information.

Edge Triggered Flip Flop

It can be of two types:

1. Positive edge triggered flip flop: Output of the flip flop changes its state at positive or rising edge of the clock *i.e.* flip flop is sensitive to the input when clock makes transition from low to high.

2. Negative edge triggered flip flop: Output of the flip flop changes its state at negative or falling edge of the clock *i.e.* flip flop is sensitive to the input when clock makes transition from high to low.

Latch

A latch is a bi-stable multivibrator with two stable states *i.e.* 0 and 1. It is similar to a flip flop and is used as a temporary storage device. As soon as the input is applied to the latch, it changes its state accordingly. The output does not depend on the triggering input clock.

S-R (Set-Reset) latch

It has two inputs Set (S) and Reset (R) with two outputs as Q and Q'. It can be constructed by using universal gates *i.e.* either NAND gate or NOR gate.

When Set input is high, output is set *i.e.* Q = 1 and Q' = 0 and when Reset input is made high, output is reset *i.e.* Q = 0 and Q' = 1. When both the inputs are low, the output remains unchanged and when both the inputs are high, output is unpredictable. The last state is not used.

Active low input $\overline{S}$-$\overline{R}$ latch

It is formed by two cross-coupled NAND gates as shown in Fig. 4.7.1.

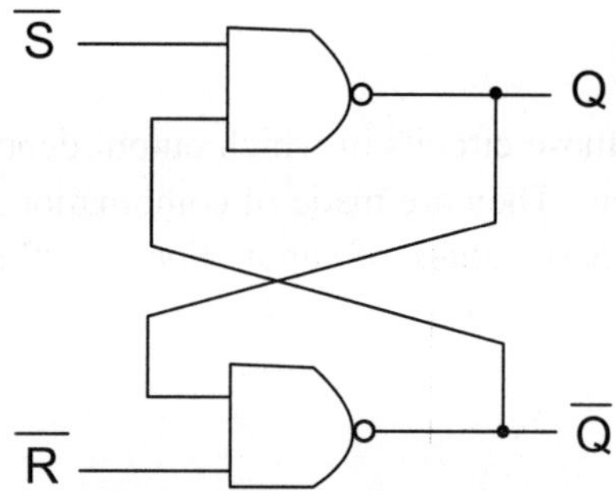

Fig. 4.7.1 Active low $\overline{S}$-$\overline{R}$ latch

Truth table

Table 4.7.1 Truth table for active low $\overline{S}$-$\overline{R}$ latch

$\overline{S}$	$\overline{R}$	Q_n	Q_{n+1}	State
0	0	0	X	Invalid
0	0	1	X	
0	1	0	1	Set
0	1	1	1	
1	0	0	0	Reset
1	0	1	0	
1	1	0	0	No change
1	1	1	1	

Active high input S-R latch

It is formed by two cross-coupled NAND gates as shown in Fig. 4.7.2.

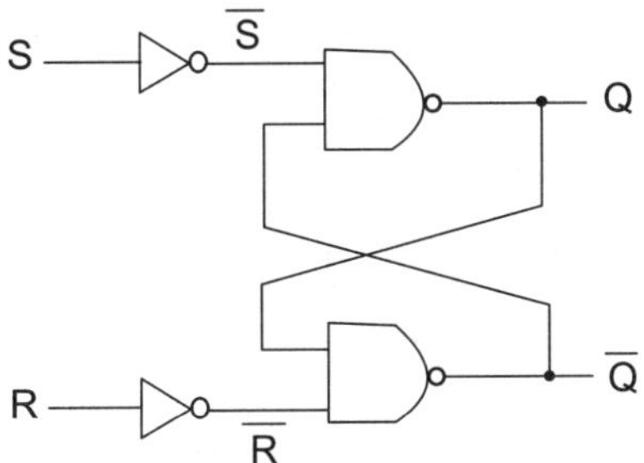

Fig. 4.7.2 Active high S-R latch

Truth table

Table 4.7.2 Truth table for active high S-R latch

S	R	Q_n	Q_{n+1}	State
0	0	0	0	No change
0	0	1	1	
0	1	0	0	Reset
0	1	1	0	
1	0	0	1	Set
1	0	1	1	
1	1	0	X	Invalid
1	1	1	X	

Clocked S-R flip flop

Clocked S-R flip flop is similar to S-R latch with an additional clock input. The two inputs set and reset will change the output of flip flop in accordance to the clock. Fig. 4.7.3 shows a clocked S-R flip flop where clock needs to be high for output to respond hence they are also called as level triggered flip flops.

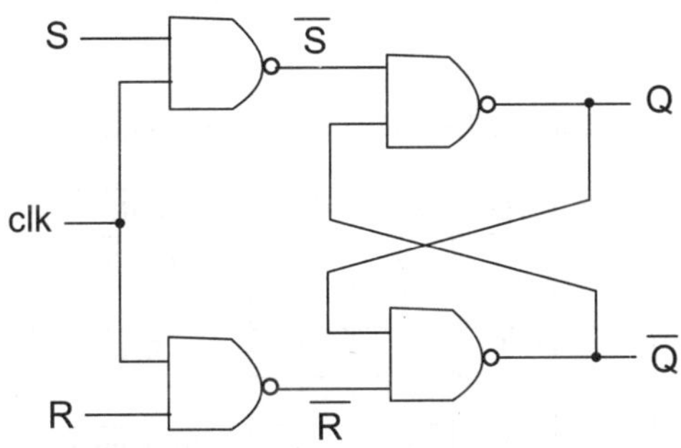

Fig. 4.7.3 Clocked S-R flip flop

Truth table

Table 4.7.3 Truth table for clocked S-R flip flop

Clk	S	R	Q_n	Q_{n+1}	State
1	0	0	0	0	No change
1	0	0	1	1	
1	0	1	0	0	Reset
1	0	1	1	0	
1	1	0	0	1	Set
1	1	0	1	1	
1	1	1	0	X	Invalid
1	1	1	1	X	
0	X	X	0	0	No change
0	X	X	1	1	

D flip flop

It is also known as delay flip flop. It has only one input 'D' and the other input is just the complement of first which can be implemented by using NOT gate. If S and R are connected together with a NOT gate in S-R flip flop, it becomes D flip flop as shown in Fig. 4.7.4.

When D = 1 (S = 1 R = 0), Q = 1 and when D = 0 (S = 0 R = 1), Q = 0 *i.e.* output follows the input only when clock is high.

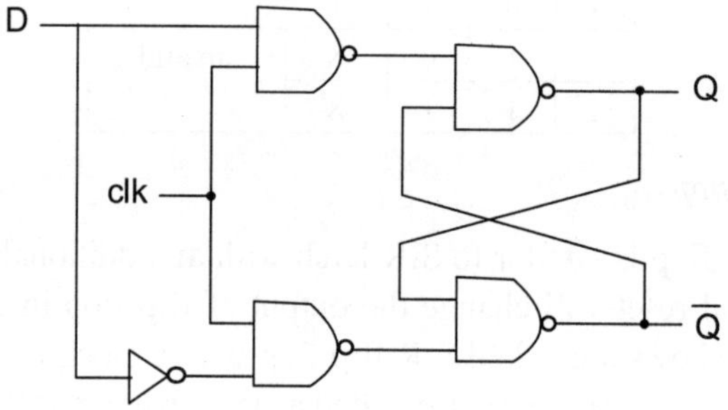

Fig. 4.7.4 D flip flop

Truth table

Table 4.7.4 Truth table for D flip flop

Clk	D	Q_n	Q_{n+1}	State
1	0	0	0	Reset
1	0	1	0	
1	1	0	1	Set
1	1	1	1	
0	X	0	0	No change
0	X	1	1	

PROCEDURE

1. Mount ICs 7404, 7400 on the breadboard.

2. Connect 14th pin of all ICs to 5V and 7th pin to ground.

3. Check gates of all the ICs before using them in the circuit.

4. For input, connect the input pin of gate to either ground (low) or 5V (high).

5. For output, connect the output pin of gate to the anode of LED and cathode to ground through 220Ω resistor. Role of resistor is to limit the current flowing through the LED.

6. Connect the circuit for S-R latch as shown in Fig. 4.7.1 and Fig. 4.7.2 on breadboard.

7. Give different combinations of input and observe the output on LEDs.

8. Record the readings in form of truth table.

9. Now connect the circuit for flip flops as shown in Fig. 4.7.3 and Fig. 4.7.4.

10. Connect function generator using probe to give the clock input.

11. Set function generator to give a square wave with frequency 1Hz.

12. Repeat steps 7 and 8.

OBSERVATIONS

Truth tables for RS, clocked RS and D type flip flop.

RESULT

S-R latch, S-R clocked flip flop, D type flip flop and their respective truth tables have been designed and verified successfully.

DISCUSSION

Flip flops are used as memory element in computers to store the binary data. One flip flop can store one bit of data. Hence, to store 'n' bit of data, 'n' flip flops are required. Group of flip flops is called as register. Flip flops can be connected together to design shift registers and counters which can count the pulses electronically. 'N' flip flops can count 2^N pulses.

Experiment 8

COUNTER

AIM

Design a Counter using D/T/JK Flip-Flop.

APPARATUS REQUIRED

Breadboard, connecting wires, 5V dc power supply, resistor - 220Ω, LED, IC-7476, function generator, probe.

THEORY

Counter

It is a set of flip flops used to count the number of clock pulses. Since each clock pulse occur at a known interval hence a counter can be used to measure period or frequency of a signal. The basic difference between a register and a counter is that a register has no specified sequence of states whereas a counter has a specified sequence of states.

Modulus of counter

A counter can count in up direction *i.e.* 0, 1, 2.....N or it can count in down direction *i.e.* N, N-1......1, 0. Each of this count is called as a state of counter and the number of states through which counter passes is called modulus of the counter. An n-bit counter has 'n' flip flops and it can count 2^n states.

Two types of counter

1. Asynchronous/ripple counters: Asynchronous counters, also known as serial counters, are easy to design and require less hardware. All the flip flops used in the counter change their state irrespective of the clock. Their speed is low as compared to synchronous counter.

2. Synchronous counters: Synchronous counters, also known as parallel counters, are complex to design and implement. Clock is applied to all the flip flops simultaneously and all the flip flops used in the counter change their state with respect to clock. Their speed is fast as compared to asynchronous counter.

J-K Flip Flop

It is the most versatile and widely used flip flop. The logic symbol is shown in Fig. 4.8.1. It is identical to R-S flip flop except that there is no invalid state as in the case of R-S flip flop.

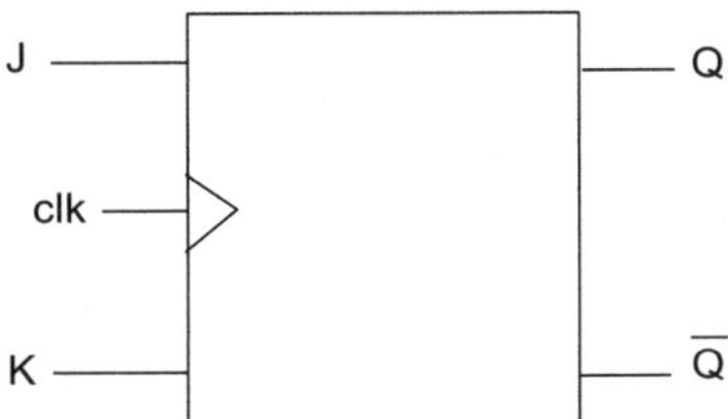

Fig. 4.8.1 Logic symbol for J-K flip flop

Truth table

Table 4.8.1 Truth table for JK flip flop

Clk	J	K	Q_{n+1}	Q'_{n+1}	State
High	0	0	Q_n	Q'_n	No change
High	0	1	0	1	Reset
High	1	0	1	0	Set
High	1	1	Q'_n	Q_n	Toggle
0	X	X	Q_n	Q'_n	No change

T Flip Flop

It is also known as Toggle flip flop. It has only one input 'T'. When input is high, output toggles with every clock pulse. If J and K are connected together, it becomes T flip flop as shown in Fig. 4.8.2.

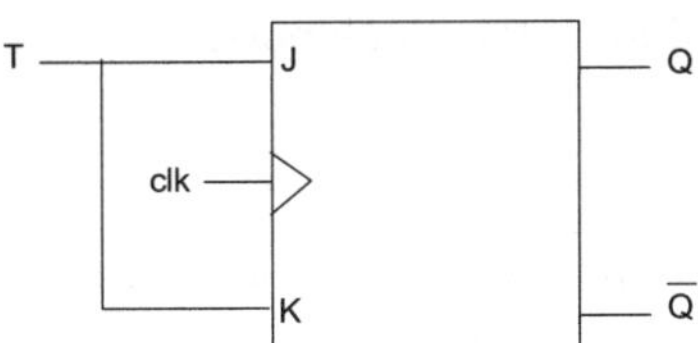

Fig. 4.8.2 Logic symbol for T flip flop

Truth table

Table 4.8.2 Truth table for T flip flop

Clk	T	Q_{n+1}	Q'_{n+1}	State
High	0	Q_n	Q'_n	No change
High	1	Q'_n	Q_n	Toggle
0	X	Q_n	Q'_n	No change

D Flip Flop

It is also known as delay flip flop. It has only one input 'D'. In S-R flip flop, when the inputs S and R are connected together with a NOT gate, it becomes D flip flop as shown in Fig. 4.8.3. When input is high, output is high and when input is low, output is also low *i.e.* output follows the input only when clock is high.

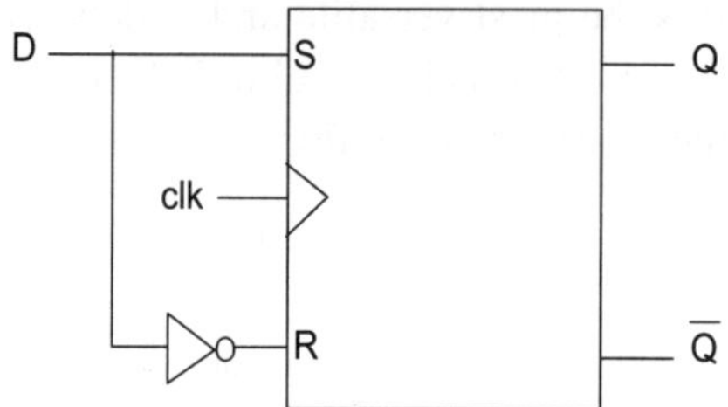

Fig. 4.8.3 Logic symbol for D flip flop

Truth table

Table 4.8.3 Truth table for D flip flop

Clk	D	Q_{n+1}	Q'_{n+1}	State
High	0	0	1	Reset
High	1	1	0	Set
0	X	Q_n	Q'_n	No change

Synchronous Inputs

S-R, D, T, J-K are called as synchronous inputs because data is transferred to the flip flop only according to the clock pulse.

Asynchronous Inputs

There are two types of asynchronous inputs: Preset and Clear. These inputs can affect the state of the flip flop irrespective of the clock. If Preset = 1, the flip flop will set to 1 and if Clear = 1, the flip flop will reset to 0. If Pr = Cr = 0, asynchronous inputs become ineffective and synchronous inputs *i.e.* JK/SR/D/T will work normally according to the clock pulse.

IC 7476

It is a 16 pin dual JK flip flop with PRESET and CLEAR pin.

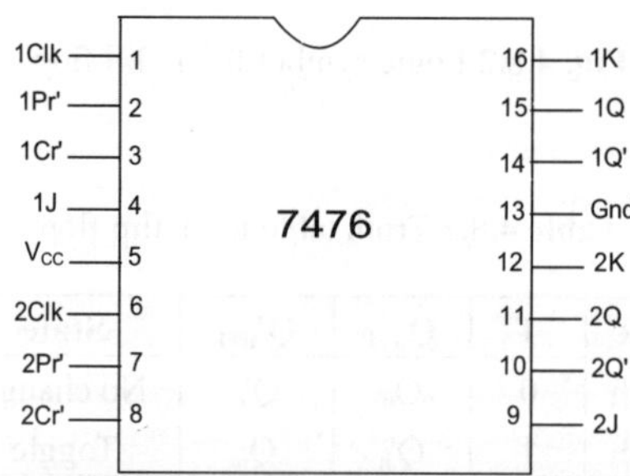

Fig. 4.8.4 Pin diagram for IC-7476

Up Counter using J-K Flip Flop

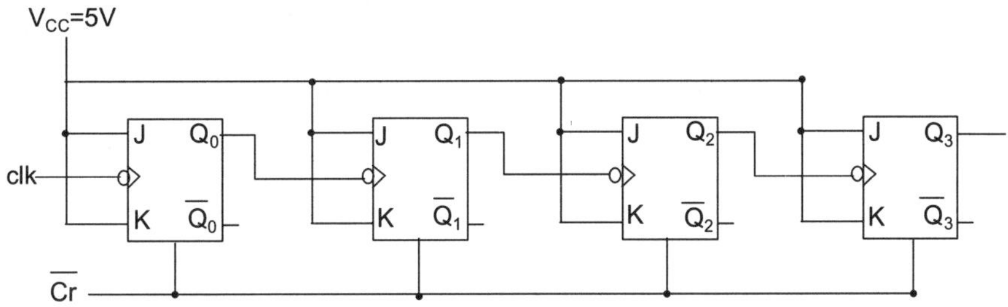

Fig. 4.8.5 Four bit up asynchronous counter

Fig. 4.8.5 shows a four bit up asynchronous counter which count 16 states *i.e.* from 0 to 15 and then again come back to 0. It is constructed with four negatively edge triggered J-K type flip flops. J and K inputs are connected together and therefore it works as a T flip-flop. Clock is connected to the first flip flop and then output of each flip flop is connected as a clock to the next flip flop. Hence, with the negative edge of every clock pulse, only first flip flop will toggle and then the output of each flip flop will change the state of next flip flop.

Let us assume initially all the flip flops are reset. As the first clock pulse arrives, the first flip flop will toggle its state from 0 to 1. With the second clock pulse, first flip flop output will again toggle its state from 1 to 0. Since Q_0 is connected as the clock to second flip flop and it is a negatively edge triggered flip flop therefore, as Q_0 goes from 1 to 0, the second flip flop output will also change its state from 0 to 1. Similarly, third and fourth flip flop will change its state when Q_1 and Q_2 go from 1 to 0 respectively.

Truth table

Table 4.8.4 Truth table for 4 bit up asynchronous counter

Clock pulse no.	Q3	Q2	Q1	Q0
0	0	0	0	0
1	0	0	0	1
2	0	0	1	0
3	0	0	1	1
4	0	1	0	0
5	0	1	0	1
6	0	1	1	0
7	0	1	1	1
8	1	0	0	0
9	1	0	0	1
10	1	0	1	0
11	1	0	1	1

12	1	1	0	0
13	1	1	0	1
14	1	1	1	0
15	1	1	1	1
16	0	0	0	0

Down Counter using JK Flip Flop

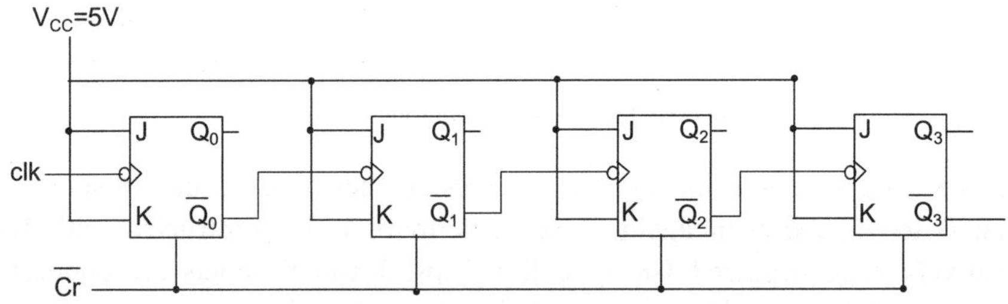

Fig. 4.8.6 Four bit down asynchronous counter

Fig. 4.8.6 shows a four bit down asynchronous counter which count 16 states *i.e.* from 15 to 0 and then again return back to 15. It is constructed with four negatively edge triggered J-K type flip flops. J and K inputs are connected together and therefore it works as a T flip-flop. Clock is connected to the first flip flop and then $\overline{Q}$ of each flip flop is connected as a clock to the next flip flop. Hence, with the negative edge of every clock pulse, only first flip flop will toggle and then the output of each flip flop will change the state of next flip flop.

Let us assume that initially all the flip flops are reset. As the first clock pulse arrives, the first flip flop will toggle its state from 0 to 1. With the second clock pulse, first flip flop output will again toggle its state from 1 to 0. Since $\overline{Q_0}$ is connected as the clock to second flip flop and it is a negatively edge triggered flip flop therefore, as $\overline{Q_0}$ goes from 1 to 0, the second flip flop output will also change its state from 0 to 1. Similarly, third and fourth flip flop will change its state when $\overline{Q_1}$ and $\overline{Q_2}$ go from 1 to 0 respectively.

Truth table

Table 4.8.5 Truth table for 4 bit down asynchronous counter

Clock pulse no.	Q3	Q2	Q1	Q0	$\overline{Q0}$
0	0	0	0	0	1
1	1	1	1	1	0
2	1	1	1	0	1
3	1	1	0	1	0
4	1	1	0	0	1
5	1	0	1	1	0
6	1	0	1	0	1

7	1	0	0	1	0
8	1	0	0	0	1
9	0	1	1	1	0
10	0	1	1	0	1
11	0	1	0	1	0
12	0	1	0	0	1
13	0	0	1	1	0
14	0	0	1	0	1
15	0	0	0	1	0
16	0	0	0	0	1

PROCEDURE

1. Mount IC 7476 on the breadboard.

2. Check the functionality of J-K flip flop before making any connections.

3. Connect the circuit for up counter as shown in Fig. 4.8.5.

4. Connect function generator to first flip flop to give the clock input using a probe.

5. Set function generator to give a square wave output with frequency 1Hz.

6. Connect anode of LED to the output pin of flip flop and cathode to the clock input of next flip flop through a 220Ω resistor.

7. Set asynchronous input Pr' to 1 by connecting it to 5V and make it ineffective.

8. Set asynchronous input Cr' to 0 by connecting it to ground and clear all the flip flops.

9. Now, set Cr' to 1 by connecting it to 5V and make it ineffective.

10. Observe the state of all flip flops with each clock cycle on the LEDs connected at their output pins.

11. Now, connect the circuit for down counter as shown in Fig. 4.8.6 and repeat the above steps.

OBSERVATIONS

Truth table for up and down counter.

RESULT

Asynchronous up and down counter have been designed successfully using J-K flip flop. These counters can count 16 different states and hence, modulus of the counter is 16.

DISCUSSION

Counters are widely used to count the number of pulses which can be used to measure time, period or frequency of a signal. They are also used as frequency divider to obtain the waveform of required frequencies; to perform the timing function in digital watches; to create time delays; to produce non-sequential binary counts and to generate pulse trains.

Experiment **9**

SHIFT REGISTERS

AIM

Design a Shift Register and Study Serial and Parallel Shifting of Data.

APPARATUS REQUIRED

Breadboard, connecting wires, 5V dc power supply, resistor - 220Ω, LED, function generator, probe, IC – 7400, 7474.

THEORY

IC – 7474

It is a 14 pin dual D flip flop IC with Preset (S') and Clear (R').

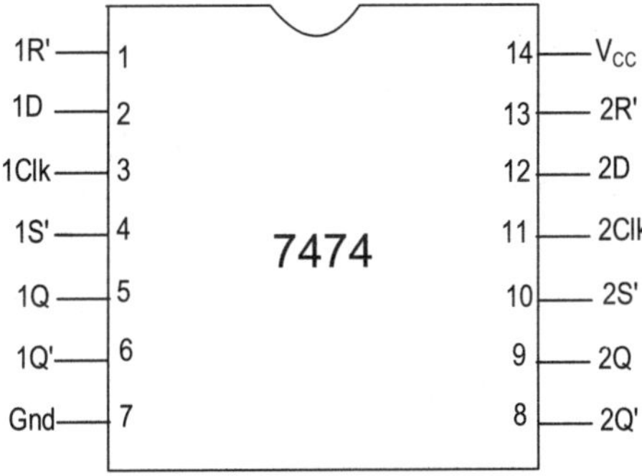

Fig. 4.9.1 Pin diagram for IC 7474

Register

A register is a group of flip flops which is capable of storing more than one bit long data. Since, a flip flop can store only one bit of data hence, it is called as a single bit register. To store 'n' bit of data, 'n' flip flops are required in a register. This storage capability of register makes them useful for a memory device. The two basic functions of register are: Data storage and Data movement

Shift Registers

A register capable of shifting the data present in binary form in one or both the directions is called as shift register. The data can be shifted into the register or out of the register. Whenever new information is transferred into a register, it is called loading of the register. This can be done in two ways.

1. Serial loading: In serial loading of data, the information is loaded into the register one bit at a time *i.e.* only one bit shifts out of the source register and enters the destination register.

2. Parallel loading: In parallel loading of data, all the data bits are loaded into the register with a common clock pulse *i.e.* at the same time.

Different types of data movement in shift registers are the following.

Serial in serial out (SISO)

When the data is shifted in or out of the registers serially *i.e.* one bit at a time then it is called as serial in serial out. The basic structure is shown in Fig. 4.9.2.

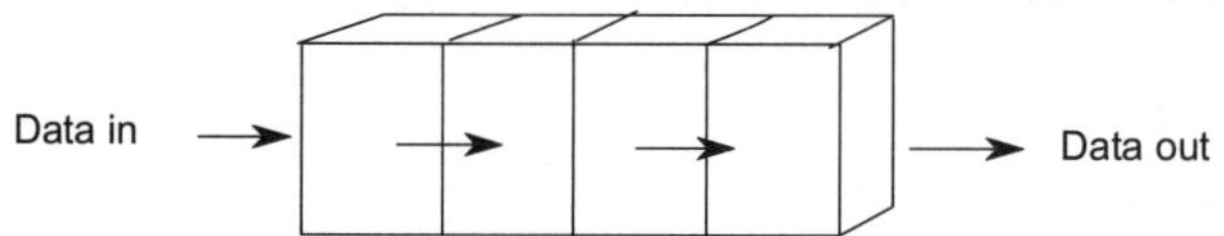

Fig. 4.9.2 Basic structure of SISO

Fig. 4.9.3 shows a four bit serial in serial out shift right register. It is constructed with four negatively edge triggered D type flip flops, all connected with a common clock. It can store 4 bit of data. The data is loaded into the first flip flop through serial in. With negative edge of every clock pulse, this data gets shifted to the next flip flop and then finally goes out through serial out pin.

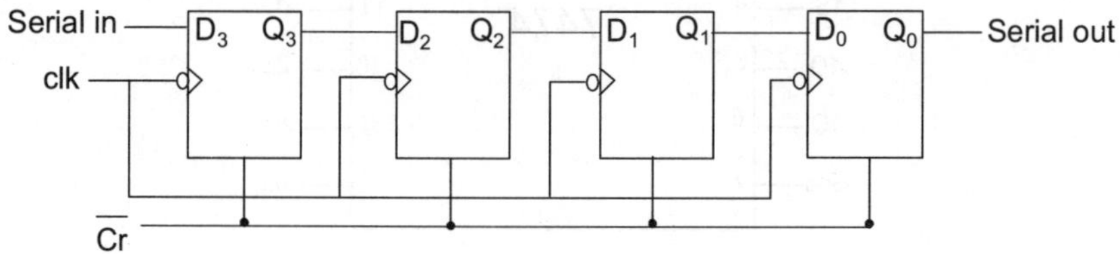

Fig. 4.9.3 Serial in serial out shift register

Truth table

Let data be 1011

Table 4.9.1 Truth table for SISO register

Clock	Serial in	D3 Q3	D2 Q2	D1 Q1	D0 Q0	Serial out
		0	0	0	0	
1	1	1	0	0	0	
2	1	1	1	0	0	
3	0	0	1	1	0	
4	1	1	0	1	1	
5		0	1	0	1	1

6		0	0	1	0	1
7		0	0	0	1	0
8		0	0	0	0	1

Serial in parallel out (SIPO)

When the data is shifted serially *i.e.* one bit at a time and all the output bits are available simultaneously *i.e.* at the same time then it is called as serial in parallel out. The basic structure is shown in Fig. 4.9.4.

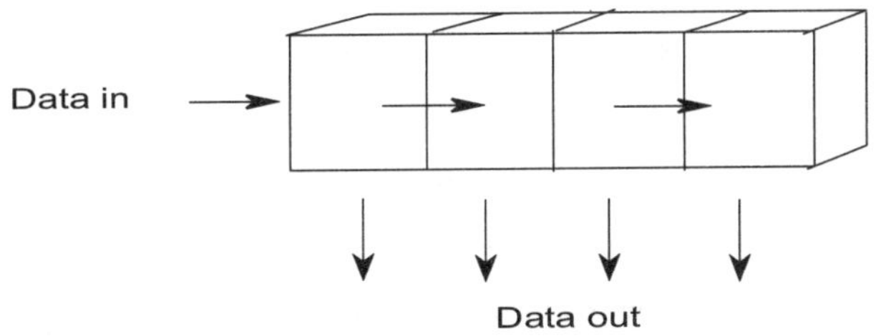

Fig. 4.9.4 Basic structure of SIPO

Fig. 4.9.5 shows a four bit serial in parallel out shift register. It is constructed with four negatively edge triggered D type flip flops, all connected with a common clock. It can store 4 bit of data. The data is first loaded into the first flip flop through serial in. Once the data is available in four bit register, the four bits are available at the same clock pulse at the output.

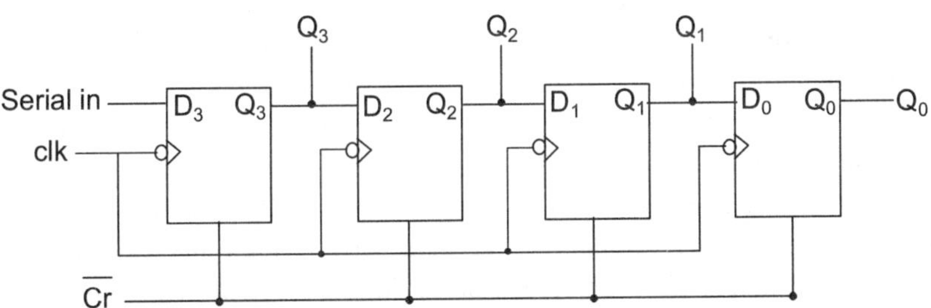

Fig. 4.9.5 Serial in parallel out shift register

Truth table

Let data be 1011

Table 4.9.2 Truth table for SIPO register

Clock	Serial in	D3 Q3	D2 Q2	D1 Q1	D0 Q0
		0	0	0	0
1	1	1	0	0	0
2	1	1	1	0	0
3	0	0	1	1	0
4	1	1	0	1	1

After the fourth clock pulse, the register has been loaded serially with the data and this data is available at the output at a same time.

Parallel in serial out (PISO)

When the data is loaded into the register simultaneously on parallel lines rather than on a single line and the output is taken serially *i.e.* one bit at a time then it is called as parallel in serial out. The basic structure is shown in Fig. 4.9.6.

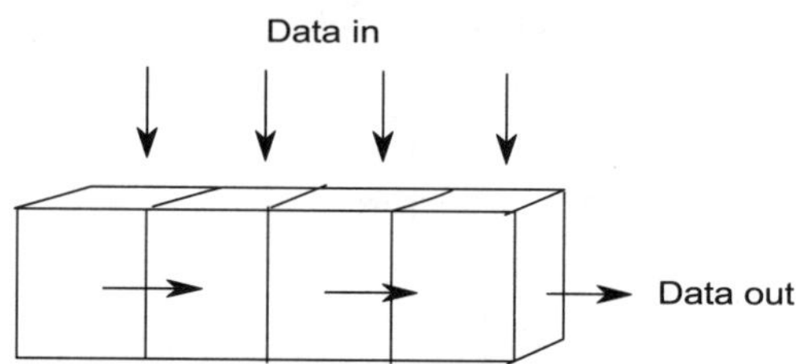

Fig. 4.9.6 Basic structure of PISO

Fig. 4.9.7 shows a four bit parallel in serial out shift register. It is constructed with four negatively edge triggered D type flip flops, all connected with a common clock. It can store 4 bit of data. The data is first loaded into all the four flip flops at a same time. Once the data is available, then with negative edge of every clock pulse, this data gets shifted to the next flip flop and then finally goes out through serial out pin.

Load line is used for parallel loading of the data. Initially all the flip flops are cleared. Now, when the load line is made high, all the NAND gates get enabled which allows parallel loading of the bits into the register. For *e.g.* load = 1, a_3 = 1, therefore, output of first NAND gate = 0. Hence Pr' = 0 *i.e.* Pr = 1. It will set the first flip flop to 1. If a_2 = 0, output of second NAND gate = 1. Hence Pr' = 1 *i.e.* Pr = 0. The flip flop will remain cleared to 0. Once the data has been loaded into the register, load line is made low which makes Preset input ineffective and then with every clock pulse data is shifted serially and is taken out at serial out pin.

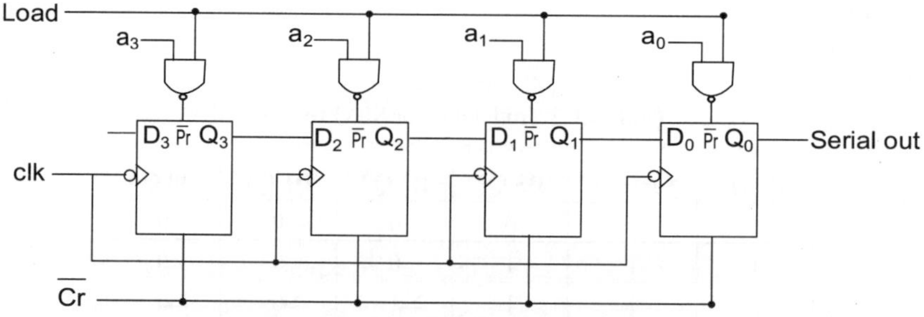

Fig. 4.9.7 Parallel in serial out shift register

Truth table

Let data be 1011 (a_3=1, a_2=0, a_1=1, a_0=1)

Table 4.9.3 Truth table for PISO register

Clock	D3 Q3	D2 Q2	D1 Q1	D0 Q0	Serial out
	1	0	1	1	
1	0	1	0	1	1
2	0	0	1	0	1
3	0	0	0	1	0
4	0	0	0	0	1

Parallel in parallel out (PIPO)

When the data is loaded into the register simultaneously on parallel lines rather than on a single line and the output bits are also available simultaneously then it is called as parallel in parallel out. The basic structure is shown in Fig. 4.9.8.

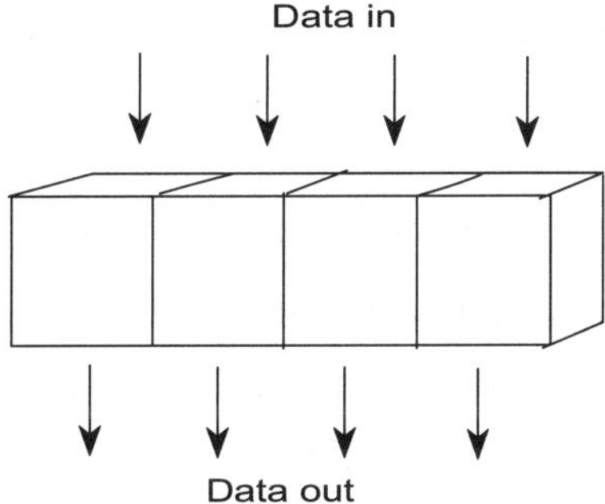

Fig. 4.9.8 Basic structure of PIPO

Figure 4.9.9 shows a four bit parallel in parallel out shift register. It is constructed with four negatively edge triggered D type flip flops, all connected with a common clock. It can store 4 bits of data. The data is first loaded into all the four flip flops at a same time. Once the data is available in four bit register, the four bits are available at the output on the same clock pulse.

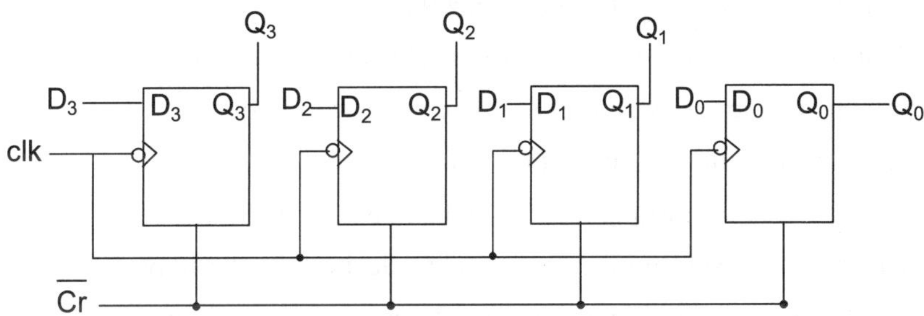

Fig. 4.9.9 Parallel in parallel out shift register

Truth table

Let data be 1011

Table 4.9.4 Truth table for PIPO register

Clock	D3 Q3	D2 Q2	D1 Q1	D0 Q0
1	1	0	1	1

PROCEDURE

Serial In Serial Out Shift Register

1. Mount IC 7474 on the breadboard.
2. Check the functionality of D flip flop before making any connections.
3. Connect the circuit as shown in Fig. 4.9.3.
4. Connect function generator to give the clock input using a probe.
5. Set function generator to give a square wave with frequency 1Hz.
6. Connect anode of LED to the serial out pin and cathode to ground through 220Ω resistor.
7. Set Pr' = 1 by connecting it to 5V and make it ineffective.
8. Set Cr' = 0 by connecting it to ground to clear all the flip flops.
9. Now, set Cr' = 1 by connecting it to 5V and make it ineffective.
10. Give input data through serial in pin.
11. Observe the output with each clock cycle at the LED connected at the serial out pin.

Serial In Parallel Out Shift Register

1. Mount IC 7474 on the breadboard.
2. Check the functionality of D flip flop before making any connections.
3. Connect the circuit as shown in Fig. 4.9.5.
4. Connect function generator to give the clock input using a probe.
5. Set function generator to give a square wave with frequency 1Hz.
6. Connect LEDs at the Q output of all four flip flops.
7. Set Pr' = 1 by connecting it to 5V and make it ineffective.
8. Set Cr' = 0 by connecting it to ground to clear all the flip flops.
9. Now, set Cr' = 1 by connecting it to 5V and make it ineffective.
10. Give input data through serial in pin.

11. After fourth clock cycle, 4 bit data is available in the register which can be seen on four LEDs.

Parallel In Serial Out Shift Register

1. Mount IC 7474 on the breadboard.

2. Check the functionality of D flip flop before making any connections.

3. Connect the circuit as shown in Fig. 4.9.7.

4. Connect function generator to give the clock input using a probe.

5. Set function generator to give a square wave output with frequency 1Hz.

6. Connect anode of LED to the serial out pin and cathode to ground through 220Ω resistor.

7. Set load line to 0 to make preset input ineffective.

8. Set Cr' = 0 by connecting it to ground to clear all the flip flops.

9. Now, set Cr' = 1 by connecting it to 5V and make it ineffective.

10. Set load line to 1 and give inputs through a_3, a_2, a_1 and a_0.

11. Set load line to 0 to make preset input ineffective.

12. Connect D_3 to ground and observe the output at the LED connected at serial out pin with every clock cycle.

Parallel In Parallel Out Shift Register

1. Mount IC 7474 on the breadboard.

2. Check the functionality of D flip flop before making any connections.

3. Connect the circuit as shown in Fig. 4.9.9.

4. Connect function generator to give the clock input using a probe.

5. Set function generator to give a square wave output with frequency 1Hz.

6. Connect LEDs at the Q output of all four flip flops.

7. Set Pr' = 1 by connecting it to 5V and make it ineffective.

8. Set Cr' = 0 by connecting it to ground to clear all the flip flops.

9. Now, set Cr' = 1 by connecting it to 5V and make it ineffective.

10. Give input data through D input of flip flops.

11. Observe the output at the LEDs connected at the output pin of flip flops.

Note

To view the change in state of all flip flops, connect LED at the Q pin of all flip flops.

OBSERVATIONS

Truth table for shift registers.

RESULT

Various types of shift registers and their respective truth tables have been studied and verified successfully.

DISCUSSION

Parallel data transmission is preferred when high speed data transmission is required. All the bits are transferred simultaneously on parallel lines which results in fast data transmission. On the other hand, serial data transmission is preferred when the amount of data to be sent is very large which becomes impractical and costly in case of parallel data transmission as the required number of parallel lines will also be very large.

Shift registers are widely used for storing and transferring the data in a digital system; designing Ring counter and Johnson counter; converting serial form of data to parallel form and vice-versa.

Computers send and receive data in parallel form but other peripheral devices which are connected to it send and receive data serially. Hence, UART *i.e.* Universal Asynchronous Receiver Transmitter is the device which is used to provide the synchronization between computer and peripheral attached to it. It contains shift registers required to convert the parallel data to serial form and vice-versa.

Experiment 10

DIGITAL to ANALOG CONVERTER

AIM

To Design a Digital to Analog Converter of Given Specification.

APPARATUS REQUIRED

Breadboard, connecting wires, $\pm$ 12V and 5 V dc power supply, resistors - 1kΩ and 2kΩ, IC-741 (op-amp), multimeter.

THEORY

Digital to Analog Converter

DAC is a device which is used to convert the data in digital form into analog form.

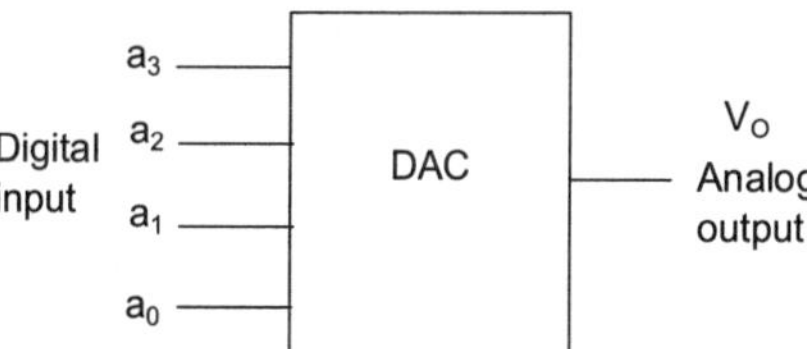

Fig. 4.10.1 Block diagram of 4 bit DAC

The digital inputs can have value either 0 or 1. Therefore, for four input lines, 16 different combinations of input can be given and a unique analog output is produced for each of the corresponding inputs.

There are two methods of conversion from digital to analog value:

1. Weighted Resistor DAC: In weighted resistor DAC, number of resistors depends on the number of bits present in the binary number. All the resistances are of different values with their value proportional to the numerical significance of the bit and hence, it is called as weighted resistor DAC. Disadvantage of this method is that if number of bits in a binary number is very large then a wide range of resistances is required and therefore, this method is not preferred.

2. R – 2R ladder type DAC: This is the most preferred method because it uses only two values of resistances *i.e.* R and 2R and the absolute value of resistances is not important but the ratio of two resistances is important. In this, number of resistances used is double the number of bits.

CIRCUIT DIAGRAM

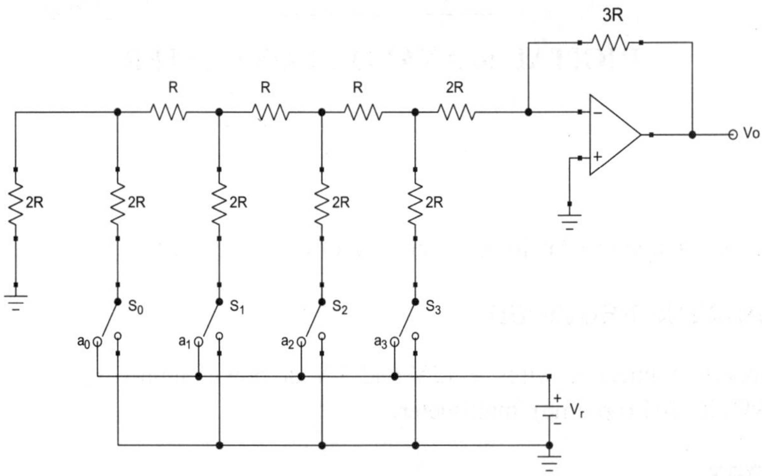

Fig. 4.10.2 Circuit for R-2R ladder type DAC

Fig. 4.10.2 shows the circuit for R-2R ladder type DAC. The ladder acts as current splitting device and consists of only two values of resistances *i.e.* R and 2R. The operational amplifier is used as voltage follower to prevent the loading of the circuit by providing very high input impedance and low output impedance.

The equivalent circuit for the ladder is shown in Fig. 4.10.3.

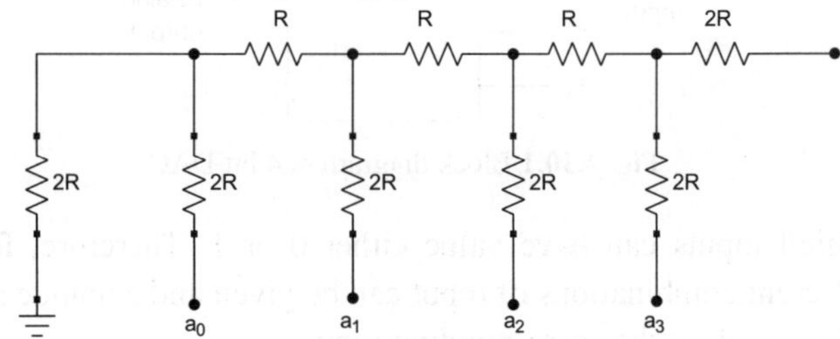

Fig. 4.10.3 Equivalent circuit for R-2R ladder

a. Let us assume $a_0 = 1$ and reset all bits to zero, then the equivalent circuit is shown in Fig. 4.10.4.

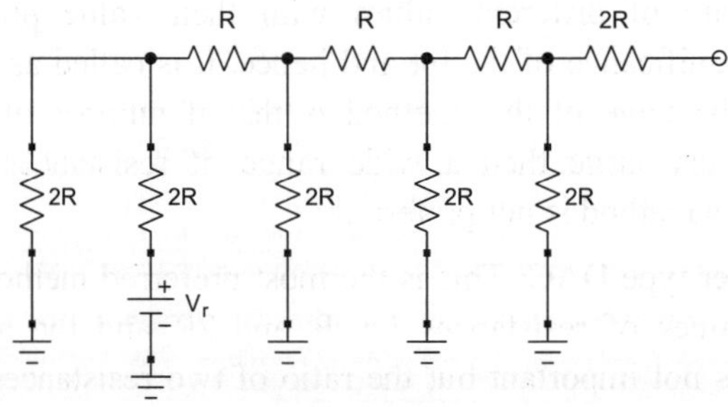

Fig. 4.10.4

Let us find the total voltage at the output of this ladder network using thevenin theorem.

Step 1:

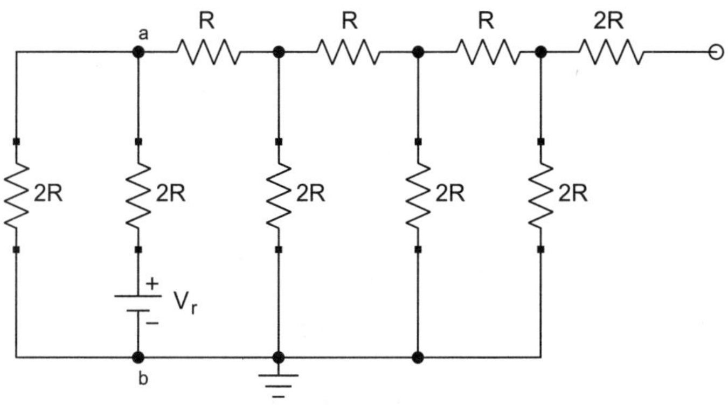

Fig. 4.10.5

Thevenin voltage and thevenin resistance at point a-b,

$$V_{ab} = \frac{V_r \times 2R}{4R} = \frac{V_r}{2}$$

$$R_{ab} = \frac{2R * 2R}{2R + 2R} = R$$

Step 2:

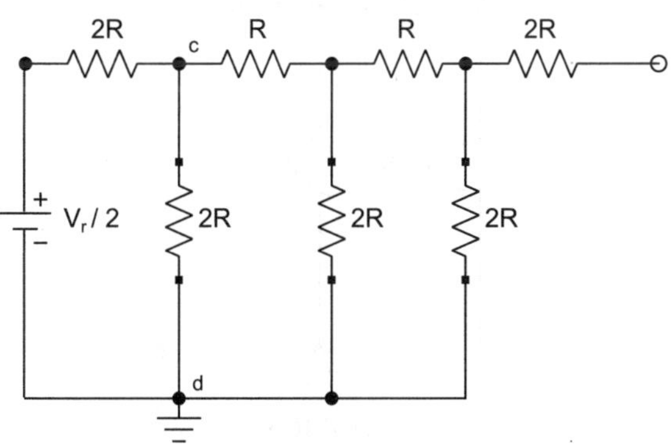

Fig. 4.10.6

Thevenin voltage and thevenin resistance at point c-d,

$$V_{cd} = \frac{V_r/2 \times 2R}{4R} = \frac{V_r}{4}$$

$$R_{cd} = \frac{2R \times 2R}{2R + 2R} = R$$

Step 3:

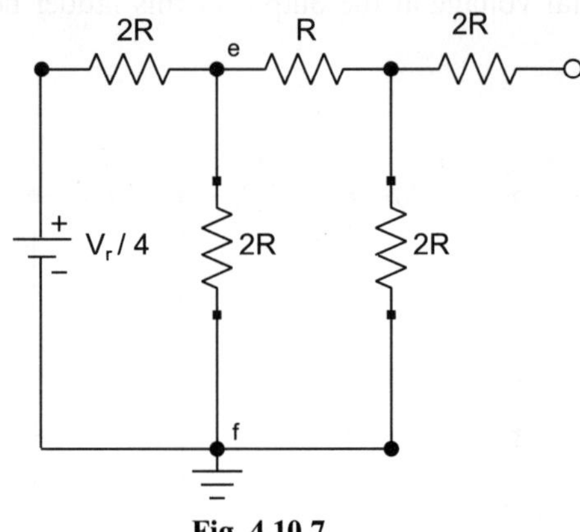

Fig. 4.10.7

Thevenin voltage and thevenin resistance at point e-f,

$$V_{ef} = \frac{V_r/4 \times 2R}{4R} = \frac{V_r}{8}$$

$$R_{ef} = \frac{2R \times 2R}{2R + 2R} = R$$

Step 4:

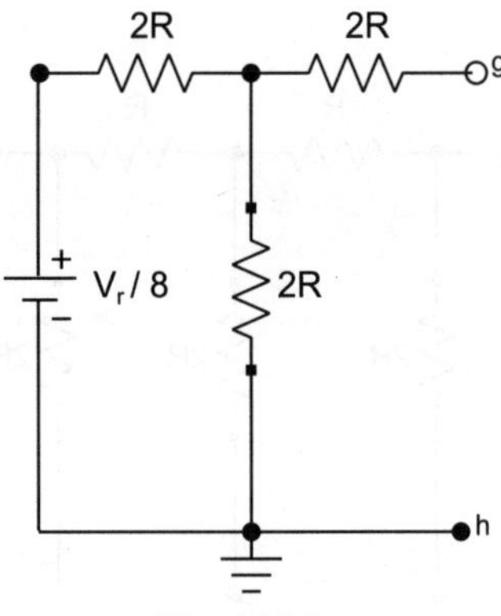

Fig. 4.10.8

Thevenin voltage and thevenin resistance at point g-h *i.e.* output terminals of the ladder network,

$$V_{ef} = \frac{V_r/8 \times 2R}{4R} = \frac{V_r}{16}$$

$$R_{ef} = \frac{2R \times 2R}{2R + 2R} = R$$

The equivalent ladder network followed by op-amp is shown in Fig. 4.10.9.

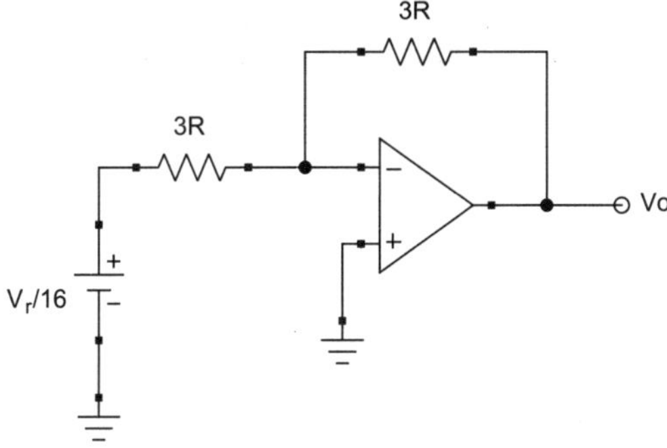

Fig. 4.10.9

Figure 4.10.9 shows that when only bit a_0 is equal to 1, then the voltage at the output of ladder is Vr/16.

b. Let us assume $a_1 = 1$ and reset all bits to zero, then the equivalent circuit will be as shown in Fig. 4.10.10.

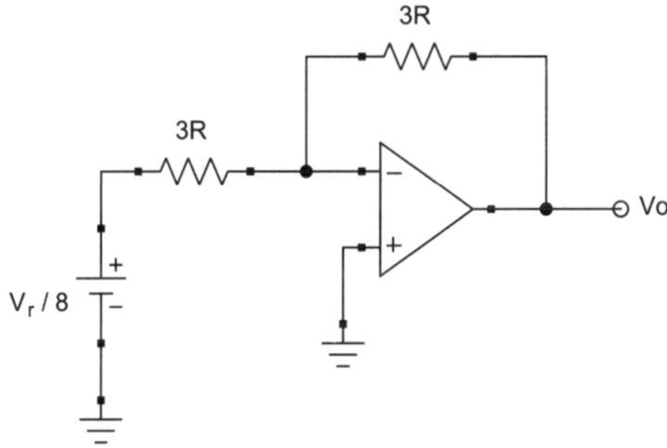

Fig. 4.10.10

Figure 4.10.10 shows that when only bit a_1 is equal to 1, then the voltage at the output of ladder is Vr/8.

c. Let us assume $a_2 = 1$ and reset all bits to zero, then the equivalent circuit will be as shown in Fig. 4.10.11.

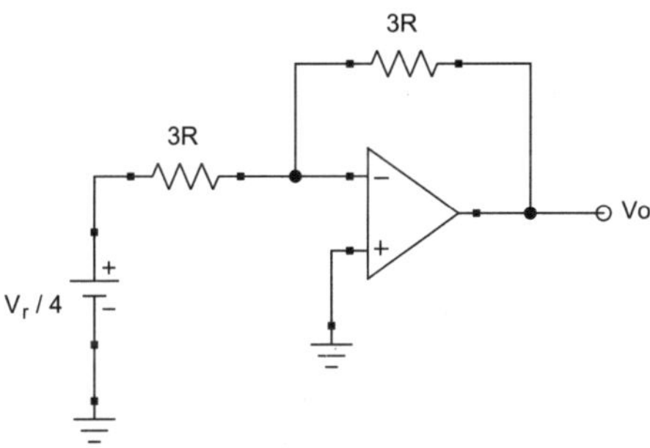

Fig. 4.10.11

Figure 4.10.11 shows that when only bit a_2 is equal to 1, then the voltage at the output of ladder is Vr/4.

d. Let us assume $a_3 = 1$ and reset all bits to zero, then the equivalent circuit will be as shown in Fig. 4.10.12.

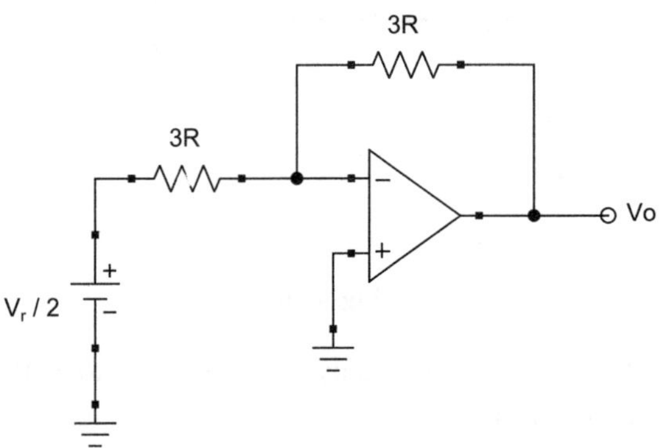

Fig. 4.10.12

Figure 4.10.12 shows that when only bit a_3 is equal to 1, then the voltage at the output of ladder is Vr/2.

Hence, taking all the bits into account, the output voltage for the ladder network becomes

$$V_0' = V_r \times \left[\frac{a_3}{2} + \frac{a_2}{2^2} + \frac{a_1}{2^3} + \frac{a_0}{2^4}\right]$$

Since, op-amp is used as an inverting amplifier with gain equal to

$$A_f = -{R_f}/{R_1} = -{3R}/{3R} = -1$$

Negative sign shows it is an inverting amplifier. Since, gain is equal to 1 hence it is acting as a voltage follower. Role of op-amp is to prevent loading of the circuit. Hence, voltage at the output terminal of op-amp becomes

$$V_0 = -\frac{3R \cdot V_r}{3R} \times \left[\frac{a_3}{2} + \frac{a_2}{2^2} + \frac{a_1}{2^3} + \frac{a_0}{2^4}\right]$$

Or we can write the above equation as

$$V_0 = -V_r/2^4 \times [a_3 2^3 + a_2 2^2 + a_1 2^1 + a_0 2^0]$$

Generalized formulae for N number of bits

$$V_0 = -V_r/2^N \times [a_{N-1} 2^{N-1} + a_{N-2} 2^{N-2} + ... + a_1 2^1 + a_0 2^0]$$

The above equation shows that the output analog voltage is proportional to input digital number.

PROCEDURE

1. Connect the circuit as shown in Fig. 4.10.2 on breadboard.

2. Take the value of R as 1kΩ.

3. Connect +12V to 7th pin of 741 IC and -12V to 4th pin of 741 IC.

4. Connect the multimeter at the 6th pin of 741 IC to measure the analog voltage.

5. Give different combinations of digital inputs by a_3 a_2 a_1 a_0 and note the corresponding analog voltage.

6. Note the readings in tabular form and compare the theoretical and practical value.

OBSERVATION TABLE

Table 4.10.1 Observation table for Digital to Analog converter

Digital input				Analog output (V)	
a_3	a_2	a_1	a_0	Theoretical	Practical
0	0	0	0		
0	0	0	1		
0	0	1	0		
0	0	1	1		
0	1	0	0		
0	1	0	1		
0	1	1	0		
0	1	1	1		
1	0	0	0		
1	0	0	1		
1	0	1	0		
1	0	1	1		
1	1	0	0		
1	1	0	1		
1	1	1	0		
1	1	1	1		

RESULT

DAC using R-2R ladder has been designed successfully and the theoretical and practical analog values for the corresponding digital inputs are found out to be almost same.

GRAPH

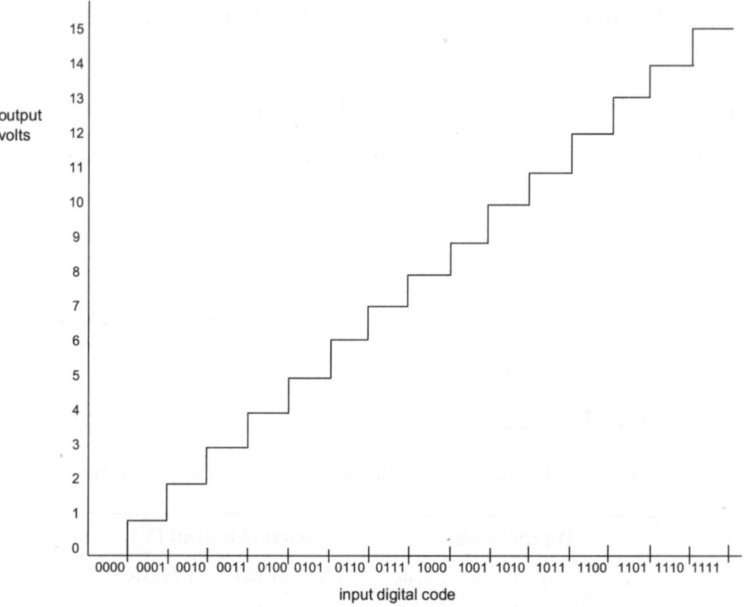

Fig. 4.10.13 Graph between digital input vs. analog output voltage

DISCUSSION

Most of the physical quantities like temperature, speed, distance, pressure *etc.* are analog or continuous in nature. These analog values represent the exact value of the quantities. If these physical quantities are to be controlled by computer then the analog value of physical quantity should be converted into digital value and vice-versa to establish the communication between the two. For *e.g.* in order to control the temperature of a room by using microprocessor, DAC and ADC are required. Since, temperature is an analog quantity and microprocessor works on digital data, so to establish the communication between the two, the analog value of temperature is first converted to digital value using Analog-to-Digital Converter and then digital value is processed by the microprocessor. The microprocessor will take the required action and gives digital output which then can be converted into analog value by Digital to Analog converter. In this way, the temperature of the room can be controlled.

DACs are also used in music players to convert digital data into analog audio signals; in televisions and mobile phones to convert digital video data into analog video signals; in military radar systems; sampling oscilloscopes.

Note:

The above practical can be performed without the use of op-amp as well. In that case, multimeter will be connected at the output terminal of the ladder network.

5

Experiments on Analog Electronics - II

LIST OF EXPERIMENTS

1. To Study Op-amp Characteristics: CMRR and Slew Rate.

2. To Design an Amplifier of Given Gain for an Inverting and Non-Inverting Configuration using an Op-amp and Study its Frequency Response.

3. To Design an Integrator using Op-amp for a Given Specification and Study its Frequency Response.

4. To Design a Differentiator using Op-amp for a Given Specification and Study its Frequency Response.

5. To Design a First Order and Second Order Low Pass Filter using Op-amp.

6. To Design a First Order and Second Order High Pass Filter using Op-amp.

7. To Design a Band Pass Filter using Op-amp.

8. To Design a Band Reject Filter using Op-amp.

9. To Design a RC Phase Shift Oscillator using Op-amp for a Given Specification.

10. To Study IC 555 as Monostable and Astable Multivibrator.

Introduction

OPERATIONAL AMPLIFIER

OPERATIONAL AMPLIFIER

An operational amplifier (abbreviated as op-amp) is a multistage, high gain and direct coupled amplifier that can amplify both ac and dc signals. As the name suggests, it is designed for performing mathematical operations such as addition, subtraction, multiplication, integration, differentiation, *etc*. It is available in the form of integrated circuit.

SCHEMATIC SYMBOL

Op-amp has two inputs and one output terminal as shown in Fig. 5a. Negative terminal is called as the inverting terminal and the positive terminal is called as the non-inverting terminal.

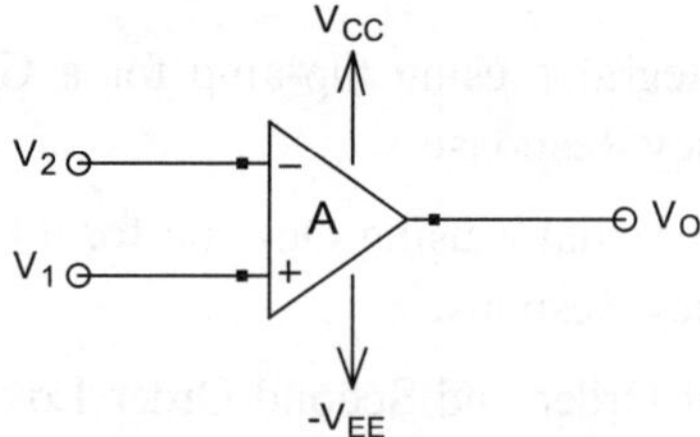

Fig. 5a Schematic symbol for op-amp

Where,

V_1: non - inverting input, V_2: inverting input, V_O: output voltage, A: large signal voltage gain.

Differential input voltage

It is defined as the difference of the two input voltages *i.e.* voltage at the inverting terminal and non-inverting terminal.

$$V_{in} = V_d = V_1 - V_2$$

Output voltage

Op-amp amplifies the difference between the two input voltages. The output voltage is given as

$$V_o = A.V_{in} = A.(V_1 - V_2)$$

The above equation shows that if voltage applied at the non-inverting terminal is greater than inverting terminal of op-amp, the output signal will be in phase with the input. Similarly, if voltage applied at the inverting terminal is greater than non-inverting terminal of op-amp, the output signal will be 180° out of phase to the input signal.

IDEAL CHARACTERISTICS OF AN OPERATIONAL AMPLIFIER

1. Infinite voltage gain, $A = \infty$.

2. Infinite input resistance, $R_{in} = \infty$.

3. Zero output resistance, $R_{out} = 0$.

4. Zero offset voltage, *i.e.* zero output voltage when input voltage is zero.

5. Infinite bandwidth, $BW = \infty$.

6. Infinite common mode rejection ratio, $CMRR = \infty$.

7. Infinite slew rate, $SR = \infty$.

PIN DIAGRAM OF 741C OPAMP

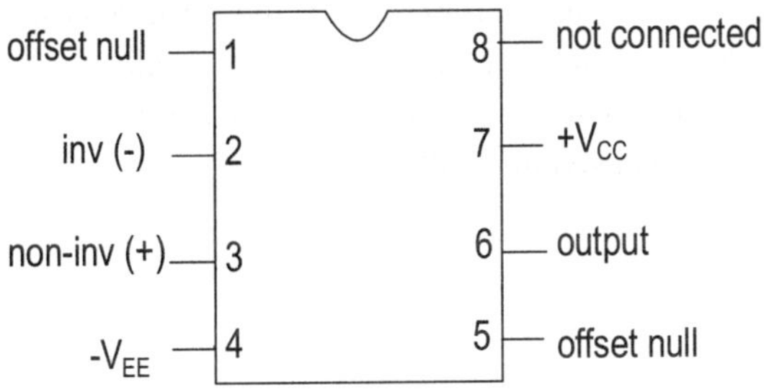

Fig. 5b Pin diagram for IC 741

CONFIGURATION

There are two basic configurations of an op-amp

1. Open loop configuration

2. Closed loop configuration

Open Loop Configuration (without feedback)

When the output is not fed back in any form to the input, the op-amp is said to be in open loop configuration. The gain of an op-amp in open loop configuration is very high (ideally infinite), therefore, even if a very small signal is applied at the input, it will get amplified to very large amplitude. If the output voltage becomes more than the saturation voltage, it gets clipped and swings between the positive and negative saturation voltage only *i.e.* ±Vsat. This action is highly undesirable in linear applications.

Closed Loop Configuration (with feedback)

In the closed loop configuration, feedback is introduced between the input and the output *i.e.* the output is either directly or via any network is feedback to the input.

Negative feedback

When a part of the output is feedback to the negative input terminal of an op-amp or in other words, if the signal fed back is of opposite polarity or out of phase by 180° with respect to input signal, the feedback is called as negative feedback. It is also known as degenerative feedback because it reduces the amplitude of output voltage and thereby the voltage gain is also reduced. The negative feedback is used in amplifiers.

Positive feedback

When a part of the output is feedback to the positive input terminal of an op-amp or in other words, if the signal fed back is of the same polarity or in phase with the input signal, the feedback is called as positive feedback. It is also known as regenerative feedback because the feedback signal adds with the input signal. The positive feedback is used in oscillators.

Experiment 1

OP-AMP CHARACTERISTICS: CMRR and SLEW RATE

AIM

To Study Op-amp Characteristics: CMRR and Slew Rate.

APPARATUS REQUIRED

Breadboard, connecting wires, $\pm$ 12V power supply, resistors – 1kΩ and 10kΩ, IC-741, cathode ray oscilloscope, function generator, probes.

THEORY

Common Mode Rejection Ratio (CMRR)

When both the input terminals of the op-amp are applied with the same input voltage then the op-amp is said to be in common mode configuration. The applied input which is common to both inverting and non-inverting terminal is called as common mode voltage V_{CM} and the corresponding output voltage is called as V_{OCM}. The input voltage can be either dc or ac or both.

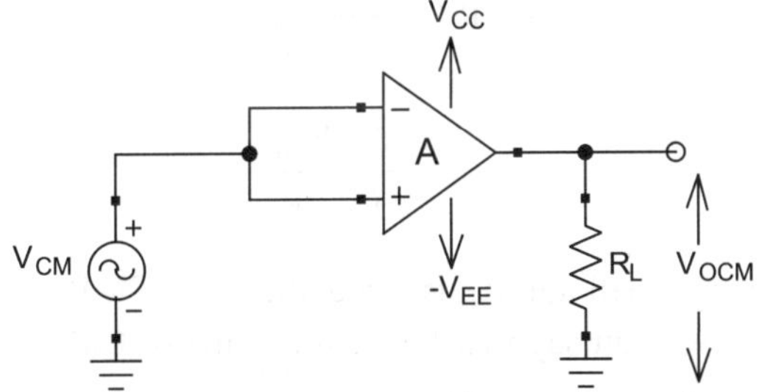

Fig. 5.1.1 Open loop op-amp connected in common mode configuration

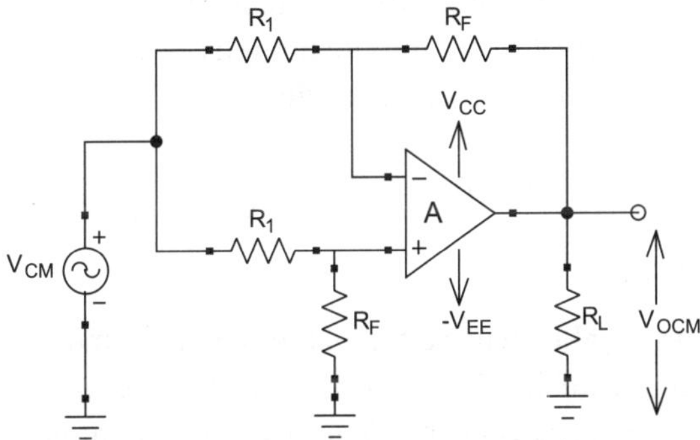

Fig. 5.1.2 Closed loop op-amp connected in common mode configuration

Since, op-amp amplifies the difference between the two input voltages, hence in common mode configuration, when the input to both terminals is same, the output V_{OCM} should be zero ideally. But practically, V_{OCM} is not equal to zero but is very small as compared to V_{CM}. Common mode voltage gain A_{CM} is given as

$$A_{CM} = {V_{OCM}}/{V_{CM}}$$

Ideally, A_{CM} is zero but practically it is smaller than 1. CMRR is defined as the ratio of differential gain A_D to the common mode gain A_{CM} *i.e.*

$$CMRR = \frac{A_D}{A_{CM}} \qquad\qquad - \text{eqn 1}$$

A_D is the open loop gain which is equal to A when the op-amp is connected in open loop configuration as shown in Fig. 5.1.1. When the op-amp is connected in closed loop configuration as shown in Fig. 5.1.2, A_D is equal to closed loop gain $A_F = R_F/R_1$. Rewriting the above equation, we get

$$\Rightarrow \qquad\qquad CMRR = \frac{A_D}{{V_{OCM}}/{V_{CM}}}$$

$$\Rightarrow \qquad\qquad CMRR = \frac{A_D \cdot V_{CM}}{V_{OCM}}$$

$$\Rightarrow \qquad\qquad V_{OCM} = \frac{A_D \cdot V_{CM}}{CMRR} \qquad\qquad - \text{eqn 2}$$

Since, A_{CM} is very small and A_D is large, therefore CMRR value is also very large (refer eqn1) or we can say that larger the value of CMRR, smaller will be V_{OCM} (refer eqn2).

An ideal op-amp has infinite CMRR so that output common mode voltage is zero but practically, CMRR is not infinite but it has a very large value. The value of CMRR is usually specified in decibels (dB).

$$CMRR \text{ (dB)} = 20 \log \left({A_D}/{A_{CM}}\right)$$

Importance of CMRR

Larger is its value, smaller will be the output common mode voltage V_{OCM} and hence, better is the matching between the two input terminals. In case of poor CMRR, we will get a large value of V_{OCM} that shows a large imperfection in the device.

Slew Rate

Slew rate is defined as maximum rate of change of the output voltage with respect to time. It is given as

$$SR = \left.\frac{dV_0}{dt}\right|_{max} \text{ V/usec.}$$

Slew rate tells how fast the output of an op-amp can change with respect to the changes in the input frequency. It is a very important factor for using op-amp with ac signals at high frequencies. If the slew rate of an op-amp is less, it cannot be used with high frequency applications. For *e.g.* slew rate equal to 0.5V/μsec means when a large step signal in applied at the input, the output can change by 0.5V in one microsecond and if the slew rate is 1000V/μsec then the op-amp is capable of providing a change of 1000V in one microseconds. This shows larger the value of slew rate is, better will be the responding time of the device. Hence, ideally the slew rate should be infinite *i.e.* output of op-amp should change simultaneously with respect to the input signal.

Effect of slew rate

Consider an op-amp connected in voltage follower configuration as shown in Fig. 5.1.3.

If the input is a square wave, then the output is also expected to be a square wave. But at a particular frequency, due to the effect of slew rate, the output wave is distorted as shown in Fig. 5.1.4.

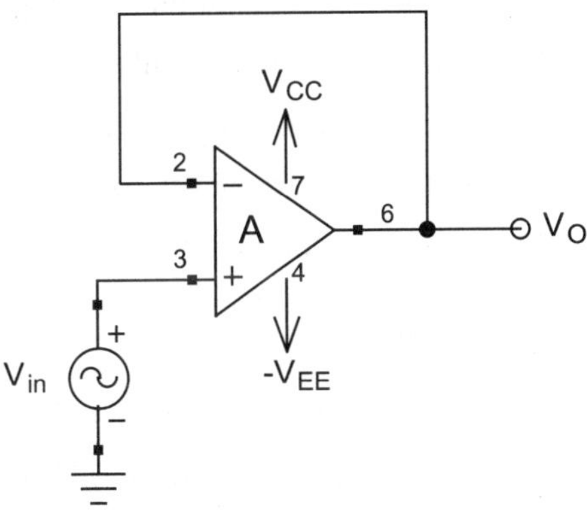

Fig. 5.1.3 Circuit diagram to measure slew rate

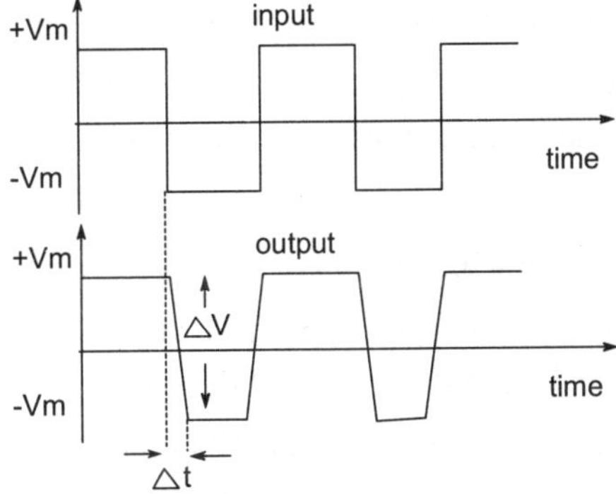

Fig. 5.1.4 Effect of slew rate

PROCEDURE

Common Mode Rejection Ratio

1. Connect the circuit shown in Fig. 5.1.1 on breadboard.
2. Take $R_L = 10k\Omega$.
3. Connect pin 7 and 4 of IC 741 to +12V and -12V respectively.
4. Connect function generator at the inverting and non-inverting terminal of op-amp using probe.
5. Set it to sine wave of frequency 2 kHz and amplitude in μV range and note its voltage as V_{CM}.
6. Connect channel 1 of CRO across the input and channel 2 across the load resistance to see the input and output waveform.
7. Measure the output voltage V_{OCM} using CRO.
8. Calculate the value of CMRR.
9. Change the value of input voltage and take the readings again.
10. Now, connect the circuit as shown in Fig. 5.1.2 on breadboard.
11. Take $R_F = 10k\Omega$, $R_1 = 1k\Omega$ and $R_L = 10k\Omega$.
12. Set the input frequency to 2 kHz and amplitude in mV range.
13. Repeat the above steps.

Slew Rate

1. Connect the circuit shown in Fig. 5.1.3 on breadboard.
2. Connect pin 7 and 4 of IC 741 to +12V and -12V respectively.
3. Connect function generator at the non-inverting terminal of an op-amp using probe.
4. Set it to square wave of 1V peak to peak amplitude and 25 kHz frequency.
5. Connect channel 1 of CRO across the input and channel 2 at the output terminal to trace the input and output waveform.
6. Output waveform will be same as shown in Fig. 5.1.4.
7. Calculate the slew rate which is equal to the slope of the output voltage.
8. Now set the frequency of input signal at 1 kHz. Adjust the input voltage to get 20V peak to peak amplitude wave at the output of op-amp.
9. Increase the frequency of input signal until output distorts.
10. Note this as the maximum frequency at which op-amp can be operated and is called the bandwidth of an op-amp.

OBSERVATION TABLE

Common Mode Rejection Ratio

Op-amp in open loop configuration

Table 5.1.1 Observation table for CMRR (open loop op-amp)

S.No.	V_{CM} (V)	V_{OCM} (V)	$A_D = A$	A_{CM} $=V_{OCM}/V_{CM}$	CMRR (dB) $20 \log \left(A_D / A_{CM} \right)$
1.	1		2×10^5		
2.	2		2×10^5		
3.	3		2×10^5		
4.	...		2×10^5		
5.	...		2×10^5		

Op-amp in closed loop configuration

Table 5.1.2 Observation table for CMRR (closed loop op-amp)

S.No.	V_{CM} (V)	V_{OCM} (V)	$A_D =$ R_F/R_1	A_{CM} $=V_{OCM}/V_{CM}$	CMRR (dB) $20 \log \left(A_D / A_{CM} \right)$
1.	1				
2.	2				
3.	3				
4.	...				
5.	...				

Slew Rate

Table 5.1.3 Observation table for slew rate

S.No.	Input Voltage	Frequency	ΔV (V)	ΔT (μsec)	SR = $\Delta V/\Delta T$
1.	1V	25kHz			

Table 5.1.4 Observation table for bandwidth

Input Voltage	Output Voltage	Frequency	BW
	$20V_{p-p}$		

OBSERVATIONS

Attach the traces for slew rate.

DISCUSSION

As the name suggests, operational amplifiers are widely used for performing mathematical operations like addition, subtraction, multiplication, *etc.* They are also used as ac and dc signals amplifiers, filters, oscillators, comparators, *etc.*

For 741 IC, value of CMRR is 90dB and slew rate is 0.5V/μsec. Larger is the value of CMRR, better is the ability of op-amp to reject common mode voltage. For *e.g.* while working in a noisy environment, the same noise appears at both the input terminals of op-amp *i.e.* noise appears as a common mode voltage. Since, op-amp amplifies the difference of the two input voltages, therefore output voltage corresponding to the noise will be zero. Ideally, the value of CMRR should be infinite.

Larger is the value of slew rate, quickly the output will respond to the fast changing inputs. Ideally, the value of slew rate should be infinite.

Experiment **2**

INVERTING and NON-INVERTING AMPLIFIER

AIM

To Design an Amplifier of Given Gain for an Inverting and Non-Inverting Configuration using an Op-amp and also Plot the Frequency Response.

APPARATUS REQUIRED

Breadboard, connecting wires, ± 12V power supply, IC 741, function generator, cathode ray oscilloscope, resistors - 1kΩ, 9kΩ, 10kΩ, probes.

THEORY

Inverting Amplifier

It is a negative feedback amplifier with input signal connected at its inverting terminal and non-inverting terminal is grounded as shown in Fig. 5.2.1. The negative feedback is achieved by connecting feedback resistance between the output terminal and the inverting terminal.

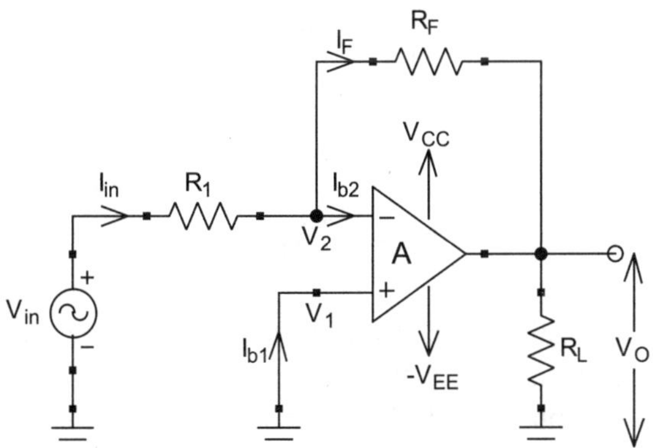

Fig. 5.2.1 Op-amp as an inverting amplifier

Where,

V_{in}: input voltage, R_1: input resistance, R_F: feedback resistance, V_O: output voltage.

To find the gain of an inverting amplifier, we apply Kirchhoff's current equation at node V_2.

$$I_{in} = I_F + I_{b2}$$

Since, input resistance of an op-amp is ideally infinite, therefore, I_{b2} is zero and hence same current will pass through resistances R_1 and R_F.

$\therefore$

$$I_{in} = I_F$$

Re-writing the above equation in terms of voltage and resistance,

$$\frac{(V_{in} - V_2)}{R_1} = \frac{(V_2 - V_O)}{R_F}$$

Since, non-inverting terminal is connected to ground, therefore $V_1 = 0$. The differential input voltage V_{id} which is the difference of V_1 and V_2 is equal to V_O/A. Since, the gain of an op-amp is very large, ideally infinite, hence V_{id} is zero. Therefore, voltage at non-inverting terminal is equal to the voltage at inverting terminal *i.e.* $V_1 = V_2$. Since, $V_1 = 0$ therefore, V_2 is also zero or we can say that the inverting terminal is virtually at the ground. Therefore, above equation can be written as

$\Rightarrow$

$$V_{in}/R_1 = -V_O/R_F$$

$\Rightarrow$

$$V_O/V_{in} = -R_F/R_1$$

$\Rightarrow$

$$\text{gain} = -R_F/R_1$$

Hence, gain of the inverting amplifier is given as

$$A_V = -R_F/R_1$$

Thus, the voltage gain of an inverting amplifier is the ratio of feedback resistor to the input resistor and the negative sign indicates that the output voltage is 180° out of phase with the input. Hence, the amplifier is called an inverting amplifier.

Non-Inverting Amplifier

It is a negative feedback amplifier with input signal connected at its non-inverting terminal and inverting terminal is grounded through resistance as shown in Fig. 5.2.2. The negative feedback is achieved by connecting feedback resistance between the output and the inverting terminal.

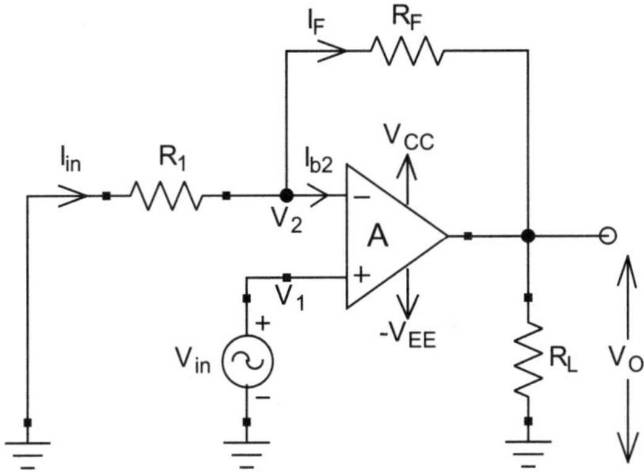

Fig. 5.2.2 Op-amp as a non-inverting amplifier

To find the gain of a non-inverting amplifier, we apply Kirchhoff's current equation at node V_2.

$$I_{in} = I_F + I_{b2}$$

Since, input resistance of an op-amp is ideally infinite, therefore, I_{b2} is zero and hence the same current would pass through R_1 and R_F.

$$\therefore \qquad I_{in} = I_F$$

Re-writing the above equation in terms of voltage and resistance,

$$\frac{V_2}{R_1} = \frac{(V_O - V_2)}{R_F}$$

Since, the gain of op-amp is infinite, therefore differential input voltage V_{id} which is equal to $V_1 - V_2 = V_O/A$ is zero. Hence, we can write

$$V_1 = V_2$$

Using this relation, we can write the above equation as

$$\frac{V_1}{R_1} = \frac{(V_O - V_1)}{R_F}$$

Since, $V_1 = V_{in}$

$$\Rightarrow \qquad \frac{V_{in}}{R_1} = \frac{(V_O - V_{in})}{R_F}$$

$$\Rightarrow \qquad V_{in}\left(\frac{1}{R_1} + \frac{1}{R_F}\right) = \frac{V_O}{R_F}$$

$$\Rightarrow \qquad \frac{V_{in}}{V_O} = \frac{R_1}{(R_1 + R_F)}$$

$$\Rightarrow \qquad \frac{V_O}{V_{in}} = 1 + \frac{R_F}{R_1}$$

$$\Rightarrow \qquad \text{gain} = 1 + \frac{R_F}{R_1}$$

Hence, gain of the non-inverting amplifier is given as

$$A_V = 1 + R_F/R_1$$

Thus, voltage gain of a non-inverting amplifier is one plus ratio of feedback resistor to the input resistor and the positive sign shows that the output is in phase with the input signal. Hence, the amplifier is called as non-inverting amplifier.

CALCULATIONS

Inverting Amplifier

Let Gain = 10

Since,

$$A_V = -R_F/R_1$$

Take $R_F = 10k\Omega$ and $R_1 = 1k\Omega$

Non-Inverting Amplifier

Let Gain = 10

Since,

$$A_V = 1 + R_F/R_1$$

Take $R_F = 9k\Omega$ and $R_1 = 1k\Omega$

PROCEDURE

Gain

1. Connect the circuit shown in Fig. 5.2.1 on breadboard.
2. Take load resistance as $10k\Omega$.
3. Connect pin 7 and 4 of IC 741 to +12V and -12V respectively.
4. Connect function generator at the inverting terminal of op-amp through resistance R_1. Set it to sine wave with amplitude 1V.
5. Connect channel 1 of CRO across the input and channel 2 across the load resistance R_L (pin 6 and ground) to view the input and output waveform.
6. Measure the input and output voltage using CRO and trace the two waveforms.

7.   Calculate the practical value of gain and compare it with the given gain.

8.   Change the value of input voltage and take the readings again.

9.   Repeat the above steps for another set of voltage gain.

Frequency response

10. Now, set the input voltage to 1V.

11. Change the frequency of input signal starting from 100 Hz to 100 kHz and measure the output voltage.

12. Note the readings in the form of a table and find the gain.

13. Plot the frequency response curve between gain and frequency.

14. Now, connect the circuit shown in Fig. 5.2.2 on breadboard.

15. Connect function generator at the non-inverting terminal of op-amp and repeat the above steps for finding gain and frequency response.

OBSERVATION TABLE

Inverting Amplifier

Gain

Table 5.2.1 Theoretical and practical values of gain for inverting amplifier

S.No.	Input voltage Vin	Output voltage Vo	Theoretical Gain $-R_F/R_1$	Practical gain Vo/Vin
1.				
2.				
3.				
4.				

Frequency response

Table 5.2.2 Frequency response of inverting amplifier

S.No.	Frequency Hz	Input voltage Vin	Output voltage Vo	Gain Vo/Vin
1.	100			
2.	200			
3.	300			
4.	...			
5.	...			
6.	...			
7.	...			
8.	100k			

Non-Inverting Amplifier

Gain

Table 5.2.3 Theoretical and practical values of gain for non-inverting amplifier

S.No.	Input voltage Vin	Output voltage Vo	Theoretical Gain $1+R_F/R_1$	Practical gain Vo/Vin
1.				
2.				
3.				
4.				

Frequency response

Table 5.2.4 Frequency response of non-inverting amplifier

S.No.	Frequency Hz	Input voltage Vin	Output voltage Vo	Gain Vo/Vin
1.	100			
2.	200			
3.	300			
4.	…			
5.	…			
6.	…			
7.	…			
8.	100k			

GRAPH

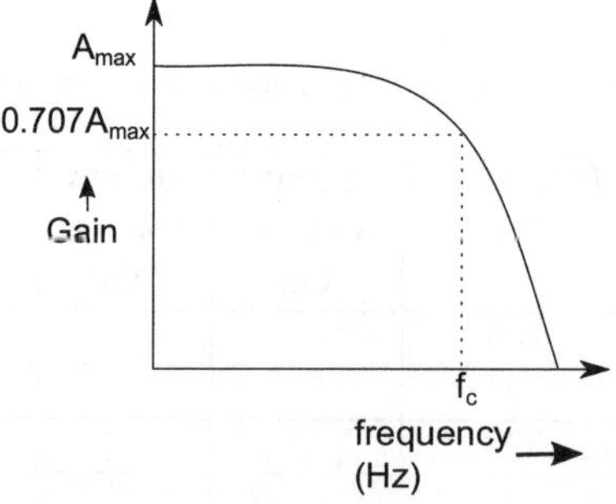

Fig. 5.2.3 Frequency response of feedback amplifier

OBSERVATIONS

Attach the traces of input and output waveforms.

Frequency Response

Theoretical values

Gain bandwidth product is given as

$$\text{Gain} \times \text{Bandwidth} = 1\text{MHz}$$

Since, Gain = 10

$$\text{Bandwidth} = \frac{1\text{MHz}}{10}$$

$$\text{Bandwidth} = 100\text{kHz}$$

Practical values

Gain = …

Bandwidth = …

RESULT

An amplifier of given gain for an inverting and non-inverting configuration using an op-amp has been designed and frequency response curve has been plotted successfully.

The calculated and observed value of gain is found to be nearly same. The output of inverting amplifier is 180° out of phase with the input while in non-inverting amplifier, it is found to be in phase with the input. As we increase the value of input voltage, the output voltage also increases. If the amplitude of output voltage becomes greater than ±12V (power supply), the op-amp goes into saturation and output gets clipped to ±12V.

The frequency response curve of feedback amplifier shows that the gain of an op-amp changes with the change in the frequency of the input signal and amplifier behaves as a low pass filter.

DISCUSSION

Disadvantages of op-amp in open loop configuration are:

1. Gain is very high, ideally infinite. Therefore, even if a small amplitude signal is applied at its input, it will get amplified by a large amount and peaks will get clipped.
2. Bandwidth is very small (~ 5Hz). Therefore, signals with very low frequency only can be amplified properly.

This limits the practical use of open loop op-amp in linear applications. By introducing a negative feedback between output and the input, the gain decreases and bandwidth increases. Op-amp with negative feedback is widely used as integrator, differentiator, summing, scaling, averaging amplifiers, *etc*.

Experiment 3

INTEGRATOR using OP-AMP

AIM

To Design an Integrator using Op-amp for a Given Specification and Study its Frequency Response.

APPARATUS REQUIRED

Breadboard, resistors, capacitors, ± 12V dc power supply, IC 741, cathode ray oscilloscope, probes, function generator, connecting wires.

THEORY

Integrator

One of the most widely used applications of op-amp is the integrator circuit. It is a circuit which performs the mathematical operation of integration *i.e.* it produces an output voltage proportional to the integral of the input voltage. The basic integrator circuit is shown in Fig. 5.3.1.

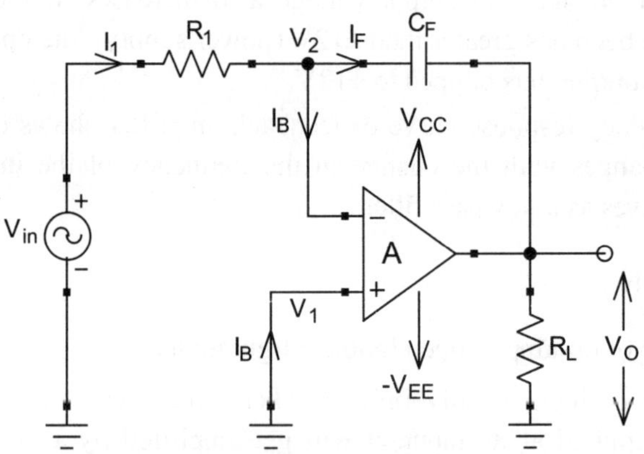

Fig. 5.3.1 Op-amp as an integrator

To find an expression for the output voltage of the integrator circuit, we apply KCL at node V_2.

$$I_1 = I_F + I_B$$

Since, input resistance of an op-amp is ideally infinite, therefore, base current $I_B = 0$. Hence, same current would pass through R_1 and C_F.

$$\therefore \qquad I_1 = I_F$$

$$\Rightarrow \qquad I_1 = C_F \frac{d(V_2 - V_O)}{dt}$$

Since, non-inverting terminal is connected to ground, therefore $V_1 = 0$. The differential input voltage V_{id} which is the difference of V_1 and V_2 is equal to V_O/A. Since, the gain of an op-amp is very large, ideally infinite, hence V_{id} is zero. Therefore, voltage at non-inverting terminal is equal to the voltage at inverting terminal *i.e.* $V_1 = V_2$. Since, $V_1 = 0$ therefore, V_2 is also zero or we can say that the inverting terminal is virtually at the ground. Therefore, above equation can be written as

$$\Rightarrow \qquad I_1 = -C_F \frac{dV_O}{dt}$$

$$\Rightarrow \qquad \frac{V_{in}}{R_1} = -C_F \frac{dV_O}{dt}$$

$$\Rightarrow \qquad \frac{dV_O}{dt} = -\frac{V_{in}}{R_1 C_F}$$

$$\Rightarrow \qquad V_O = -\frac{1}{R_1 C_F} \int V_{in} \cdot dt + C$$

Where, C is an integration constant and is proportional to the output voltage at time t=0 seconds.

The above expression for the output voltage shows that it is directly proportional to the integral of input voltage and inversely proportional to the time constant $R_1 C_F$.

Limitation of basic integrator circuit

If the frequency of the applied input signal is very low, the capacitor acts as an open circuit. Due to this, the closed loop configuration changes to open loop configuration. As a result, the gain of an op-amp becomes infinite and output voltage swings between the saturation voltages which results in instability. Hence, this circuit is not used.

Practical integrator

The limitation of basic integrator is overcome by introducing a feedback resistor R_F in parallel with the capacitor C_F as shown in Fig. 5.3.2. This R_F limits the low frequency gain and hence, minimizes the variations in the output voltage.

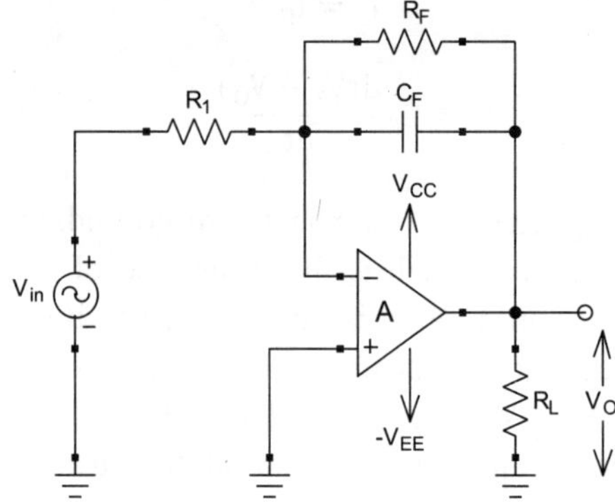

Fig. 5.3.2 Practical integrator

Frequency response

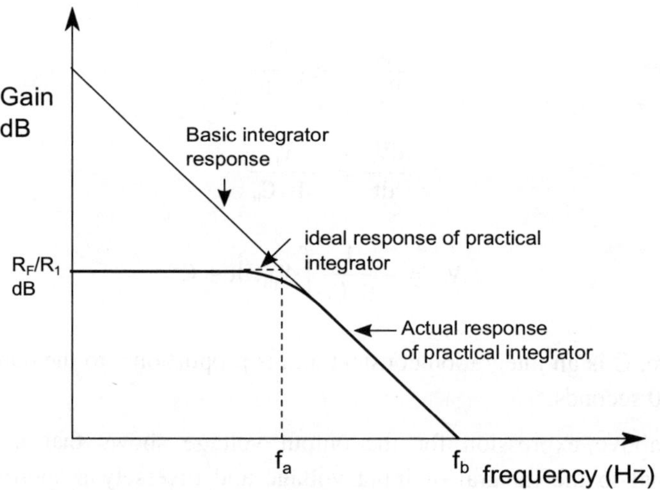

Fig. 5.3.3 Frequency response of basic and practical integrator

Frequency response of basic and practical integrator is shown in Fig. 5.3.3. The figure shows that f_b is the frequency at which gain is 0dB and is given as

$$f_b = \frac{1}{2\Pi R_1 C_F} \qquad - \text{eqn 1}$$

The gain R_F/R_1 is constant for the frequencies below f_a where f_a is given as

$$f_a = \frac{1}{2\Pi R_F C_F} \qquad - \text{eqn 2}$$

The gain decreases at a rate of 20dB/decade as the frequency of input signal becomes above f_a and the circuit works as an integrator for the frequencies between f_a and f_b.

CALCULATION

1. Choose the cut off frequency f_a for which the integrator is to be designed.
2. Choose a particular value of C_F.
3. Calculate R_F by using equation 2.
4. Choose the value of f_b such that $f_b = 10\, f_a$.
5. Calculate R_1 by using equation 1.

PROCEDURE

Integrator

1. Design the integrator circuit by calculating the values of R_1, R_F and C_F as mentioned above.
2. Connect the circuit as shown in Fig. 5.3.2 on breadboard.
3. Take load resistance as $10k\Omega$.
4. Connect function generator to the inverting terminal of op-amp through resistance R_1 and set it to sine wave.
5. Give the input signal of frequency such that (a) $f < f_a$ (b) $f_a < f < f_b$ (c) $f > f_b$.
6. Connect channel 1 of CRO at the input terminal and channel 2 across the load resistance to observe the input and output waveforms.
7. Take the traces of the input and output signals for all three cases.
8. Change the input signal to square wave and triangular wave and take traces.

Frequency Response

9. Now, set the input voltage to sine wave with amplitude of 1V.
10. Change the frequency of input signal starting from 100 Hz to 100 kHz and measure the output voltage.
11. Note the readings in the form of a table and find the gain.
12. Plot the frequency response curve between gain and frequency.

INPUT AND OUTPUT WAVEFORMS

1. Square wave

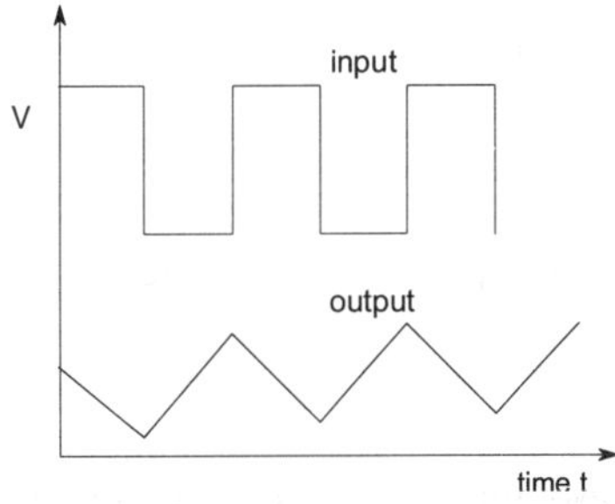

Fig. 5.3.4 Input and output waveform for square wave input

2.   Triangular wave

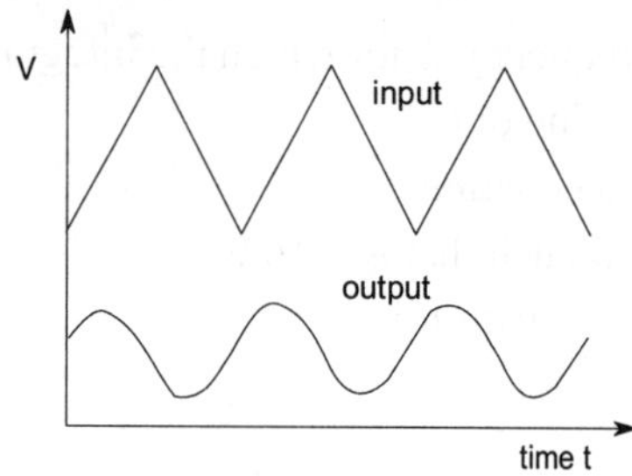

Fig. 5.3.5 Input and output waveform for triangular wave input

3.   Sine wave

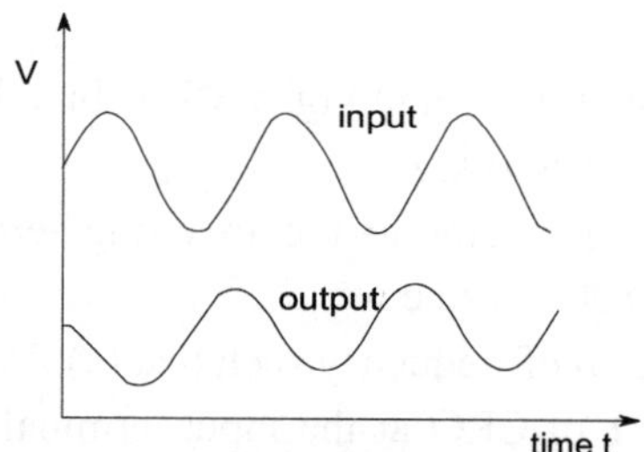

Fig. 5.3.6 Input and output waveform for sine wave input

OBSERVATION TABLE

Table 5.3.1 Frequency response of integrator

S.No.	Frequency Hz	Input voltage Vin	Output voltage Vo	Gain Vo/Vin
1.	100			
2.	200			
3.	300			
4.	…			
5.	…			
6.	…			
7.	…			
8.	100k			

OBSERVATIONS

Attach the traces for input and output waveforms.

RESULT

Integrator circuit using op-amp has been designed and frequency response curve has been plotted successfully. Traces for input and output signals have been taken and it is found that the circuit behaves as an integrator between frequency $f_a = \dots$ Hz and $f_b = \dots$ Hz.

DISCUSSION

The integrator circuit produces cosine wave when the input is sine wave, triangular wave when the input is square wave and a parabola when the input is a triangular wave. The circuit is widely used as analog to digital converters, wave shaping circuits, active low pass filters, *etc*.

Experiment 4

DIFFERENTIATOR using OP-AMP

AIM

To Design a Differentiator using Op-amp for a Given Specification and Study its Frequency Response.

APPARATUS REQUIRED

Breadboard, resistors, capacitors, ± 12V dc power supply, IC 741, cathode ray oscilloscope, probes, function generator, connecting wires.

THEORY

Differentiator

It is a circuit which performs the mathematical operation of differentiation *i.e.* its output voltage is proportional to the rate of change of its input voltage. An active differentiator circuit consists of an operational amplifier, feedback resistor and capacitor as shown in Fig. 5.4.1.

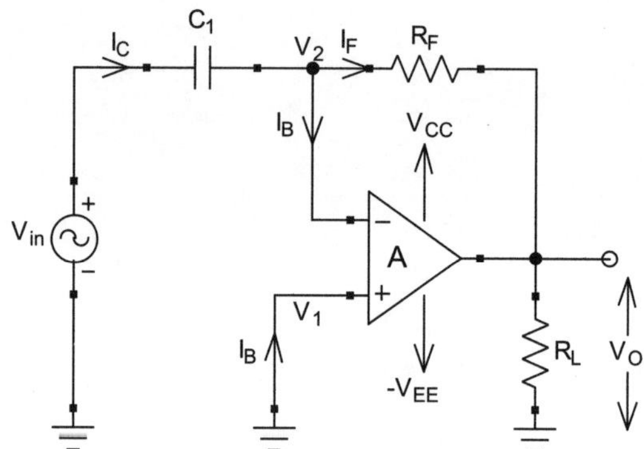

Fig. 5.4.1 Op-amp as a differentiator

To find an expression for the output voltage of the differentiator circuit, we apply KCL at node V_2.

$$I_C = I_F + I_B$$

Since, input resistance of an op-amp is ideally infinite, therefore, base current I_B = 0. Hence, same current would pass through R_F and C_1.

$\therefore$ $$I_C = I_F$$

$\Rightarrow$ $$C_1 \frac{d(V_{in} - V_2)}{dt} = \frac{V_2 - V_O}{R_F}$$

Since, non-inverting terminal is connected to ground, therefore $V_1 = 0$. The differential input voltage V_{id} which is the difference of V_1 and V_2 is equal to V_O/A. Since, the gain of an op-amp is very large, ideally infinite, hence V_{id} is zero. Therefore, voltage at non-inverting terminal is equal to the voltage at inverting terminal *i.e.* $V_1 = V_2$. Since, $V_1 = 0$ therefore, V_2 is also zero or we can say that the inverting terminal is virtually at the ground. Therefore, above equation can be written as

$\Rightarrow$ $$C_1 \frac{dV_{in}}{dt} = -\frac{V_O}{R_F}$$

$\Rightarrow$ $$V_O = -R_F C_1 \frac{dV_{in}}{dt}$$

The above expression for the output voltage shows that it is directly proportional to $R_F C_1$ times the negative rate of change of the input voltage V_{in} at that instant of time.

Limitation of basic differentiator circuit

As the frequency of the input signal increases, X_{C1} decreases. This results in the following problems.

- Gain of the basic circuit *i.e.* R_F/X_{C1} increases which as a result makes the circuit unstable.

- The circuit becomes very susceptible to high frequency noise. During amplification, this noise can completely override the differentiated output signal.

Therefore, the basic differentiator circuit is not used for practical applications.

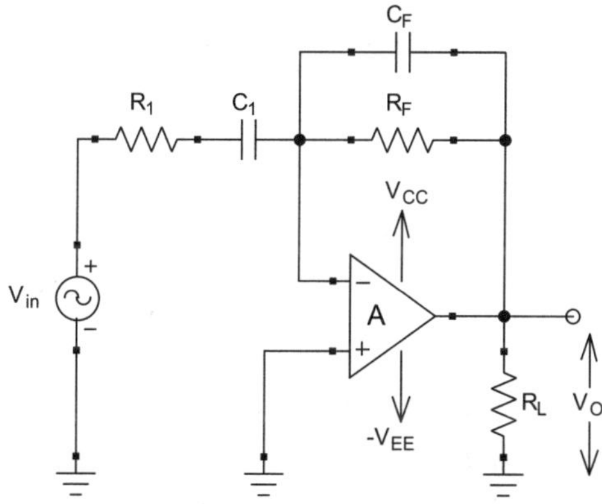

Fig. 5.4.2 Practical differentiator

Practical differentiator

Both these limitations can be overcome by adding two components R_1 and C_F. The modified circuit is called as practical differentiator as shown in Fig. 5.4.2.

Frequency response

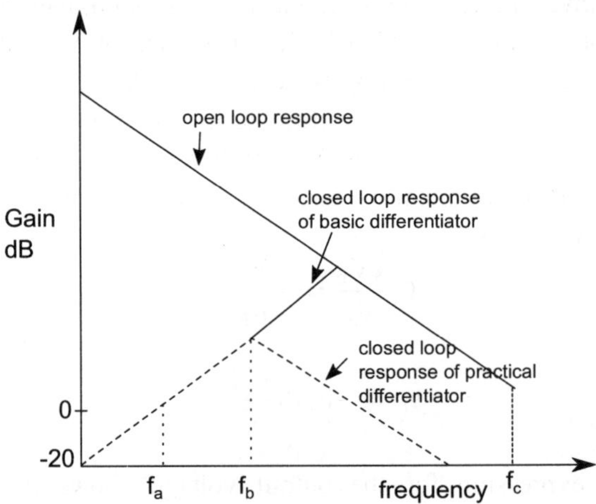

Fig. 5.4.3 Frequency response of basic and practical differentiator

Frequency response of basic and practical differentiator is shown in Fig. 5.4.3. Frequency f_C is the unity gain-bandwidth of the op-amp and f_a is the frequency at which gain is 0dB which is given as:

$$f_a = \frac{1}{2\Pi R_F C_1} \qquad - \text{eqn 1}$$

As shown in figure, the gain increases with frequency for basic differentiator which makes the circuit unstable. For practical differentiator, gain increases till frequency f_b at a rate of 20dB per decade and after that gain starts decreasing at 20dB per decade. This decrease in gain occurs due to the presence of $R_1 C_1$ and $R_F C_F$ combinations in practical differentiator.

Frequency f_b is called as the gain limiting frequency and is given as

$$f_b = \frac{1}{2\Pi R_1 C_1} \qquad - \text{eqn 2}$$

Where,

$$R_1 C_1 = R_F C_F \qquad - \text{eqn 3}$$

CALCULATION

1. Choose the highest frequency f_a of an input signal to be differentiated.

2. Assuming the value of C_1 less than 1µf, calculate the value of feedback resistor R_F by using equation 1.

3. Choose $f_b = 20f_a$. Calculate the value of R_1 by using equation 2.

4. Calculate the value of C_F by using equation 3.

PROCEDURE

Differentiator

1. Design the differentiator circuit by calculating the values of R_1, R_F and C_F as mentioned above.

2. Connect the circuit as shown in Fig. 5.4.2 on breadboard.

3. Take load resistance as $10k\Omega$.

4. Connect function generator to the inverting terminal of op-amp through R_1 and C_1 and set it to sine wave.

5. Give the input signal of frequency such that (a) $f < f_a$ (b) $f_a < f < f_b$ (c) $f > f_b$.

6. Connect channel 1 of CRO at the input terminal and channel 2 across the load resistance to observe the input and output waveforms.

7. Take the traces of the input and output signals for all three cases.

8. Change the input signal to square wave and triangular wave and take traces.

Frequency Response

9. Now, set the input voltage to sine wave with amplitude of 1V.

10. Change the frequency of input signal starting from 100 Hz to 100 kHz and measure the output voltage.

11. Note the readings in the form of a table and find the gain.

12. Plot the frequency response curve between gain and frequency.

INPUT AND OUTPUT WAVEFORMS

1. Square wave

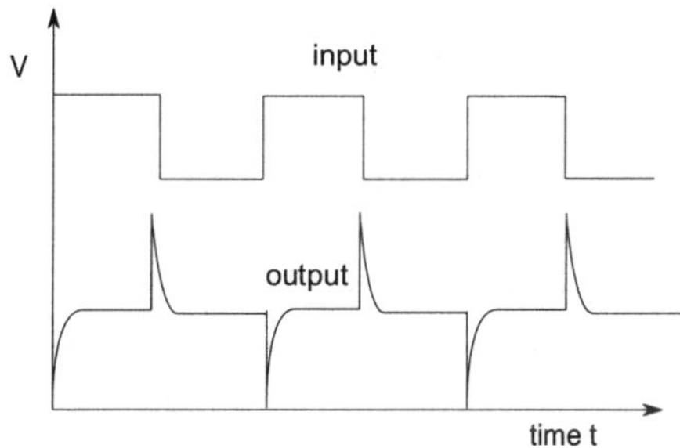

Fig. 5.4.4 Input and output waveform for square wave input

2.   Triangular wave

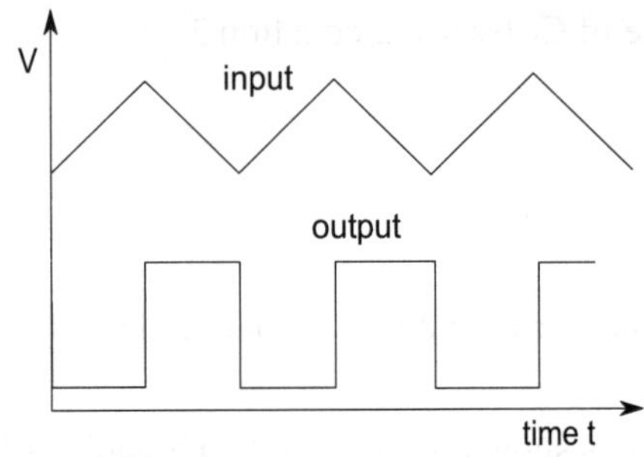

Fig. 5.4.5 Input and output waveform for triangular wave input

3.   Sine wave

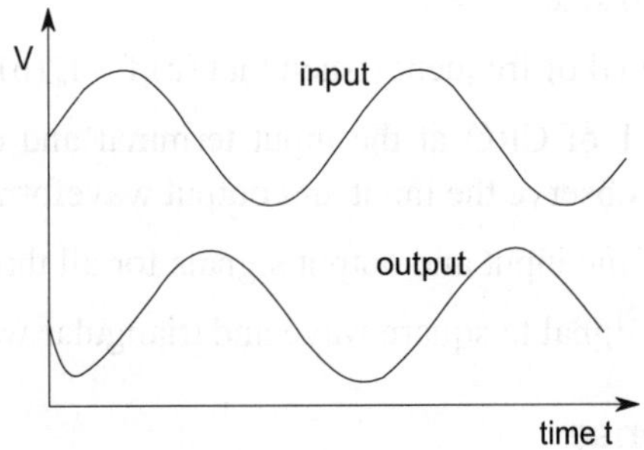

Fig. 5.4.6 Input and output waveform for sine wave input

OBSERVATION TABLE

Table 5.4.1 Frequency response of differentiator

S.No.	Frequency Hz	Input voltage Vin	Output voltage Vo	Gain Vo/Vin
1.	100			
2.	200			
3.	300			
4.	…			
5.	…			
6.	…			
7.	…			
8.	100k			

OBSERVATIONS

Attach the traces for input and output waveforms.

RESULT

Differentiator circuit using op-amp has been designed and frequency response curve has been plotted successfully. Traces for input and output signals have been taken and it is found that the circuit behaves as a differentiator for the frequencies below f_a = ... Hz.

DISCUSSION

The differentiator circuit produces cosine wave when the input is sine wave, square wave when the input is triangular wave and spikes when the input is a square wave.

The differentiator circuit is used in wave shaping circuits to detect high frequency components in an input signal and also used as a rate-of-change detector in FM modulators. Differentiators are also an important part in analog PID controllers and analog computers to differentiate the input analog voltage.

Experiment 5

LOW PASS FILTER

AIM

To Design a First and Second Order Low Pass Filter using Op-amp.

APPARATUS REQUIRED

Breadboard, resistors, capacitors, ±12V dc power supply, IC 741, cathode ray oscilloscope, probes, function generator, connecting wires.

THEORY

Filters

These are the electronic circuits which are used to pass the wanted frequency components and reject the unwanted one.

Passive and active filters

- Passive filters use passive elements like resistors, capacitors and inductors. They do not depend on any external power supply for their operation.

- Active filters use amplifying components like operational amplifiers, transistors in addition to resistors, capacitors and inductors. They need an external power source for their operation.

Advantage of active filter over passive filter

- Active filters use amplifying device and hence the output signal is not attenuated in pass band as in the case of passive filters.

- Op-amp used in active filters has high input impedance and low output impedance and therefore, it does not cause loading of the source or load.

- Active filters are more economical as compared to passive filters.

Low Pass Filter

A low pass filter is a frequency selective network that passes signals of frequency lower than a specified frequency called as cut off frequency and attenuates signals of frequency higher than the cut off frequency.

In low pass filter, as the frequency of input signal increases from 0 Hz, the gain of the circuit remains constant. At a certain frequency called cut-off frequency, the magnitude of the gain becomes 0.707 times its maximum value. As the input frequency increases beyond the cut off frequency, the signals get attenuated and gain decreases with a particular rate called as roll off rate. Therefore, the band from 0 Hz to cut off frequency f_C is called as pass-band and beyond this frequency, the band is called as stop-band. It is to be noted that the attenuation provided by any filter also depends on the order of the filter that is used. The roll off rate or the rate at which the gain of the filter in the stop-band decreases is calculated by using the formula $-n \times 20$ db/decade, where n is the order of the filter.

For first order low pass filter, the attenuation is of -20 db/ decade. Likewise, second order low pass filter, attenuation is of -40db/decade.

Ideal low pass filter

An ideal low pass filter is a network that passes a specified band of frequencies, known as pass-band and blocks all the signals of frequencies outside this band, known as stop-band. In other words, an ideal low pass filter has zero loss in its pass-band and infinite loss in its stop-band. But due to practical limitations of linear networks, ideal response of a low pass filter cannot be obtained. The response of such a filter is shown in Fig. 5.5.1.

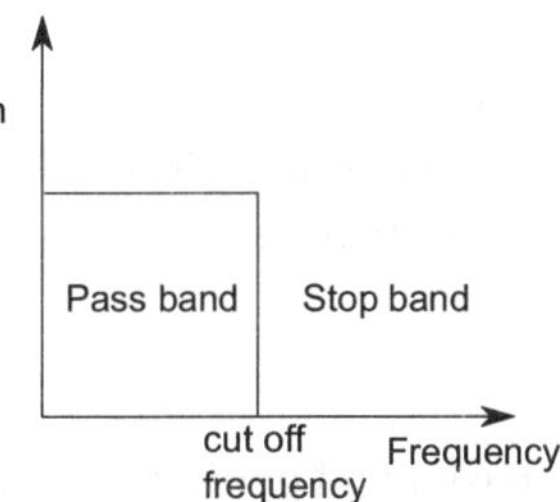

Fig. 5.5.1 Frequency response of ideal low pass filter

First Order Low Pass Filter with Op-amp

It consists of a RC network and an op-amp in non-inverting configuration as shown in Fig. 5.5.2.

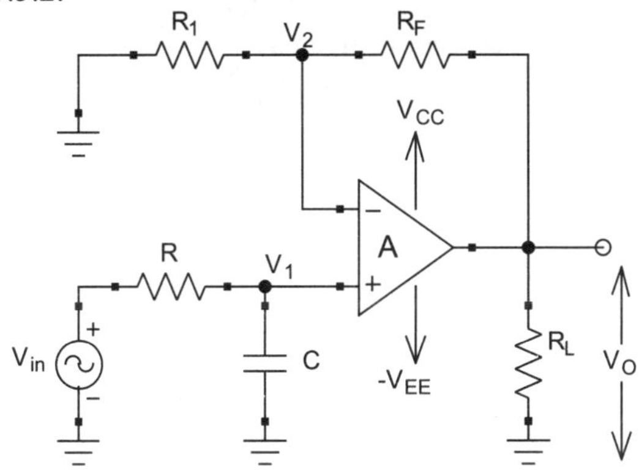

Fig. 5.5.2 First order low pass filter

The voltage at the non-inverting terminal *i.e.* across capacitor C is given as:

$$V_1 = \frac{-jX_C}{R - jX_C} V_{in}$$

Since,

$$X_C = \frac{1}{2\Pi fC}$$

Putting this value in above equation, we get

$$\Rightarrow \qquad V_1 = \frac{V_{in}}{1 + j2\Pi fRC}$$

Output voltage for non-inverting amplifier is given as

$$V_O = \left(1 + \frac{R_F}{R_1}\right) V_1$$

$$\Rightarrow \qquad V_O = \left(1 + \frac{R_F}{R_1}\right) \frac{V_{in}}{1 + j2\Pi fRC}$$

$$\Rightarrow \qquad \frac{V_O}{V_{in}} = \frac{A_F}{1 + j\left(f/f_C\right)}$$

Where, V_O/V_{in}: gain of filter, f: frequency of input signal, fc: cut off frequency which is given as

$$f_C = \frac{1}{2\Pi RC}$$

A_F: Pass-band gain of the filter and is given as

$$A_F = 1 + \frac{R_F}{R_1}$$

The magnitude of gain is given as

$$\left|\frac{V_O}{V_{in}}\right| = \frac{A_F}{\sqrt{1 + \left(f/f_C\right)^2}} \qquad\qquad - \text{eqn 1}$$

The phase angle of gain (in degrees) is given as

$$\phi = -\tan^{-1}\left(f/f_C\right)$$

Frequency response

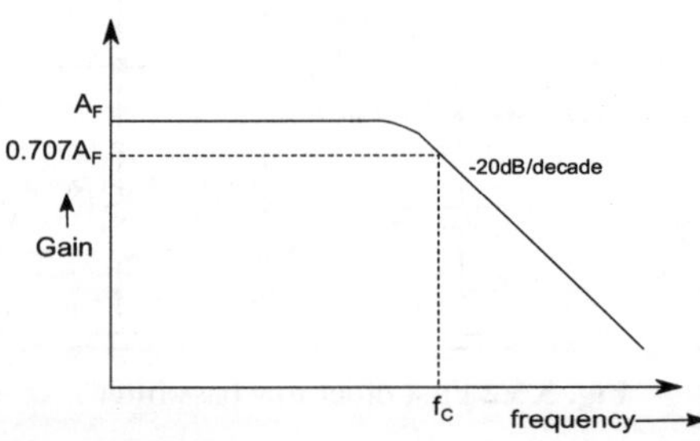

Fig. 5.5.3 Frequency response of first order low pass filter

Explanation

1. For frequencies below cut-off frequency *i.e.* $f < f_C$

Since, $f < fc$, hence $f/fc < 1$, we can write equation 1 as

$$\left|\frac{V_O}{V_{in}}\right| \approx A_F$$

Therefore, for frequencies below cut-off frequency, the signal is passed without any attenuation and gain is maximum which remains constant.

2. At cut-off frequency *i.e.* $f = fc$

Putting this value in equation 1, we get

$$\left|\frac{V_O}{V_{in}}\right| = \frac{A_F}{\sqrt{1 + \left(\frac{f_c}{f_c}\right)^2}}$$

$\Rightarrow$
$$\left|\frac{V_O}{V_{in}}\right| = \frac{A_F}{\sqrt{2}}$$

$\Rightarrow$
$$\left|\frac{V_O}{V_{in}}\right| = 0.707\, A_F$$

Therefore, at cut-off frequency the gain is 0.707 times the maximum gain or in other words, gain is 3dB (= 20log 0.707) down the maximum gain.

3. For frequencies above cut-off frequency *i.e.* $f > f_C$

Since, $f > fc$, hence $f/fc > 1$, we can write equation 1 as

$$\left|\frac{V_O}{V_{in}}\right| < A_F$$

Therefore, for frequencies above cut-off frequency, the signal is attenuated and as the frequency increases, gain decreases.

Second Order Low Pass Filter

It consists of two RC network and op-amp in non-inverting configuration as shown in Fig. 5.5.4.

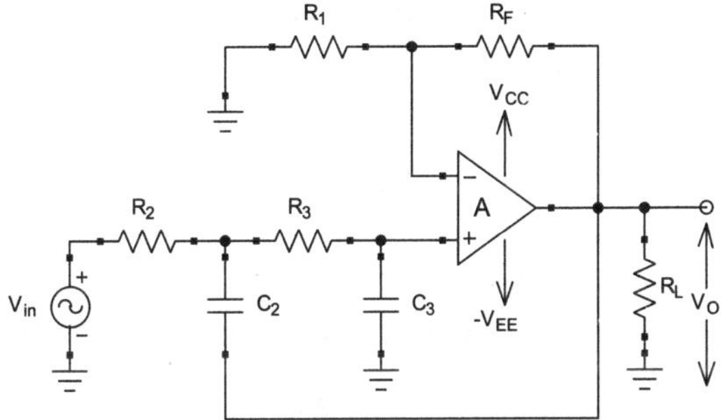

Fig. 5.5.4 Second order low pass filter

The magnitude of gain is given as

$$\left|\frac{V_O}{V_{in}}\right| = \frac{A_F}{\sqrt{1 + \left(f/f_C\right)^4}} \qquad - \text{eqn 2}$$

Where,

V_O/V_{in}: gain of filter, f: frequency of input signal, f_C: cut off frequency which is given as

$$f_C = \frac{1}{2\Pi\sqrt{R_2 R_3 C_2 C_3}}$$

A_F: Pass-band gain of the filter and is given as

$$A_F = 1 + \frac{R_F}{R_1}$$

Frequency response

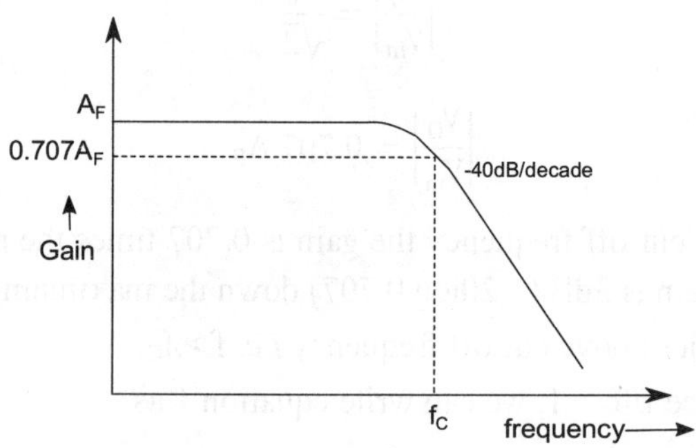

Fig. 5.5.5 Frequency response of second order low pass filter

The frequency response of second order low pass filter shows that the roll-off rate is 40 db per decade in the stop-band.

CALCULATION

First Order Low Pass Filter

1. Choose a particular value of cut-off frequency f_C.
2. Select the value of capacitor $C \leq 1\ \mu F$.
3. Calculate the value of resistor by using formula for cut-off frequency *i.e.*

$$f_C = \frac{1}{2\pi RC}$$

4. Choose the desired value of pass-band gain *i.e.* A_F.
5. Calculate the values of resistors R_1 and R_F by using the formula

$$A_F = 1 + \frac{R_F}{R_1}$$

Second Order Low Pass Filter

1. Choose a particular value of cut-off frequency f_C.

2. Take $R_2 = R_3$ (R) and $C_2 = C_3$ (C).

3. Select the value of capacitor $C \leq 1\ \mu F$.

4. Calculate the value of resistor by using formula for cut-off frequency *i.e.*

$$f_C = \frac{1}{2\Pi\sqrt{R_2 R_3 C_2 C_3}}$$

$\Rightarrow$
$$f_C = \frac{1}{2\Pi RC}$$

5. For $R_2 = R_3$ and $C_2 = C_3$, gain A_F is 1.586. Select $R_1 \leq 100k\Omega$ and calculate the values of resistor R_F by using the formula

$$A_F = 1 + \frac{R_F}{R_1}$$

$\Rightarrow$
$$R_F = 0.586R_1$$

PROCEDURE

1. Design the circuits according to the method as mentioned above.

2. Connect the circuit as shown in Fig. 5.5.2 on breadboard.

3. Take load resistance $R_L = 10k\Omega$.

4. Connect function generator to the input of circuit using a probe. Set it to sine wave with fixed amplitude.

5. Connect channel 1 of CRO at the input and channel 2 across R_L to view the input and output waveforms.

6. Vary the frequency of input signal from 100Hz to 1MHz.

7. Measure the input and output voltage through CRO and note the readings in a tabular form.

8. Plot the graph on semi-log graph paper between gain and input frequency.

9. Find the pass-band gain and cut-off frequency graphically where gain becomes 0.707 times of its maximum value.

10. Compare the theoretical and practical value of pass-band gain and cut-off frequency.

11. Now, connect the circuit as shown in Fig. 5.5.4 and repeat the above steps.

OBSERVATION TABLE

Table 5.5.1 Observation table for first order low pass filter

S.No.	Frequency (Hz)	Input voltage V_{in}	Output voltage V_O	Gain V_O / V_{in}
1.	100			
2.	200			
3.	300			
4.	400			
5.	500			
...	...			

Table 5.5.2 Observation table for second order low pass filter

S.No.	Frequency (Hz)	Input voltage V_{in}	Output voltage V_O	Gain V_O / V_{in}
1.	100			
2.	200			
3.	300			
4.	400			
5.	500			
...	...			

OBSERVATIONS

Table 5.5.3 Observations for cut-off frequency and pass-band gain

	First order low pass filter	Second order low pass filter
f_C (theoretical)		
f_C (practical)		
A_F (theoretical)		
A_F (practical)		

RESULT

The first and second order low pass filter has been designed and the frequency response has been drawn successfully.

DISCUSSION

An active low pass filter is extensively used in audio amplifiers, equalizers and speaker systems. It is used to reduce any high frequency noise, also known as hiss type distortion, present in the signal. It also directs the lower frequency bass signals to the larger bass speakers. Active low pass filters are also know as bass boost filter when they are used in audio applications.

They are also used in analog to digital converters as anti aliasing filters; in radio transmitters to block harmonic emissions *etc*.

Experiment 6

HIGH PASS FILTER

AIM

To Design a First and Second Order High Pass Filter using Op-amp.

APPARATUS REQUIRED

Breadboard, resistors, capacitors, ±12V dc power supply, IC 741, cathode ray oscilloscope, probes, function generator, connecting wires.

THEORY

Filters

These are the electronic circuits which are used to pass the wanted frequency components and reject the unwanted one.

Passive and active filters

- Passive filters use passive elements like resistors, capacitors and inductors. They do not depend on any external power supply for their operation.

- Active filters use amplifying components like operational amplifiers, transistors in addition to resistors, capacitors and inductors. They need an external power source for their operation.

Advantage of active filter over passive filter

- Active filters use amplifying device and hence the output signal is not attenuated in pass band as in the case of passive filters.

- Op-amp used in active filters has high input impedance and low output impedance and therefore, it does not cause loading of the source or load.

- Active filters are more economical as compared to passive filters.

High Pass Filter

A high pass filter is a frequency selective circuit that blocks the band of signals having frequency lower than a specified frequency called as cut-off frequency and passes the signals of frequency higher than the cut off frequency.

In high pass filter circuit, as the frequency of input signals increases from 0 Hz, the gain of the circuit also increases. At a certain frequency called cut-off frequency, the magnitude of the gain becomes 0.707 times its maximum value. As the input frequency increases beyond the cut off frequency, the circuit produces a constant gain. Therefore, the band from 0 Hz to cut off frequency f_C is called as stop-band and beyond this frequency, the band is called as pass-band. Below f_C, the signals get attenuated and gain decreases with a particular rate called as roll off rate. It is to be noted that the attenuation provided by any filter also depends on the order of the filter that is used. The roll off rate or the rate at which the gain of the filter in the stop-band decreases is calculated by using the formula $-n \times 20$ db/decade, where n is the order of the filter.

For first order high pass filter, the attenuation is of -20 db/ decade. Likewise, second order high pass filter, attenuation is of -40db/decade.

Ideal high pass filter

An ideal high pass filter is a network that passes a specified band of frequencies, known as pass-band and blocks all the signals of frequencies outside this band, known as stop-band. In other words, an ideal high pass filter has zero loss in its pass-band and infinite loss in its stop-band. But, due to practical limitations of linear networks, ideal response of high pass filter cannot be obtained. The response of such a filter is shown in Fig. 5.6.1.

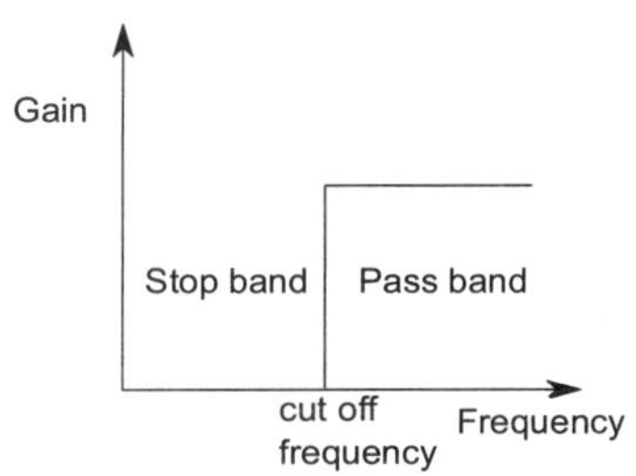

Fig. 5.6.1 Frequency response of ideal high pass filter

First Order High Pass Filter with Op-amp

It consists of an RC network and op-amp in non-inverting configuration as shown in Fig. 5.6.2.

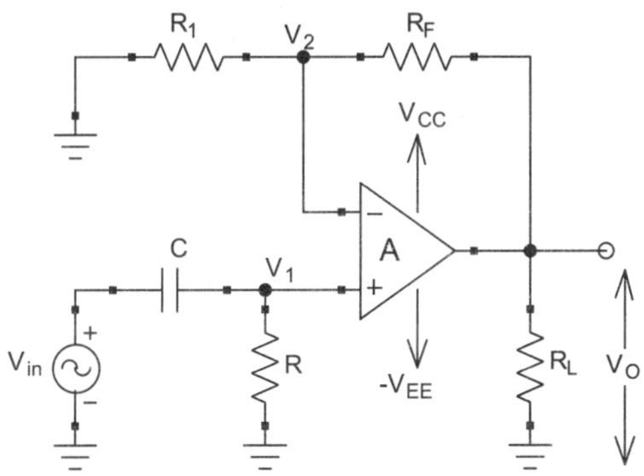

Fig. 5.6.2 First order high pass filter

The voltage at the non-inverting terminal *i.e.* across resistor R is given as:

$$V_1 = \frac{R}{R - jX_C} V_{in}$$

Since,

$$X_C = \frac{1}{2\Pi fC}$$

Putting this value in above equation, we get

$$\Rightarrow \qquad V_1 = \frac{j\,2\Pi\,f\,R\,C}{1 + j\,2\Pi\,f\,R\,C} V_{in}$$

Output voltage for non-inverting amplifier is given as

$$V_O = \left(1 + \frac{R_F}{R_1}\right) V_1$$

$$\Rightarrow \qquad V_O = \left(1 + \frac{R_F}{R_1}\right) \frac{j\,2\Pi\,f\,R\,C}{1 + j\,2\,\Pi\,f\,R\,C} V_{in}$$

$$\Rightarrow \qquad \frac{V_O}{V_{in}} = A_F \frac{j\left(f/f_C\right)}{1 + j\left(f/f_C\right)}$$

Where, V_O/V_{in}: gain of filter, f: frequency of input signal, f_C: cut off frequency which is given as

$$f_C = \frac{1}{2\Pi RC}$$

A_F: Pass-band gain of the filter and is given as

$$A_F = 1 + \frac{R_F}{R_1}$$

The magnitude of gain is given as

$$\left|\frac{V_O}{V_{in}}\right| = \frac{A_F\left(f/f_C\right)}{\sqrt{1 + \left(f/f_C\right)^2}} \qquad\qquad -\text{eqn 1}$$

Frequency response

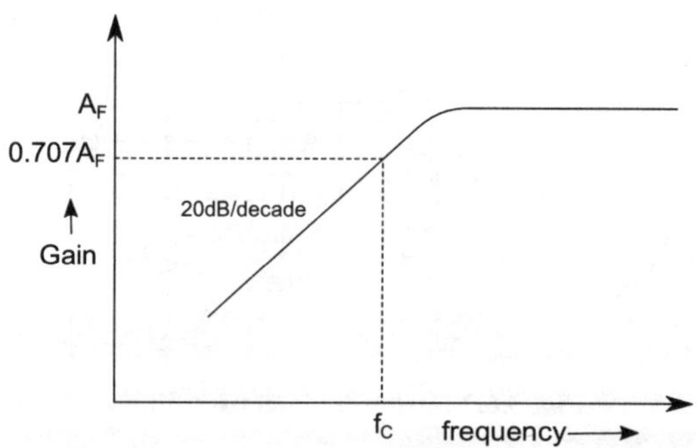

Fig. 5.6.3 Frequency response of first order high pass filter

Explanation

1.   For frequencies below cut-off frequency *i.e.* $f < f_C$

Since, $f < f_C$, hence $f/f_C < 1$, we can write equation 1 as

$$\left| \frac{V_o}{V_{in}} \right| < A_F$$

Therefore, for frequencies below cut-off frequency, the signal is attenuated and as the frequency increases, gain increases.

2.   At cut-off frequency *i.e.* $f = f_C$

Putting this value in equation 1, we get

$$\left| \frac{V_o}{V_{in}} \right| = \frac{A_F \left(\frac{f_C}{f_C} \right)}{\sqrt{1 + \left(\frac{f_C}{f_C} \right)^2}}$$

$$\Rightarrow \qquad \left| \frac{V_o}{V_{in}} \right| = \frac{A_F}{\sqrt{2}}$$

$$\Rightarrow \qquad \left| \frac{V_o}{V_{in}} \right| = 0.707 \, A_F$$

Therefore, at cut-off frequency the gain is 0.707 times the maximum gain or in other words, gain is 3dB ($= 20\log 0.707$) down the maximum gain.

3.   For frequencies above cut-off frequency *i.e.* $f > f_C$

Since, $f > f_C$, hence $f/f_C > 1$, we can write equation 1 as

$$\left| \frac{V_o}{V_{in}} \right| \approx A_F$$

Therefore, for frequencies above cut-off frequency, the gain is maximum which remains constant.

Second Order High Pass Filter

It consists of two RC network and op-amp in non-inverting configuration as shown in Fig. 5.6.4.

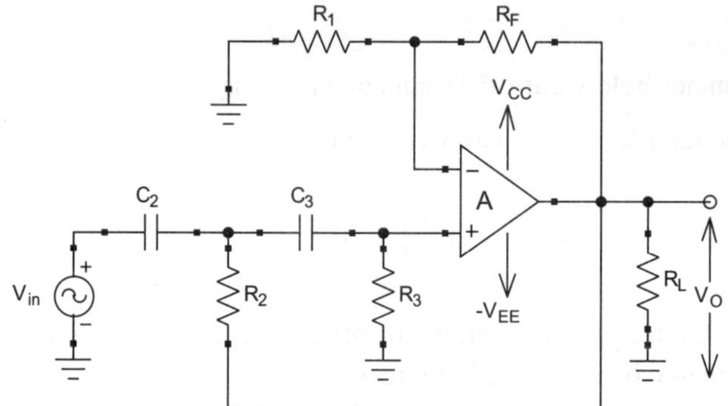

Fig. 5.6.4 Second order high pass filter

The magnitude of gain is given as

$$\left|\frac{V_O}{V_{in}}\right| = \frac{A_F}{\sqrt{1 + \left(f_C/f\right)^4}} \qquad - \text{eqn 2}$$

Where,

V_O/V_{in}: gain of filter, f: frequency of input signal, f_C: cut off frequency which is given as

$$f_C = \frac{1}{2\Pi\sqrt{R_2 R_3 C_2 C_3}}$$

A_F: Pass-band gain of the filter and is given as

$$A_F = 1 + \frac{R_F}{R_1}$$

Frequency response

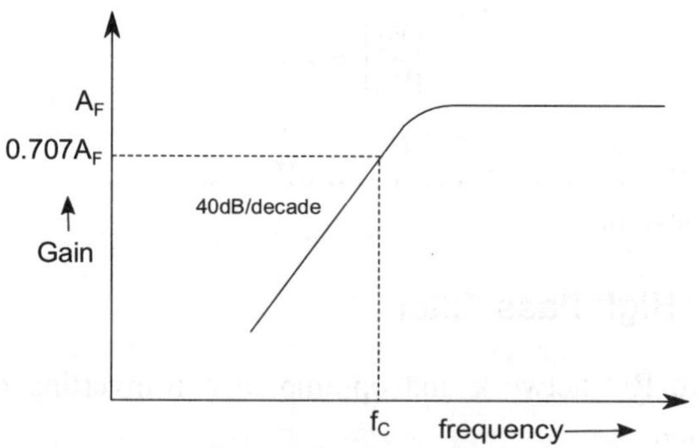

Fig. 5.6.5 Frequency response of second order high pass filter

The frequency response of second order high pass filter shows that the roll-off rate is 40 db per decade in the stop-band.

CALCULATION

First Order High Pass Filter

1. Choose a particular value of cut-off frequency f_C.
2. Select the value of capacitor $C \leq 1\ \mu F$.
3. Calculate the value of resistor by using formula for cut-off frequency *i.e.*

$$f_C = \frac{1}{2\pi RC}$$

4. Choose the desired value of pass-band gain *i.e.* A_F.
5. Calculate the values of resistors R_1 and R_F by using the formula

$$A_F = 1 + \frac{R_F}{R_1}$$

Second Order High Pass Filter

1. Choose a particular value of cut-off frequency f_C.
2. Take $R_2 = R_3$ (R) and $C_2 = C_3$ (C).
3. Select the value of capacitor $C \leq 1\ \mu F$.
4. Calculate the value of resistor by using the formula for cut-off frequency *i.e.*

$$f_C = \frac{1}{2\Pi\sqrt{R_2 R_3 C_2 C_3}}$$

$$\Rightarrow \qquad f_C = \frac{1}{2\Pi RC}$$

5. For $R_2 = R_3$ and $C_2 = C_3$, gain A_F is 1.586. Select $R_1 \leq 100\text{k}\Omega$ and calculate the values of resistor R_F by using the formula

$$A_F = 1 + \frac{R_F}{R_1}$$

$$\Rightarrow \qquad R_F = 0.586 R_1$$

PROCEDURE

1. Design the circuits according to the method as mentioned above.
2. Connect the circuit as shown in Fig. 5.6.2 on breadboard.
3. Take load resistance $R_L = 10\text{k}\Omega$.

4.   Connect function generator to the input of circuit using a probe. Set it to sine wave with fixed amplitude.

5.   Connect channel 1 of CRO at the input and channel 2 across R_L to view the input and output waveforms.

6.   Vary the frequency of input signal from 100Hz to 1MHz.

7.   Measure the input and output voltage through CRO and note the readings in a tabular form.

8.   Plot the graph on semi-log graph paper between gain and input frequency.

9.   Find the pass-band gain and cut-off frequency graphically where gain becomes 0.707 times of its maximum value.

10. Compare the theoretical and practical value of pass-band gain and cut-off frequency.

11. Now, connect the circuit as shown in Fig. 5.6.4 and repeat the above steps.

OBSERVATION TABLE

Table 5.6.1 Observation table for first order high pass filter

S.No.	Frequency (Hz)	Input voltage V_{in}	Output voltage V_O	Gain V_O / V_{in}
1.	100			
2.	200			
3.	300			
4.	400			
5.	500			
...	...			

Table 5.6.2 Observation table for second order high pass filter

S.No.	Frequency (Hz)	Input voltage V_{in}	Output voltage V_O	Gain V_O / V_{in}
1.	100			
2.	200			
3.	300			
4.	400			
5.	500			
...	...			

OBSERVATIONS

Table 5.6.3 Observations for cut-off frequency and pass-band gain

	First order high pass filter	Second order high pass filter
f_C (theoretical)		
f_C (practical)		
A_F (theoretical)		
A_F (practical)		

RESULT

The first and second order high pass filter has been designed and the frequency response has been drawn successfully.

DISCUSSION

An active high pass filter is extensively used in audio amplifiers, equalizers and speaker systems. It is used to reduce any low frequency noise, also known as rumble type distortion, present in the signal. It also directs the high frequency signals to the small tweeter speakers. Active high pass filters are also known as treble boost filter when they are used in audio applications.

Experiment 7

BAND PASS FILTER

AIM

To Design a Band Pass Filter using Op-amp.

APPARATUS REQUIRED

Breadboard, resistors, capacitors, ± 12V dc power supply, IC 741, cathode ray oscilloscope, probes, function generator, connecting wires.

THEORY

Filters

These are the electronic circuits which are used to pass the wanted frequency components and reject the unwanted one.

Passive and active filters

- Passive filters use passive elements like resistors, capacitors and inductors. They do not depend on any external power supply for their operation.

- Active filters use amplifying components like operational amplifiers, transistors in addition to resistors, capacitors and inductors. They need an external power source for their operation.

Advantage of active filter over passive filter

- Active filters use amplifying device and hence the output signal is not attenuated in pass band as in the case of passive filters.

- Op-amp used in active filters has high input impedance and low output impedance and therefore, it does not cause loading of the source or load.

- Active filters are more economical as compared to passive filters.

Band Pass Filter

Band pass filter is a circuit designed to pass the signals of specified band of frequencies while attenuating all the signals that lie outside this band. In other words, the pass-band of a band pass filter lies between two cut-off frequencies *i.e.* f_L and f_H such that $f_H > f_L$ and the stop-band consists of the input frequencies that fall outside this pass-band. Band pass filter can be of two types.

- Narrow band pass filter with Quality factor or figure of merit Q > 10.
- Wide band pass filter with Quality factor or figure of merit Q < 10.

The quality factor Q is a measure of selectivity. If the value of Q is more, filter will be more selective and bandwidth will be narrower.

Ideal band pass filter

An ideal band pass filter is a network that has zero loss in its pass-band and infinite loss in its stop-band. But, due to practical limitations of linear networks, ideal response of band pass filter cannot be obtained. The response of such a filter is shown in Fig. 5.7.1.

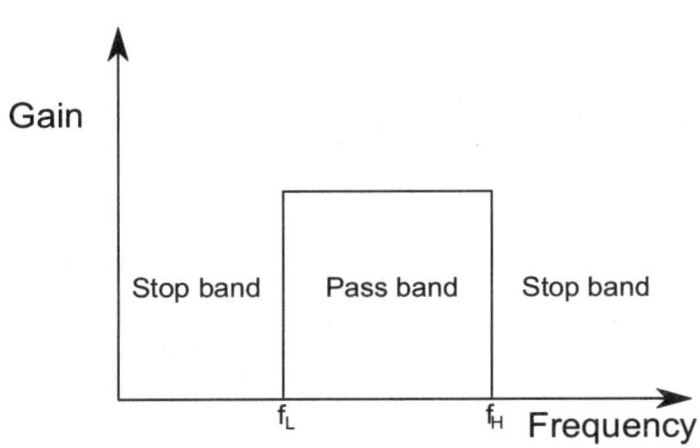

Fig. 5.7.1 Frequency response of ideal band pass filter

Wide Band Pass Filter with Op-amp

A wide band pass filter is a network that is formed by cascading high pass and low pass filter as shown in Fig. 5.7.2.

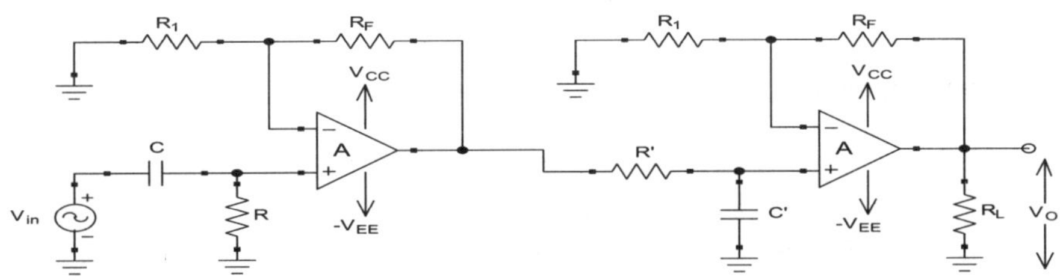

Fig. 5.7.2 First order wide band pass filter

The order of the band pass filter depends on the order of the high pass and low pass filter which are cascaded together. For *e.g.* if the first order high pass and first order low pass filter are cascaded together, it will form a first order ±20dB/decade band pass filter. Similarly, if the second order high pass and second order low pass filter are cascaded together, it will form a second order ±40dB/decade band pass filter. The total pass band gain of the wide band pass filter is given as

$$A_F = A_{F1} \cdot A_{F2}$$

Where,

A_{F1}: gain of high pass filter, A_{F2}: gain of low pass filter.

The magnitude of voltage gain is given as

$$\left|\frac{V_O}{V_{in}}\right| = \left|\frac{V_O}{V_{in}}\right|_{\text{High pass filter}} \cdot \left|\frac{V_O}{V_{in}}\right|_{\text{Low pass filter}}$$

$$\Rightarrow \qquad \left|\frac{V_O}{V_{in}}\right| = \frac{A_{F1}\left(f/f_L\right)}{\sqrt{1+\left(f/f_L\right)^2}} \cdot \frac{A_{F2}}{\sqrt{1+\left(f/f_H\right)^2}}$$

$$\Rightarrow \qquad \left|\frac{V_O}{V_{in}}\right| = \frac{A_F\left(f/f_L\right)}{\sqrt{\left[1+\left(f/f_L\right)^2\right]\left[1+\left(f/f_H\right)^2\right]}}$$

The expression for the center frequency f_C is defined as

$$f_C = \sqrt{f_H \cdot f_L}$$

Where,

f_H: high cut off frequency and f_L: low cut off frequency.

The quality factor Q is related to the bandwidth as

$$Q = \frac{f_C}{BW} = \frac{f_C}{f_H - f_L}$$

Frequency response

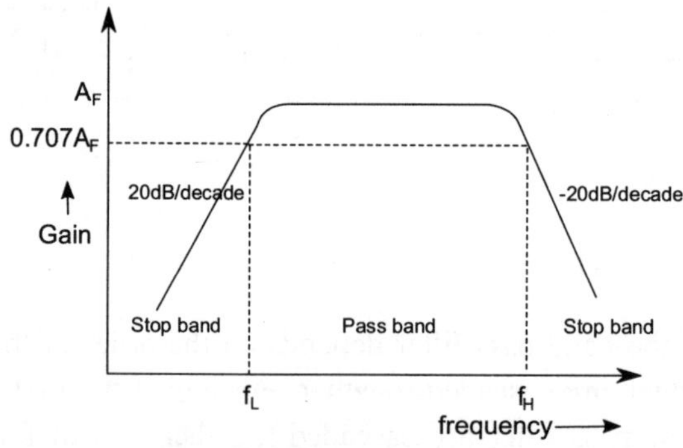

Fig. 5.7.3 Frequency response of first order wide band pass filter

CALCULATION

The wide band pass filter can be designed for a given pass band gain A_F and cut-off frequencies as per the steps mentioned below.

High Pass Filter

1. Choose a particular value of cut-off frequency f_L.

2. Select the value of capacitor $C \leq 1\ \mu F$.

3. Calculate the value of resistor by using formula for cut-off frequency *i.e.*

$$f_L = \frac{1}{2\pi RC}$$

Low Pass Filter

4. Choose a particular value of cut-off frequency f_H such that $f_H > f_L$.

5. Select the value of capacitor $C' \leq 1\ \mu F$.

6. Calculate the value of resistor R' by using formula for cut-off frequency *i.e.*

$$f_H = \frac{1}{2\pi R'C'}$$

7. Choose the desired value of pass-band gain and keep it same for both low pass and high pass filter i.e. $A_{F1} = A_{F2}$.

8. Calculate the values of resistors R_1 and R_F by using the formula

$$A_{F1} = A_{F2} = 1 + \frac{R_F}{R_1}$$

PROCEDURE

1. Design the band pass filter according to the method as mentioned above.

2. Connect the circuit as shown in Fig. 5.7.2 on breadboard.

3. Take load resistance $R_L = 10k\Omega$.

4. Connect function generator to the input of circuit using a probe. Set it to sine wave with fixed amplitude.

5. Connect channel 1 of CRO at the input and channel 2 across R_L to view the input and output waveforms.

6. Vary the frequency of input signal from 100Hz to 1MHz.

7. Measure the input and output voltage through CRO and note the readings in a tabular form.

8. Plot the graph on semi-log graph paper between gain and input frequency.

9. Find the pass-band gain and lower and higher cut-off frequency graphically where gain becomes 0.707 times of its maximum value.

10. Compare the theoretical and practical value of pass-band gain and cut-off frequencies.

OBSERVATION TABLE

Table 5.7.1 Observation table for wide band pass filter

S.No.	Frequency (Hz)	Input voltage V_{in}	Output voltage V_O	Gain V_O / V_{in}
1.	100			
2.	200			
3.	300			
4.	...			

OBSERVATIONS

Table 5.7.2 Comparison of theoretical and practical values

S.No.	Parameter	Theoretical value	Practical value
1.	High cut-off frequency f_H		
2.	Low cut-off frequency f_L		
3.	Pass band gain A_F	$A_F = A_{F1} . A_{F2}$ $= ...$	
4.	Center frequency f_C	$f_C = \sqrt{f_H . f_L}$ $= ...$	
5.	Quality factor Q	$Q = \dfrac{f_C}{f_H - f_L}$ $= ....$	

RESULT

The first order wide band pass filter has been designed and the frequency response has been drawn successfully.

DISCUSSION

Band pass filters are widely used in wireless transmitters and receivers. In transmitters, it is used to limit the bandwidth of the signal according to the band allocated for transmission. This helps in reducing the interference from other stations. In receivers, it is used to pass only a selected range of frequencies while rejecting the unwanted signals.

Experiment 8

BAND REJECT FILTER

AIM

To Design a Band-Reject Filter using Op-amp.

APPARATUS REQUIRED

Breadboard, resistors, capacitors, ±12V dc power supply, IC 741, cathode ray oscilloscope, probes, function generator, connecting wires.

THEORY

Filters

These are the electronic circuits which are used to pass the wanted frequency components and reject the unwanted one.

Passive and active filters

- Passive filters use passive elements like resistors, capacitors and inductors. They do not depend on any external power supply for their operation.

- Active filters use amplifying components like operational amplifiers, transistors in addition to resistors, capacitors and inductors. They need an external power source for their operation.

Advantage of active filter over passive filter

- Active filters use amplifying device and hence the output signal is not attenuated in pass band as in the case of passive filters.

- Op-amp used in active filters has high input impedance and low output impedance and therefore, it does not cause loading of the source or load.

- Active filters are more economical as compared to passive filters.

Band Reject Filter

Band reject filter is a circuit designed to reject the signals of specified band of frequencies while passing all the signals outside this band. This filter is also known as band stop or band eliminator filter. In other words, the stop-band or reject band of a band reject filter lies between two cut-off frequencies *i.e.* f_H and f_L such that $f_L > f_H$ and the pass-band consists of the input frequencies that fall outside this reject-band. Band reject filter can be of two types.

- Narrow band reject or notch filter with Quality factor or figure of merit Q > 10.

- Wide band reject filter with Quality factor or figure of merit Q < 10.

The quality factor Q is a measure of selectivity. If the value of Q is more, filter will be more selective and bandwidth will be narrower.

Ideal band reject filter

An ideal band reject filter is a network that has zero loss in its pass-band and infinite loss in its stop-band. But, due to practical limitations of linear networks, ideal response of band reject filter cannot be obtained. The response of such a filter is shown in Fig. 5.8.1.

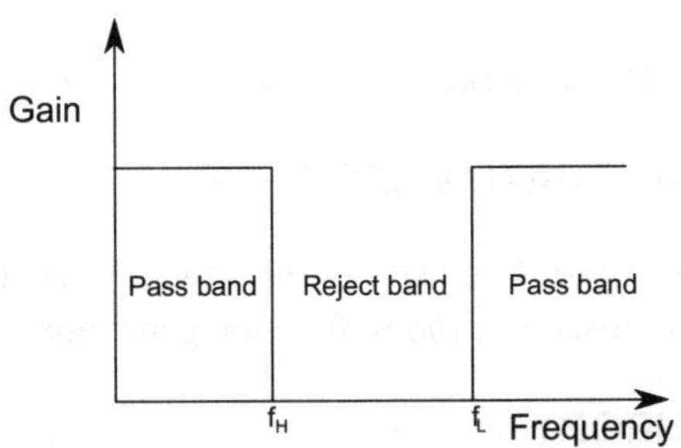

Fig. 5.8.1 Frequency response of ideal band reject filter

Wide Band Reject Filter with Op-amp

A wide band reject filter is a circuit formed by using a low pass filter, high pass filter and summing amplifier as shown in Fig. 5.8.2.

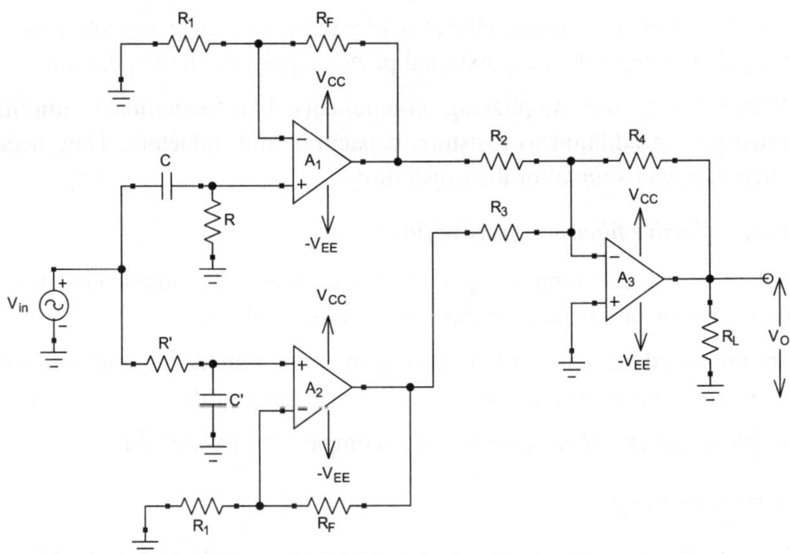

Fig. 5.8.2 First order wide band reject filter

The output voltage of summing amplifier is given as

$$V_O = -\left(\frac{R_4}{R_3} \cdot V_{\text{LPF}} + \frac{R_4}{R_2} \cdot V_{\text{HPF}}\right)$$

Taking $R_3 = R_2 = R$

$$\Rightarrow \qquad V_O = -\frac{R_4}{R}(V_{HPF} + V_{LPF})$$

The above equation shows that the output voltage of summing amplifier is equal to the negative times the gain of the circuit *i.e.* $-R_4/R$ to the sum of all the input voltages *i.e.* output voltage of high pass filter and low pass filter. Taking $R_4 = R$, the above equation becomes

$$\Rightarrow \qquad V_O = -(V_{HPF} + V_{LPF})$$

The band reject filter is designed by taking the gain of both high pass and low pass filters to be same *i.e.*

$$A_F = A_{F1} = A_{F2}$$

Where, A_{F1}: gain of low pass filter, A_{F2}: gain of high pass filter.

The total pass band gain of the wide band reject filter is given as

$$A_{FT} = A_F \cdot A_{F3}$$

Where, A_{F3}: gain of summing amplifier.

The expression for the center frequency f_C is defined as

$$f_C = \sqrt{f_H \cdot f_L}$$

Where, f_H: high cut off frequency of low pass filter, f_L: low cut off frequency of high pass filter.

The quality factor Q is related to the bandwidth as

$$Q = \frac{f_C}{BW} = \frac{f_C}{f_H - f_L}$$

Frequency response

The frequency response shows that the low cut off frequency f_L of high pass filter should be greater than the high cut off frequency f_H of low pass filter.

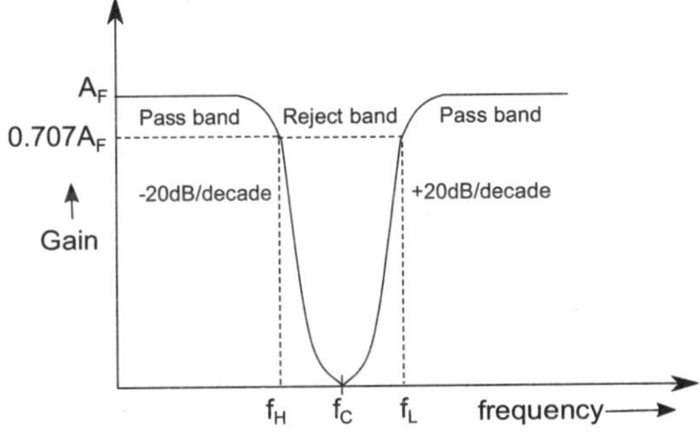

Fig. 5.8.3 Frequency response of first order wide band reject filter

CALCULATION

The wide band reject filter can be designed for a given pass band gain A_F and cut-off frequencies as per the steps mentioned below.

High Pass Filter

1. Choose a particular value of cut-off frequency f_L such that $f_L > f_H$.
2. Select the value of capacitor $C \leq 1~\mu F$.
3. Calculate the value of resistor by using formula for cut-off frequency *i.e.*

$$f_L = \frac{1}{2\pi RC}$$

Low Pass Filter

4. Choose a particular value of cut-off frequency f_H.
5. Select the value of capacitor $C' \leq 1~\mu F$.
6. Calculate the value of resistor R' by using formula for cut-off frequency *i.e.*

$$f_H = \frac{1}{2\pi R'C'}$$

7. Choose the desired value of pass-band gain *i.e.* A_F and keep it same for both low pass and high pass filter.
8. Calculate the values of resistors R_1 and R_F by using the formula

$$A_F = 1 + \frac{R_F}{R_1}$$

9. Set the gain of summing amplifier A_{F3} to be 1 and thus, take the value of R_2, R_3 and R_4 such that all are same *i.e.* $R_2 = R_3 = R_4$.

PROCEDURE

1. Design the band reject filter according to the method as mentioned above.
2. Connect the circuit as shown in Fig. 5.8.2 on breadboard.
3. Take load resistance $R_L = 10k\Omega$.
4. Connect function generator to the input of circuit using a probe. Set it to sine wave with fixed amplitude.
5. Connect channel 1 of CRO at the input and channel 2 across R_L to view the input and output waveforms.
6. Vary the frequency of input signal from 100Hz to 1MHz.
7. Measure the input and output voltage through CRO and note the readings in a tabular form.
8. Plot the graph on semi-log graph paper between gain and input frequency.

9. Find the pass-band gain and lower and higher cut-off frequency graphically where gain becomes 0.707 times its maximum value.

10. Compare the theoretical and practical value of pass-band gain and cut-off frequencies.

OBSERVATION TABLE

Table 5.8.1 Observation table for wide band reject filter

S.No.	Frequency (Hz)	Input voltage V_{in}	Output voltage V_{out}	Gain V_{out} / V_{in}
1.	100			
2.	200			
3.	300			
4.	400			
5.	500			
...	...			

OBSERVATIONS

Table 5.8.2 Comparison of theoretical and practical values

S.No.	Parameter	Theoretical value	Practical value
1.	Cut off frequency f_L		
2.	Cut off frequency f_H		
3.	Pass band gain A_{FT}	$A_{FT} = A_F \cdot A_{F3}$ $= ...$	
4.	Center frequency f_C	$f_C = \sqrt{f_H \cdot f_L}$ $= ...$	
5.	Quality factor Q	$Q = \dfrac{f_C}{f_H - f_L}$ $= ...$	

RESULT

The first order wide band reject filter has been designed and the frequency response has been drawn successfully.

DISCUSSION

Band reject filters are widely used in biomedical instruments like ECG to remove the line noise; as anti hum filters to filter out the main hum from the power line; in electric guitar amplifiers to reduce the hum at 60Hz frequency produced by electric guitar; in communication electronics to reduce the effect of unwanted harmonics on the desired signals; in telephone technology to reduce the telephone line noise; in optical communication to reduce the effect of spurious frequencies of light at the end of the optical fiber which can distort the main light beam; in high quality audio applications.

Experiment 9

PHASE SHIFT OSCILLATOR

AIM

To Design a RC Phase Shift Oscillator using Op-amp

APPARATUS REQUIRED

Breadboard, resistors, capacitors, ± 12V dc power supply, IC - 741, cathode ray oscilloscope, probes, connecting wires.

THEORY

Oscillator

An oscillator is a circuit that is used to generate a repetitive current or voltage waveform of specific amplitude and frequency without any external input signal.

Types of oscillators

1. Depending on the type of waveform generated

    *   Sinusoidal oscillator which generates pure sinusoidal waveform.

    *   Non sinusoidal or relaxation oscillator which generates square, pulses, triangular and saw tooth waveforms.

2. Depending on the type of components used

    *   RC oscillator

    *   LC oscillator

    *   Crystal oscillator

3. Depending on the frequency of oscillations

    *   Audio frequency oscillator

    *   Radio frequency oscillator

Principle

Oscillator is a positive feedback amplifier that is capable of producing an ac output signal without any ac input signal. The word 'feedback' means that the output is fed back to its input via a feedback circuit and the word 'positive' means that the output that is fed back is in phase with the input.

An oscillator needs a starting voltage to produce oscillations. This starting voltage is provided by the noise voltage produced by the random motion of electrons in resistors or active device. Noise, which is random in nature having all the frequencies, is amplified by an amplifier and is fed to the feedback network. The feedback network provides maximum gain and 180° phase shift only for a single frequency. Hence, only a single frequency present in the noise is fed back with proper phase to the amplifier. Therefore, a positive feedback amplifier *i.e.* an oscillator generates oscillations only at a single frequency as decided by the feedback network.

Block diagram of oscillator

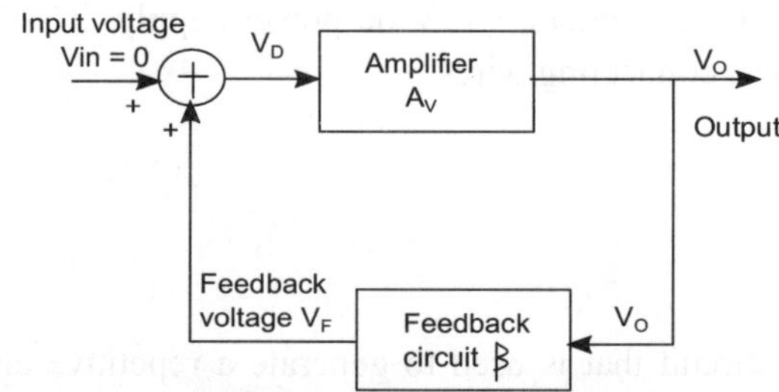

Fig. 5.9.1 Block diagram of oscillator

As shown in the Fig. 5.9.1, the closed loop gain of the amplifier and feedback circuits are A_V and β respectively. The voltage going as input to the amplifier is given as

$$V_D = V_F + V_{in}$$

The output voltage is given as

$$V_O = A_V V_D$$

$\Rightarrow$
$$V_O = A_V(V_F + V_{in}) \qquad\qquad - \text{eqn 1}$$

Since,

$$V_F = \beta V_O$$

Using this relation in equation 1, we get

$\Rightarrow$
$$V_O = A_V \beta V_O + A_V V_{in}$$

$\Rightarrow$
$$\frac{V_O}{V_{in}} = \frac{A_V}{1 - A_V\beta}$$

Since, input voltage is zero *i.e.* $V_{in} = 0$ and $V_O \neq 0$, we can say

$$A_V \beta = 1$$

or
$$A_V \beta = 1\angle 0° \text{ or } 360°$$

When $A_V\beta = 1$, V_o/V_{in} *i.e.* gain with positive feedback A_F becomes infinite *i.e.* we get the output without any input. This is the condition to start the oscillations.

Two conditions necessary to be fulfilled for the oscillations

- The magnitude of loop gain $A_V\beta$ must be ≥ 1. The value of $A_V\beta$ must be 1 for oscillations to start and for the sustained oscillations $A_V\beta > 1$, otherwise, the oscillations would die out. $A_V\beta = 1$ is called as the Barkhausen criterion of oscillation.

- Total phase shift of the loop gain must be 0° or 360°.

Phase Shift Oscillator

Phase shift oscillator is a RC type of an oscillator as shown in Fig. 5.9.2.

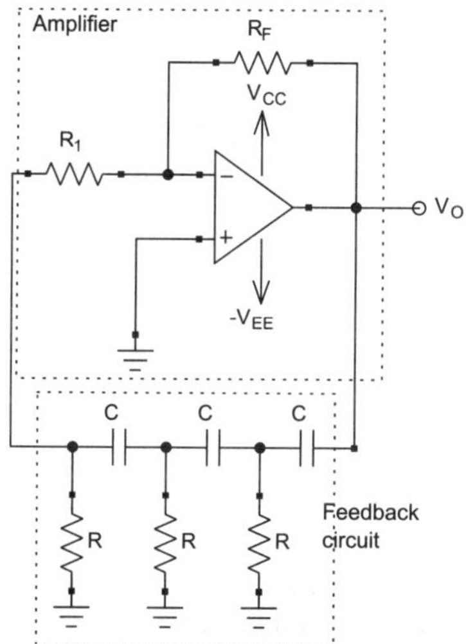

Fig. 5.9.2 Phase shift oscillator using op-amp

It consists of an op-amp and three RC networks cascaded together. Op-amp is used in inverting mode and hence, it amplifies the input signal as well as provides a phase shift of 180°. Another 180° phase shift is provided by RC network. One RC section is capable of producing a maximum phase shift of 60° practically (90° ideally). Hence, 3 RC sections are cascaded together to provide a total of 180° phase shift. Thus, the total phase shift around the loop is 360°.

Therefore, the circuit oscillates at some specific frequency only where the phase shift provided by the cascaded RC networks is exactly 180° and gain of the amplifier is sufficiently large. This frequency of oscillation is given by

$$f = \frac{1}{2\Pi RC\sqrt{6}} \qquad\qquad - \text{eqn 2}$$

At this frequency, the gain A_V must be at least 29 or we can say

$$\frac{R_F}{R_1} = 29 \text{ or } R_F = 29R_1 \qquad\qquad - \text{eqn 3}$$

CALCULATION

1. Choose the desired frequency of oscillations.
2. Choose a particular value of capacitor C.
3. Find the value of resistor R by using equation 2.
4. Find R_1 such that $R_1 = 10R$ (to prevent loading of amplifier due to RC network).
5. Find R_F by using equation 3.

PROCEDURE

1. Design phase shift oscillator according to the method as mentioned above.
2. Connect the circuit as shown in Fig. 5.9.2 on breadboard.
3. Connect channel 1 of CRO using probe at the output of op-amp.
4. Take the trace of output waveform and measure its frequency and amplitude.
5. Compare the theoretical and practical value of the frequency.
6. Change the value of R and C for different set of frequency of oscillations and repeat the above steps.

OBSERVATION TABLE

Table 5.9.1 Observation table for phase shift oscillator

S.No.	Resistance R	Capacitor C	Frequency of oscillations	
			Th.	Pr.
1.				
2.				
3.				
4.				

OBSERVATIONS

Attach the traces for output waveform.

RESULT

The phase shift oscillator has been designed successfully and the theoretical and practical value of frequency of oscillations is found to be nearly same.

DISCUSSION

Op-amp IC 741 can be used for generating low frequencies only *i.e.* for frequencies less than 1 kHz. However, op-amp ICs such as LM318 and LF351 can be used for generating higher frequencies due to its better slew rate.

As the temperature or dc power supply changes, the value of external components as well as internal components such as resistances and junction capacitances of op-amp changes. This changes the frequency of oscillations. Hence, in order to attain frequency stability, variations in the components can be controlled by appropriate designing of the circuit, by using regulated power supply and by controlling the temperature.

Phase shift oscillators are the RC oscillators which are used for generating audio frequencies. They have a wide frequency range from few Hz to several hundred kHz with good frequency stability. Circuitry is simple and cheap and it does not require inductors for their operation.

They cannot be used for high frequencies because at high frequency, capacitor will offer very low impedance and hence acts as a short. Therefore, resistances will come in parallel to each other and hence RC combinations will lose its action of producing 180° phase shift.

Experiment 10

MONOSTABLE and ASTABLE MULTIVIBRATOR

AIM

To Study IC 555 as Monostable and Astable Multivibrator.

APPARATUS REQUIRED

Breadboard, resistors, capacitors, IC - 555, cathode ray oscilloscope, probes, function generator, +5V dc power supply, connecting wires.

THEORY

555 Timer

The 555 timer is a linear integrated circuit that is used in applications of producing accurate and highly stable time delays or oscillations. Its prominent features are: operating voltage from +5V to +18V; adjustable duty cycle; timing is from microseconds through hours; high temperature stability; high current output.

Pin Configuration of 555 Timer

The 555 IC is commonly used as 8 pin DIP (Dual-in-package) as shown in Fig. 5.10.1.

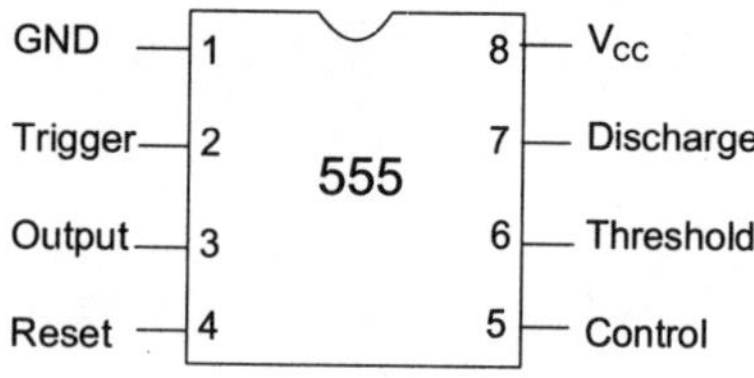

Fig. 5.10.1 Pin diagram for 555 timer

Pin 1: Ground

All the voltages are measured with respect to Pin 1.

Pin 2: Trigger

The output of the timer is controlled by amplitude of the signal applied at this terminal. This pin is connected to an inverting input of a comparator which in turn is responsible for the transition of state of flip flop.

Pin 3: Output

This pin is designated for taking the output of the timer.

Pin 4: Reset

Output of the timer can be reset by applying a negative pulse at this pin. To avoid any false triggering, this terminal must be connected to $+V_{CC}$.

Pin 5: Control voltage

The voltage at this terminal controls the variation in pulse width of the output waveform. To prevent any noise problems, this pin when unused should be connected to ground through a $0.01\,\mu f$ capacitor.

Pin 6: Threshold

This pin is connected to the non-inverting input terminal of comparator1. When the voltage at this pin is greater than $2/3\,V_{CC}$, the output of the timer will be low.

Pin 7: Discharge

This pin is connected to the collector of transistor Q_1. When output of the timer is low, transistor goes into saturation which results in discharging of capacitor.

Pin 8: $+V_{CC}$

A supply voltage ranging from $+5V$ to $+18V$ is connected to this pin.

Block Diagram of 555 Timer

A 555 timer basically consists of two op-amps working as comparators, a RS flip flop, two transistors and a resistive voltage divider network as shown in Fig. 5.10.2.

The three resistors used are of $5k\Omega$ value and hence it is called as 555 timer. Comparator 1 compares the threshold voltage with a control voltage of $2/3\ V_{CC}$ while comparator 2 compares the trigger voltage with a voltage of $1/3\ V_{CC}$. The outputs of both the comparators are connected to the input of RS latch whose input states decide the overall output of the timer.

There are two transistors in the internal circuitry of 555 timer. The collector of transistor Q_1 is connected to discharge pin *i.e.* pin 7 of 555 timer. When this transistor goes into the saturation mode, the capacitor connected to it starts discharging through it which in turn changes the output of timer to low state. The base of other transistor Q_2 is connected to reset pin *i.e.* pin 4 where a pulse when applied will reset the output of timer irrespective of state of any other input.

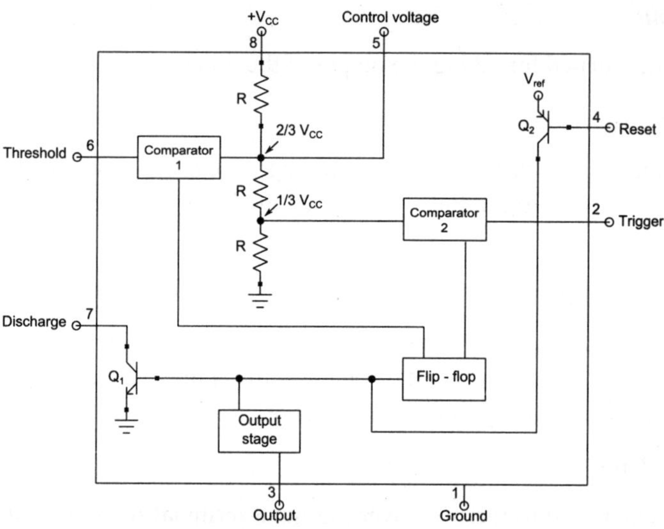

Fig. 5.10.2 Block diagram of 555 timer

Mono-stable Multivibrator

A mono-stable multivibrator, also called as one shot multivibrator, is a pulse generating circuit with only one stable state. The duration of the pulse is determined by the network consisting of resistor and capacitor connected externally to 555 timer as shown in Fig. 5.10.3.

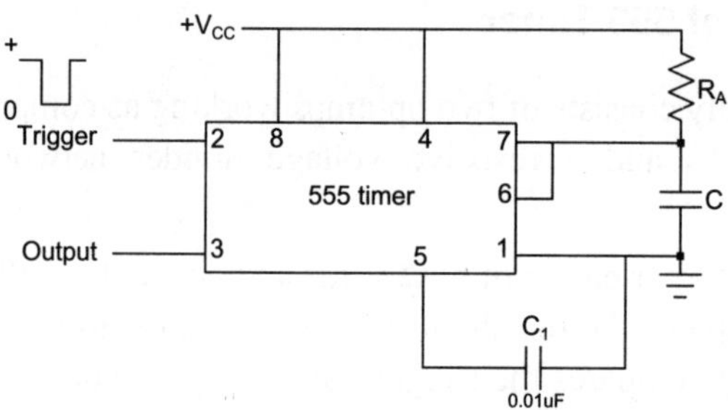

Fig. 5.10.3 555 timer as mono-stable multivibrator

The output of multivibrator is initially low and as soon as the triggering pulse is applied, the output is forced to go high. The external network R_A and C decides the duration for which output will remain high and at the end of this time interval, the output again comes back to its stable low state.

Working

Consider the block diagram shown in Fig. 5.10.4. Initially, the mono-stable multivibrator is at its stable state *i.e.* output is low which turns on the transistor Q_1 and capacitor C connected at pin 7 is shorted to ground. However, when the negative trigger pulse is applied at pin 2 of the timer and as the amplitude of this pulse goes below 1/3 V_{CC}, comparator 2 sets the flip flop and output of timer is changed to high state. At the same time, output of the flip flop turns Q_1 off and the capacitor starts charging to $+V_{CC}$ through resistance R_A. As the capacitor voltage becomes 2/3V_{CC}, the output of comparator 1 goes high and it resets the output of flip flop and the output of timer is changed to low state. At the same time, output of flip flop turns Q_1 on and hence, capacitor C starts discharging again. The output of mono-stable remains low until a trigger pulse is applied again and then the same cycle repeats.

Decoupling capacitor C' is used to avoid any unwanted voltage spike in the output waveform.

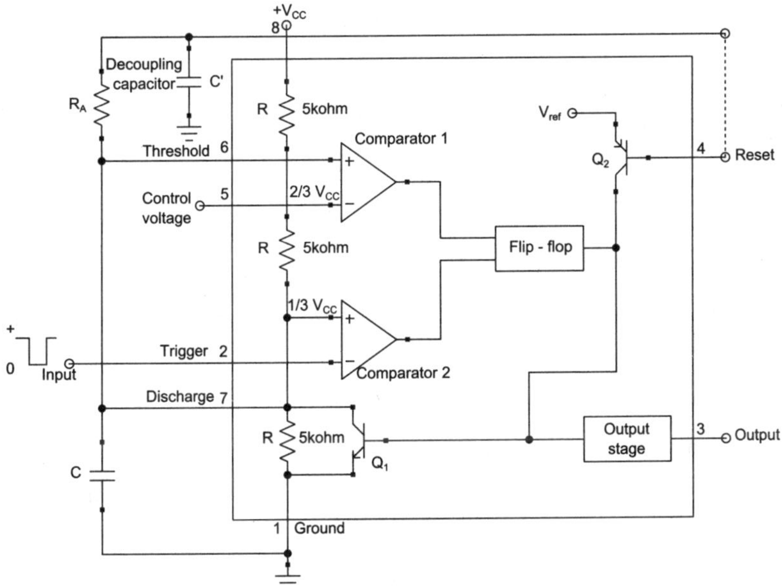

Fig. 5.10.4 Block diagram for 555 timer as mono-stable multivibrator

In order to avoid any false triggering of timer on positive pulse edges, a wave shaping circuit consisting of R', C_2 and diode D can be connected between terminals 2 and 8 as shown in Fig. 5.10.5. The values of R and C_2 should be chosen such that the time constant $R'C_2$ is smaller than the on time of the timer.

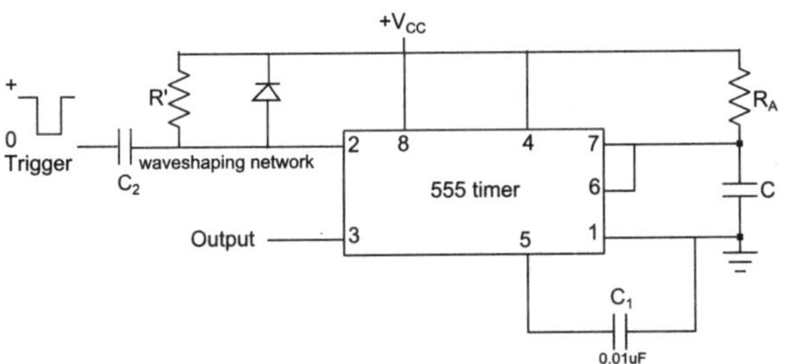

Fig. 5.10.5 555 timer as mono-stable multivibrator with wave shaping network

Time for which the output is high and timer is on is given by

$$t_{ON} = 1.1\, R_A C$$

Where, R_A (in ohms) and C (in Farad) are resistor and capacitor respectively connected externally to the timer.

Input and output waveforms

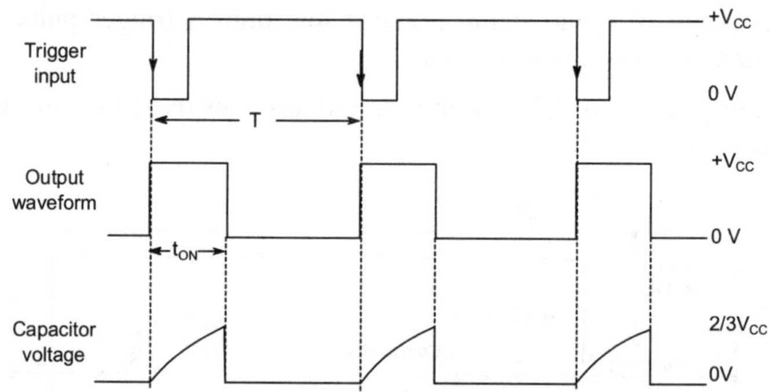

Fig. 5.10.6 Input and output waveforms

Designing

A mono-stable multivibrator can be designed for a given on time of the output waveform by the following steps:

1. Choose the on time *i.e.* t_{ON} of 555 timer.

2. Fix any value of capacitor C.

3. Calculate value of external resistance R_A by using the given relation.

$$t_{ON} = 1.1\, R_A C$$

$$\Rightarrow \qquad R_A = \frac{t_{ON}}{1.1 \times C}$$

Astable Multivibrator

An astable multivibrator is also known as free-running multivibrator as it does not require an external trigger pulse to change the output of timer. It is a circuit used to generate rectangular waveform. The off and on time of the circuit is determined by the resistors R_A, R_B and capacitor C which are connected externally to the timer.

Working

555 timer as astable multivibrator is shown in Fig. 5.10.7.

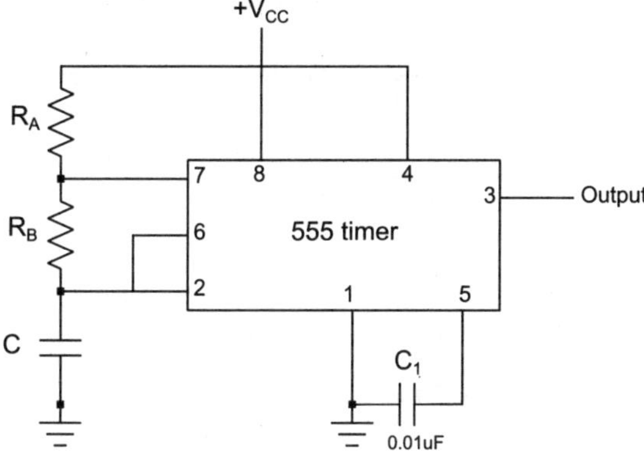

Fig. 5.10.7 555 timer as astable multivibrator

Initially, the output of timer is high which drive transistor Q_1 into cut-off region *i.e.* Q_1 is off. As a result, the capacitor C starts charging towards voltage $+V_{CC}$ through the resistive network R_A and R_B. As the capacitor voltage reaches to $2/3V_{CC}$, the output of comparator 1 goes high and it resets the flip flop and thereby, changes the output of timer from high to low state. At the same time, output of the flip flop turns Q_1 on and hence, capacitor C starts discharging through Q_1 and resistance R_B. As the capacitor voltage reaches to $1/3V_{CC}$, the comparator 2 sets the flip flop which in turn changes the output of timer from low to high state again and the same cycle repeats.

Time during which the output is high is given by

$$t_{ON} = 0.69(R_A + R_B)C$$

Time during which the output is low is given by

$$t_{OFF} = 0.69R_B C$$

Total time period of output of timer is

$$T = t_{ON} + t_{OFF} = 0.69(R_A + 2R_B)C$$

Frequency of oscillations is given by

$$f_0 = \frac{1}{T} = \frac{1.45}{(R_A + 2R_B)C}$$

Duty cycle is another parameter which is defined as the ratio of time for which the output of timer remains high to the total time period T.

$$\%\text{Duty cycle} = \frac{t_{ON}}{T} \times 100 = \left(\frac{R_A + R_B}{R_A + 2R_B}\right) \times 100\%$$

Output waveform

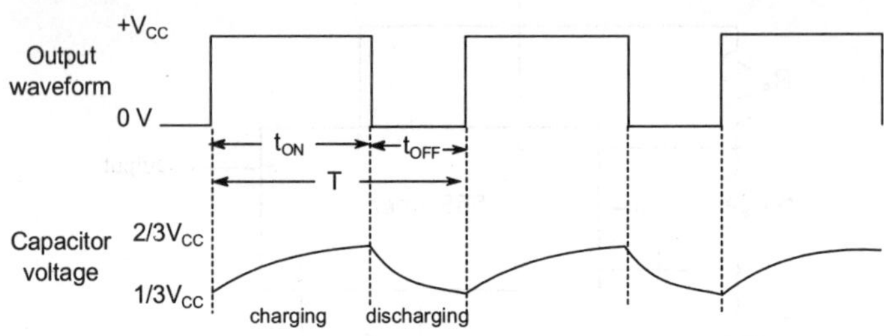

Fig. 5.10.8 Output waveforms

Designing

An astable multivibrator can be designed for a given time period of output waveform and duty cycle by following the steps mentioned below.

1. Choose the time period of 555 timer *say* 10msec.

$$T = 10 \text{ msec}$$

2. For the given duty cycle *say* 80%, find on time of the timer t_{ON} by using the given equation.

$$\%\text{Duty cycle} = \frac{t_{ON}}{T} \times 100$$

$\Rightarrow$
$$t_{ON} = \frac{\text{Duty Cycle} \times T}{100}$$

3. Use the values of total time period T and on time of 555 timer t_{ON} to calculate the off time t_{OFF}.

$$t_{OFF} = T - t_{ON}$$

4. Fix any value of capacitor C say 1μf, calculate value of external resistances R_A and R_B by using the given relation.

$$t_{ON} = 0.69(R_A + R_B)C$$

$$t_{OFF} = 0.69R_B C$$

For the above chosen values, resistances R_A and R_B are found out to be approximately 8.7kΩ and 2.9kΩ respectively.

Note

To produce a waveform with duty cycle of 50% or less, a diode is connected across R_B with its anode connected at pin 7 and cathode connected at pin 2. Due to this, the capacitor charges through resistor R_A and diode and discharges through resistor R_B. Therefore, the duty cycle becomes $R_A/(R_A+R_B)$. If R_B is made equal to R_A plus forward resistance of diode, the duty cycle becomes 50% and a symmetrical square wave can be generated.

PROCEDURE

Mono-stable Multivibrator

1. Determine the values of R_A and C depending on the desired value of on time of 555 timer.

2. Connect the circuit as shown in Fig. 5.10.3 on breadboard.

3. Connect function generator using a probe to pin 2 of 555 timer to provide the triggering pulses. Set it to square wave.

4. Connect both the channels of CRO using probes to pin 2 and pin 3 of 555 timer to observe the trigger pulse and output waveforms.

5. Take the traces of trigger pulse and output waveform and measure the on time period.

6. Compare the theoretical and practical values of t_{ON}.

7. Connect one channel of CRO across the capacitor and take the trace of voltage across it.

Astable Multivibrator

1. Determine the values of R_A, R_B and C using the given values of total time period and duty cycle of the desired output waveform.

2. Connect the circuit as shown in Fig. 5.10.7 on breadboard.

3. Connect channel 1 of CRO to the output pin of 555 timer and channel 2 across capacitor C.

4. Take the traces and measure the time period and duty cycle of output waveform.

5. Compare the practical and theoretical values.

OBSERVATION TABLE

Mono-stable Multivibrator

Table 5.10.1 Observation table for mono-stable multivibrator

S.No.	Resistance R_A	Capacitor C	ON time	
			Th.	Pr.
1.				
2.				

Astable Multivibrator

Table 5.10.2 Observation table for astable multivibrator

S.No.	R_A	R_B	C	t_{ON}		t_{OFF}		Time period T		Duty Cycle%	
				Th.	Pr.	Th.	Pr.	Th.	Pr.	Th.	Pr.
1.											
2.											

OBSERVATIONS

Attach the traces for output waveforms.

RESULT

The mono-stable and astable multivibrator has been designed and studied successfully using 555 timer.

DISCUSSION

The 555 timer is an economical and popular timing device which is used to generate pulses of varying duty cycles. It is used as mono-stable multivibrator, astable multivibrator, dc-dc converters, digital logic probes, waveform generators, analog frequency meters and tachometers, temperature measurement and control devices, infrared transmitters, burglar and toxic gas alarms, voltage regulators, electric eyes and many stable time delays or oscillation.

Its use as a monostable multivibrator finds a variety of applications such as timers, switches, frequency divider, pulse width modulator, pulse stretcher, *etc.* Besides this, the astable multivibrator also finds applications in LED and flashers, logic clocks, security alarms, square wave oscillator, free running ramp generator, pulse position modulators, *etc.*

<table><tr><td>**6**</td><td># Experiments on Signals and Systems</td></tr></table>

LIST OF EXPERIMENTS

1. Generation of Signals: Continuous Time.

2. Generation of Signals: Discrete Time.

3. Convolution of Signals.

4. Solution of Difference Equations.

5. Introduction to XCOS/SIMULINK and Calculation of Output of Systems Represented by Block Diagrams.

6. Fourier Series Representation of Continuous Time Signals.

7. Discrete Time Fourier Analysis.

8. Fourier Transform of Continuous Time Signals.

Experiment 1

CONTINUOUS TIME SIGNALS

AIM

Generation of Continuous Time Signals.

RAMP SIGNAL

```
clear;                                      // clear all variables stored in memory
clc;                                        // clear the display screen
close;                                      // close all graphic windows
upper_limit=7;                              // upper limit
t= -upper_limit:upper_limit;               // t varies from -7 to 7 in step size of 1
x= [zeros(1,upper_limit),0:upper_limit];    // x = [ 0 0 0 0 0 0 0 0 1 2 3 4 5 6 7 ]
a=gca();                                    // get the current axes
a.y_location = 'middle';                    // y axis in middle, can see –x & +x axis
plot2d(t,x);                                // plot the graph between x and t
xtitle('Ramp Signal', 'time', 'x(t) ');    // label title, x axis and y axis
```

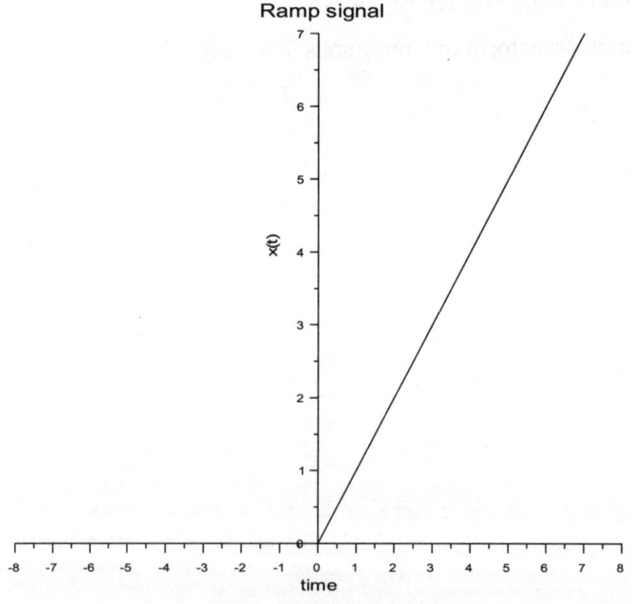

Fig. 6.1.1 Ramp signal

DECREASING FUNCTION X(T) = (0.6)^T

```
clear;                                  // clear all variables stored in memory
clc;                                    // clear the display screen
close;                                  // close all graphic windows
t= 0:15                                 // t varies from 0 to 15 in step size of 1
x= (0.6)^t;                             // defining the function
plot2d(t,x);                            // plot the graph between x and t
xtitle('Decreasing Function','t','x(t)');   // label title, x axis and y axis
```

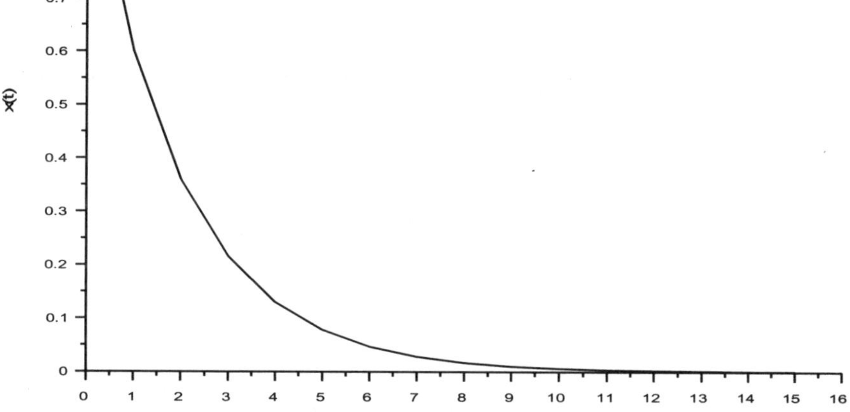

Fig. 6.1.2 Decreasing function

SINUSOIDAL SIGNALS

```
clear;                                  // clear all variables stored in memory
clc;                                    // clear the display screen
close;                                  // close all graphic windows
t=0:%pi/100:2*%pi;                      // t = 0 to 2pi in step size of pi/100
y1=sin(t);                              // sin function
subplot(2,1,1);                         // divides graphical window in 2 rows 1 column
plot2d(t,y1);                           // plot in 1st part of graphical window
```

```
xtitle('Sinusoidal Signals','t','sin t');        // label title, x axis and y axis
y2=cos(t);                                        // cosine function
subplot(2,1,2);
plot2d(t,y2);                                     // plot in 2nd part of graphical window
xlabel('t');                                      // label x axis
ylabel('cos t');                                  // label y axis
```

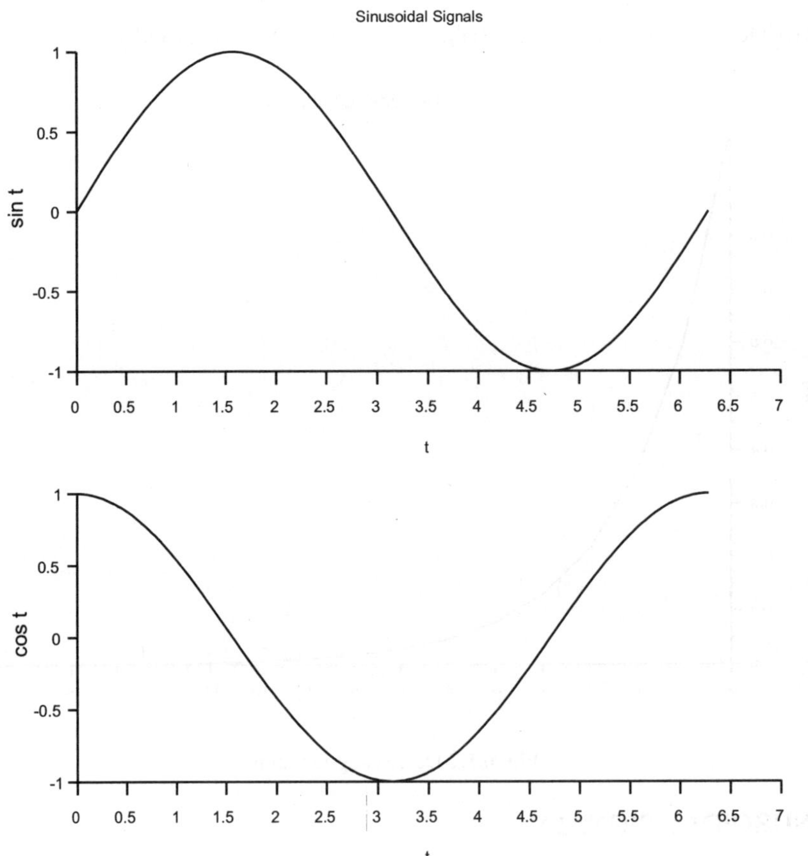

Fig. 6.1.3 Sinusoidal signals

SINC AND EXPONENTIAL FUNCTIONS

```
clear;                                            // clear all variables stored in memory
clc;                                              // clear the display screen
close;                                            // close all graphic windows
x=-50:0.01:50;                                    // x = -50 to +50 in step size of 0.01
```

```
y1=sinc(x);                          // defining sinc function
subplot(3,1,1);                      // divides graphical window in 3 rows 1 column
plot2d(x,y1);                        // plot in first part of graphical window
xtitle('Sinc and Exponential Functions', 'x', 'sinc x');
t=0:0.1:10;                          // t = 0 to 10 in step size of 0.1
y2=exp(t);                           // exponential increasing function
subplot(3,1,2);                      // plot in 2nd part of graphical window
plot2d(t,y2);
xlabel('t');                         // label x axis
ylabel('exp(t) ');                   // label y axis
y3=exp(-t);                          // exponential decreasing function
subplot(3,1,3);                      // plot in third part of graphical window
plot2d(t,y3);
xlabel('t');                         // label x axis
ylabel('exp(-t) ');                  // label y axis
```

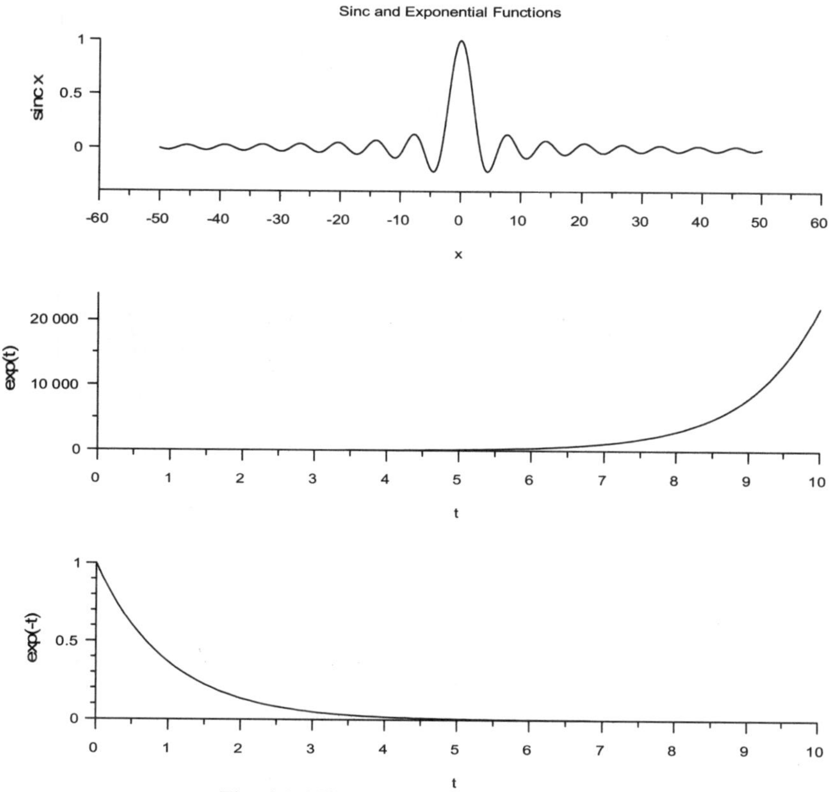

Fig. 6.1.4 Sinc and exponential functions

APERIODIC SIGNAL

```
clear;                          // clear all variables stored in memory
clc;                            // clear the display screen
close;                          // close all graphic windows
t1=0:-%pi/100:-2*%pi;           // t1 = 0 to -2pi in step size of -pi/100
y1=cos(t1);                     // defining x(t)=cos(t) for t<0
t2=0:%pi/100:2*%pi;             // t2 = 0 to 2pi in step size of pi/100
y2=sin(t2);                     // defining x(t)=sin(t) for t>0
plot(t2,y2,t1,y1);              // plot y2 w.r.t. t2 and y1 w.r.t. t1
a=gca();                        // get the current axes
a.y_location = 'middle';        // y axis in middle, can see -x & +x axis
a.x_location = "middle";        // x axis in middle, can see -y & +y axis
xtitle('Aperiodic signal: x(t) = cost for t<0 and sint for t>0')
```

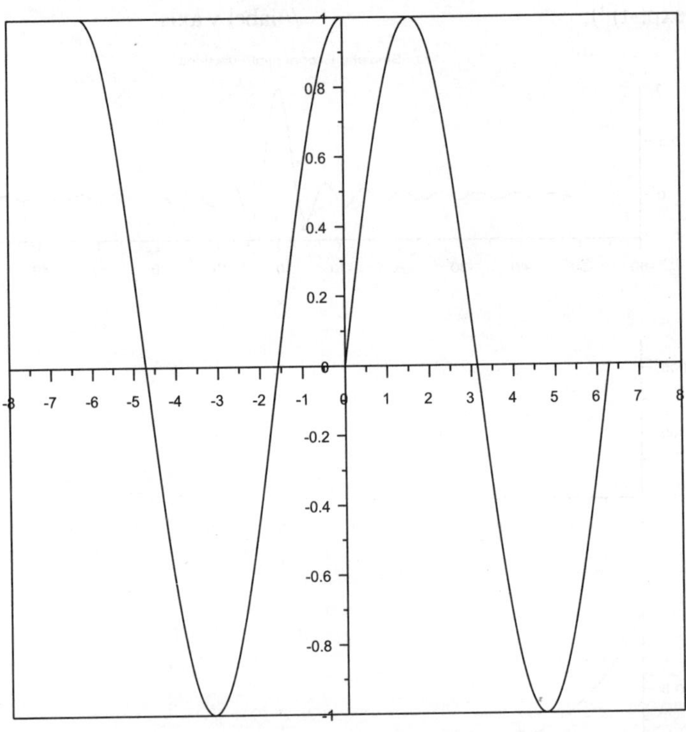

Fig. 6.1.5 Aperiodic signal

TIME SHIFTING

```
clear;                          // clear all variables stored in memory
clc;                            // clear the display screen
close;                          // close all graphic windows
t=0:%pi/100:2*%pi;              // t = 0 to 2pi in step size of pi/100
y1=sin(t);
y2=sin(t-5);                    // time shifted signal
y3=sin(t+10);                   // time shifted signal
plot(t,y1);                     // plot first signal
plot(t,y2);                     // plot signals in same graphical window
plot(t,y3);
xtitle ( 'Representation of time shifted signals x(t), x(t-5), x(t+10)');
```

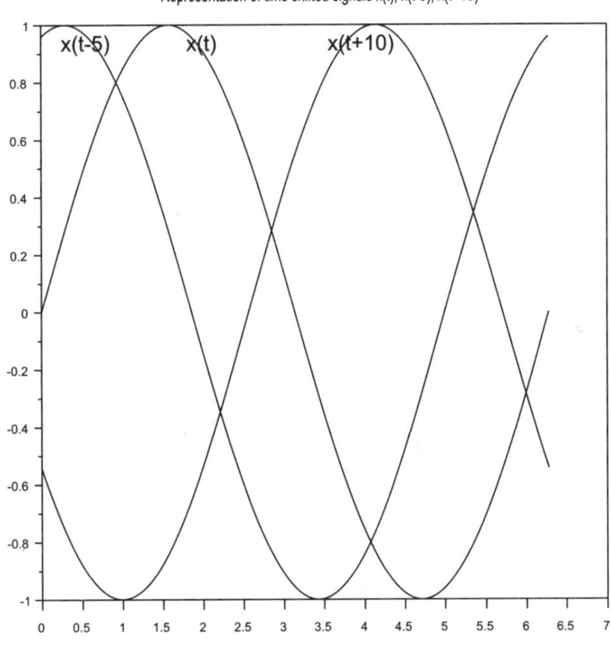

Fig. 6.1.6 Time shifted signals

TIME SCALING

```
clear;                          // clear all variables stored in memory
clc;                            // clear the display screen
close;                          // close all graphic windows
```

```
F=2.0;                              // scaling factor
t=-2*%pi:1/100:2*%pi;              // t = -2pi to 2pi with step size of 1/100
x1=sin(t);                         // original signal
x2=sin(F*t);                       // compressed signal
x3=sin(t/F);                       // expanded signal
xtitle('Representation of time scaling function x(t), x(2t), x(t/2)')
subplot(3,1,1);                    // divides graphical window in 3 rows 1 column
plot(t,x1);                        // plot in first part of graphical window
ylabel('x(t)');                    // label y axis
subplot(3,1,2);
plot(t,x2);                        // plot in 2nd part of graphical window
ylabel('x(2t)');                   // label y axis
subplot(3,1,3);
plot(t,x3);                        // plot in 3rd part of graphical window
ylabel('x(t/2)');                  // label y axis
```

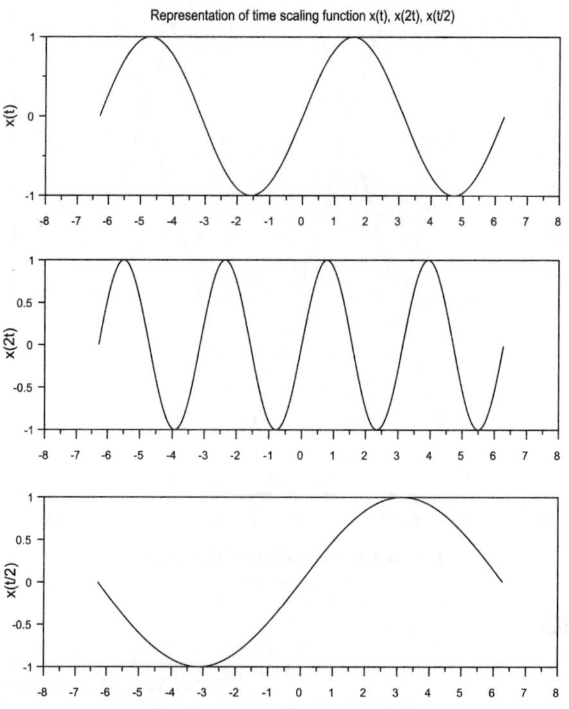

Fig. 6.1.7 Time scaling signals

Experiment 2

DISCRETE TIME SIGNALS

AIM

Generation of Discrete Time Signals.

RAMP SIGNAL

```
clear;                              // clear all variables stored in memory
clc;                                // clear the display screen
close;                              // close all graphic windows
max_value=5;                        // maximum value
n=-max_value:max_value;             // n varies from -5 to 5 in step size of 1
x=[zeros(1,max_value),0:max_value]; // x = [ 0 0 0 0 0 0 1 2 3 4 5]
a=gca();                            // get the current axes
a.y_location = 'origin';            // y axis in origin, can see -x and +x axis
plot2d3(n,x);                       // plot discrete time signal
title('discrete time ramp signal','fontsize',3)// label title with font size 3
xlabel('n','fontsize',2);           // label x axis with font size 2
ylabel('x[n]','fontsize',2);        // label y axis with font size 2
```

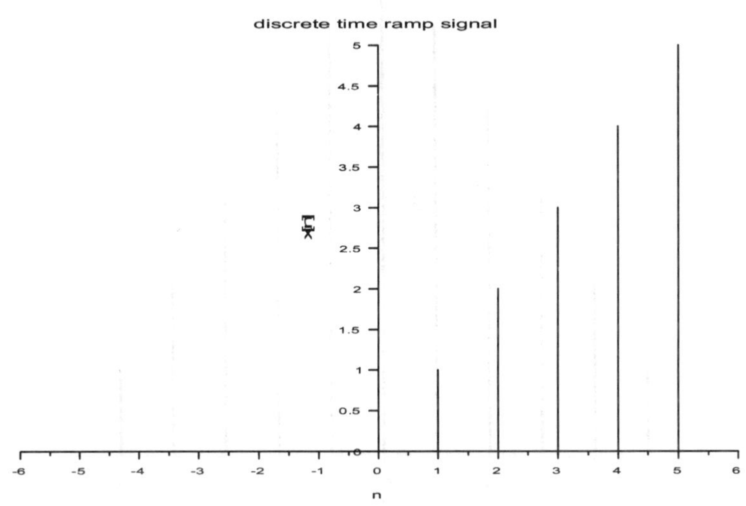

Fig. 6.2.1 Ramp signal

EVEN SIGNAL

```
clear;                                      // clear all variables stored in memory
clc;                                        // clear the display screen
close;                                      // close all graphic windows
max_value=6;
n=-max_value:max_value;                     // n varies from -6 to 6 in step size of 1
for i=1:max_value
   x1(i)=i-1;                               // loop creates x1 matrix with 1 column
end                                         // and 6 rows
x1=x1';                                     // take transpose, x1=[0 1 2 3 4 5]
x=[x1,max_value,x1(length(x1):-1:1)];       // x = [0 1 2 3 4 5, 6, 5 4 3 2 1 0]
a=gca();                                     // get the current axis
a.y_location = 'middle';                    // y axis in middle
plot2d3(n,x);                               // plot discrete time signal
a.children(1).children(1).thickness=2;      // increases thickness of plotted values
title('Even signal','fontsize',3);          // label title with font size 3
xlabel('n','fontsize',2);                   // label x axis with font size 2
ylabel('x[n]','fontsize',2);                // label y axis with font size 2
```

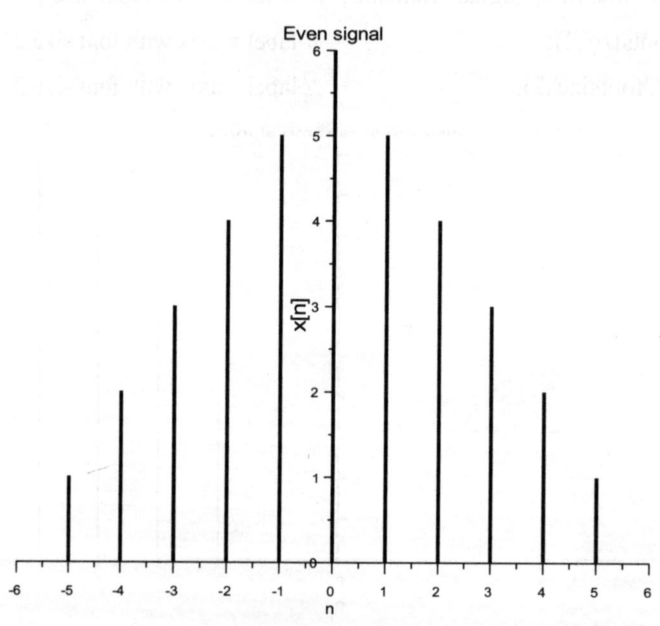

Fig. 6.2.2 Even signal

ODD SIGNAL

```
clear;                          // clear all variables stored in memory
clc;                            // clear the display screen
close;                          // close all graphic windows
max_value=5;
n=-max_value:max_value;         // n varies from -5 to 5 in step size of 1
for i=1:max_value
    x1(i)=i;                    // loop creates x1 matrix with 1 column
end                             // and 5 rows
x1=x1';                         // take transpose, x1=[1 2 3 4 5]
x=[-x1($:-1:1),0,x1];           // x = [-5 -4 -3 -2 -1, 0, 1 2 3 4 5]
a=gca();                        // get the current axis
a.thickness=2;                  // thickness of whole graph is 2
a.y_location='middle';          // y axis in middle
a.x_location='middle';          // x axis in middle
plot2d3(n,x);                   // plot discrete time signal
title('Odd signal','fontsize',3);   // label title with font size 3
xlabel('n','fontsize',2);       // label x axis with font size 2
ylabel('x[n]','fontsize',2);    // label y axis with font size 2
```

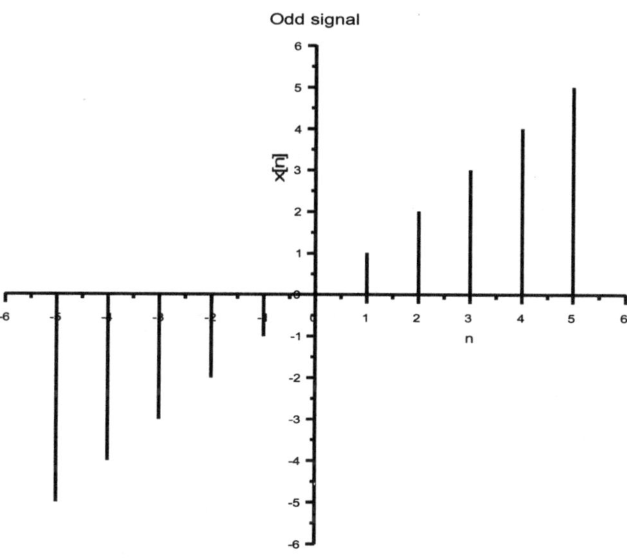

Fig. 6.2.3 Odd signal

TIME SHIFTING

```
clear;                                    // clear all variables stored in memory
clc;                                      // clear the display screen
close;                                    // close all graphic windows
t=0:%pi/100:2*%pi;                        // t = 0 to 2pi in step size of pi/100
y1=sin(t);
y2=sin(t-5);                              // time shifted signal
y3=sin(t+10);                             // time shifted signal
plot(t,y1);                               // plot first signal
plot(t,y2);                               // plot signals in same graphical window
plot(t,y3);
xtitle ( 'Representation of time shifted signals x(t), x(t-5), x(t+10)');
```

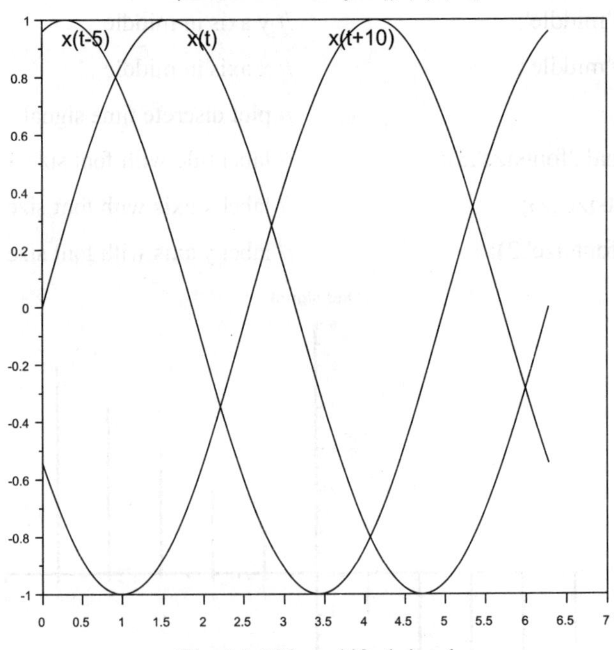

Fig. 6.1.6 Time shifted signals

TIME SCALING

```
clear;                                    // clear all variables stored in memory
clc;                                      // clear the display screen
close;                                    // close all graphic windows
```

UNIT STEP AND EXPONENTIAL FUNCTIONS

```
clear;                                  // clear all variables stored in memory
clc;                                    // clear the display screen
close;                                  // close all graphic windows
max_value=5;
n=-max_value:max_value;                 // n varies from -5 to 5 in step size of 1
u=[zeros(1,max_value),ones(1,max_value+1)];     // u = [ 0 0 0 0 0 1 1 1 1 1 1]
a=gca();                                // get the current axes
a.y_location = 'middle';                // y axis in middle, can see -x and +x
                                        // axis
subplot(3,1,1);                         // divides graphical window in 3 rows 1
                                        // column
plot2d3(n,u);                           // plot unit step signal in first part of
                                        // graphical window
title('Unit step function','fontsize',3);       // label title with font size 3
xlabel('n','fontsize',2);               // label x axis with font size 2
ylabel('u[n]','fontsize',2);            // label y axis with font size 2
n1=0:0.2:2;                             // n1 = 0 to 2 in step size of 0.2
y=exp(n1);                              // exponential increasing function
subplot(3,1,2);                         // plot signal in second part of graphical
                                        // window
plot2d3(n1,y);
title('Exponential increasing function','fontsize',3);
xlabel('n','fontsize',2);               // label x axis with font size 2
ylabel('exp[n]','fontsize',2);          // label y axis with font size 2
z=exp(-n1);                             // exponential decreasing function
subplot(3,1,3);                         // plot signal in third part of graphical
                                        // window
plot2d3(n1,z);
title('Exponential decreasing function','fontsize',3);
xlabel('n','fontsize',2);               // label x axis with font size 2
ylabel('exp[-n]','fontsize',2);         // label y axis with font size 2
```

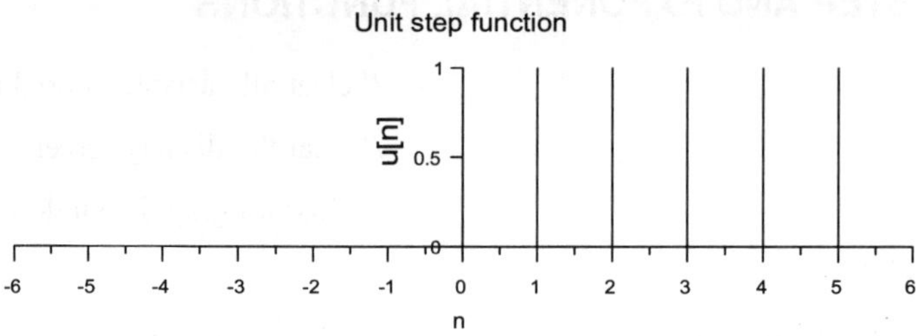

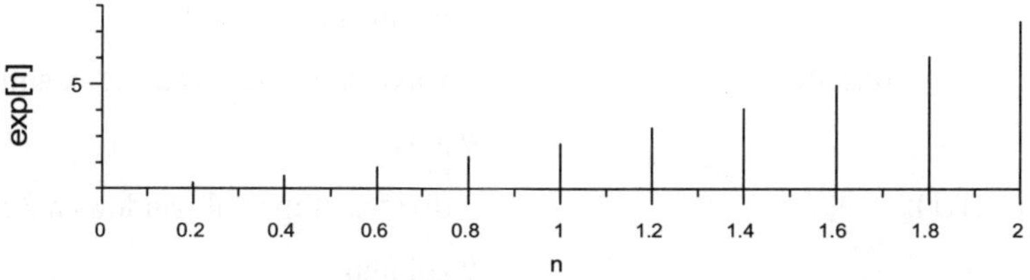

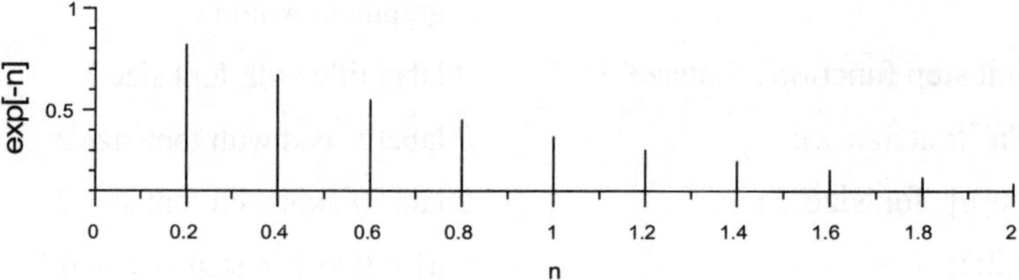

Fig. 6.2.5 Unit step and exponential functions

Experiment 3

CONVOLUTION

AIM

Convolution of Signals.

CONVOLUTION SUM

1. Using Convol Function

```
clear;                                  // clear all variables stored in memory
clc;                                    // clear the display screen
close;                                  // close all graphic windows
x=input('enter x sequence: ');          // enter the input signal
h=input('enter h sequence: ');          // enter the impulse response
y=convol(x,h);                          // convolution using inbuilt function
disp(y,'convolution sum: ');            // display convolution sum
```

Result

enter x sequence: [1 1 1]
enter h sequence: [1 0]

convolution sum:

    1.   1.   1.   0.

2. Without Convol Function

```
clear;                                  // clear all variables stored in memory
clc;                                    // clear the display screen
close;                                  // close all graphic windows
x=input('enter x sequence: ');          // enter the input signal
h=input('enter h sequence: ');          // enter the impulse response
x_length=length(x);                     // length of x
h_length=length(h);                     // length of h
convol_sum=[zeros(1,x_length+h_length-1)];// convolution sum with zero values
```

```
for i=1:x_length                            // loop to find convolution
    temp=[x(i)*h,zeros(1,x_length-i)];      // multiply one value of input with
                                            // impulse response

    shifted_h=[zeros(1,1),h];               // shifting h for next multiplication
    h=shifted_h;
    convol_sum=convol_sum+temp;             // summing up all multiplied terms
end                                         // end of for loop
disp(convol_sum,'convolution sum: ');       // display convolution sum
```

Result

enter x sequence: [1 1 1]

enter h sequence: [1 0]

convolution sum:

   1.   1.   1.   0.

3. x[n] = 2^n.u[-n] and h[n] = u[n]

```
clear;                                      // clear all variables stored in memory
clc;                                        // clear the display screen
close;                                      // close all graphic windows
max_limit=10;
n1=0:-1:-(max_limit-1);                     // n1=0 -1 -2 -3 -4 -5 -6 -7 -8 -9
x=(2)^n1;                                   // x=1 0.5 0.25 0.125....0.0039 0.0019
x1=x($:-1:1);                               // x1=0.0019 0.0039...0.125 0.25 0.5 1
h=ones(1,max_limit);                        // h=1 1 1 1 1 1 1 1 1 1
n2=0:length(h)-1;                           // n2=0 1 2 3 4 5 6 7 8 9
y=convol(x1,h);                             // convolution sum
n=-length(x)+1:length(h)-1;                 // n=-9 to 9
subplot(3,1,1);                             // divides graphical window in 3 rows 1 column
a=gca();                                    // get the current axes
a.y_location="origin";                      // y axis in origin
title('Input signal','fontsize',3);         // label title with font size 3
xlabel('n','fontsize',2);                   // label x axis with font size 2
ylabel('x[n]','fontsize',2);                // label y axis with font size 2
```

plot2d3(n1,x);	// plot input signal
subplot(3,1,2);	// plot in 2nd part of graphical window
plot2d3(n2,h);	// plot impulse response
title('Impulse response','fontsize',3);	// label title with font size 3
xlabel('n','fontsize',2);	// label x axis with font size 2
ylabel('h[n]','fontsize',2);	// label y axis with font size 2
subplot(3,1,3);	// plot in 3rd part of graphical window
plot2d3(n,y);	// plot output
a=gca();	// get the current axis
a.y_location ="origin";	// bring y axis in origin
title('Output','fontsize',3);	// label title with font size 3
xlabel('n','fontsize',2);	// label x axis with font size 2
ylabel('y[n]','fontsize',2);	// label y axis with font size 2

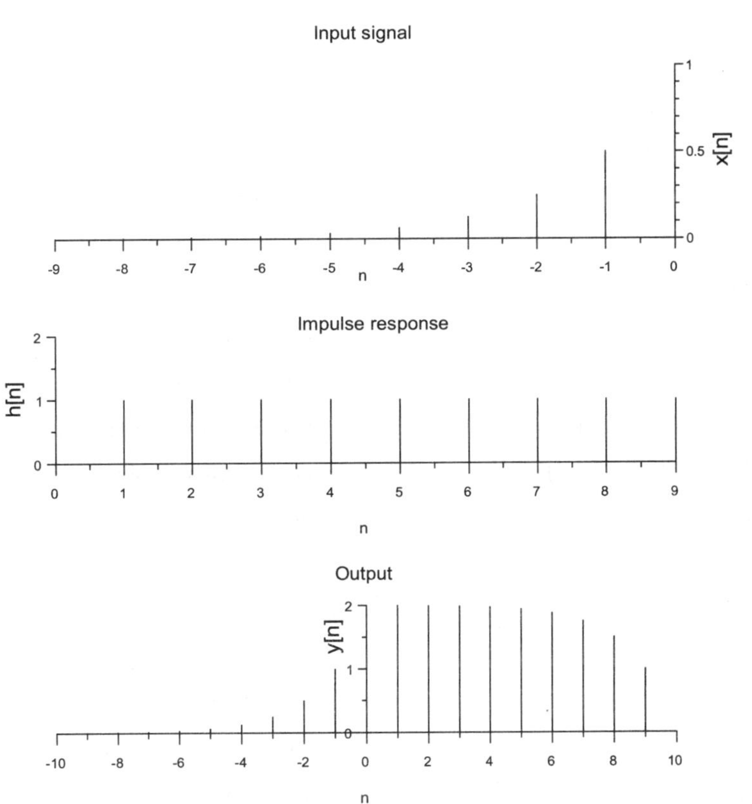

Fig. 6.3.1 Convolution sum for $x[n] = 2^n.u[-n]$ and $h[n] = u[n]$

4. $x[n] = 0.5^n u[n]$ and $h[n] = u[n]$

```
clear;                              // clear all variables stored in memory
clc;                                // clear the display screen
close;                              // close all graphic windows
max_limit=10;
n1=0:1:max_limit-1;                 // n1=0 1 2 3 4 5 6 7 8 9
x=(0.5)^n1;                         // x=1 0.5 0.25 0.125....0.0039 0.0019
h=ones(1,max_limit);                // h=1 1 1 1 1 1 1 1 1 1
n2=0:length(h)-1;                   // n2=0 1 2 3 4 5 6 7 8 9
y=convol(x,h);                      // convolution sum
n=0:length(x)+length(h)-2;          // n = 0 1 2 3... 18
subplot(3,1,1);                     // divides graphical window in 3 rows 1
                                    // column

a=gca();                            // get the current axes
a.y_location="origin";              // y axis in origin
title('Input signal','fontsize',3); // label title with font size 3
xlabel('n','fontsize',2);           // label x axis with font size 2
ylabel('x[n]','fontsize',2);        // label y axis with font size 2
plot2d3(n1,x);                      // plot input signal
subplot(3,1,2);                     // plot in 2nd part of graphical window
plot2d3(n2,h);                      // plot impulse response
title('Impulse response','fontsize',3); // label title with font size 3
xlabel('n','fontsize',2);           // label x axis with font size 2
ylabel('h[n]','fontsize',2);        // label y axis with font size 2
subplot(3,1,3);                     // plot in 3rd part of graphical window
plot2d3(n,y);                       // plot output
a=gca();                            // get the current axis
a.y_location ="origin";             // bring y axis in origin
title('Output','fontsize',3);       // label title with font size 3
xlabel('n','fontsize',2);           // label x axis with font size 2
ylabel('y[n]','fontsize',2);        // label y axis with font size 2
```

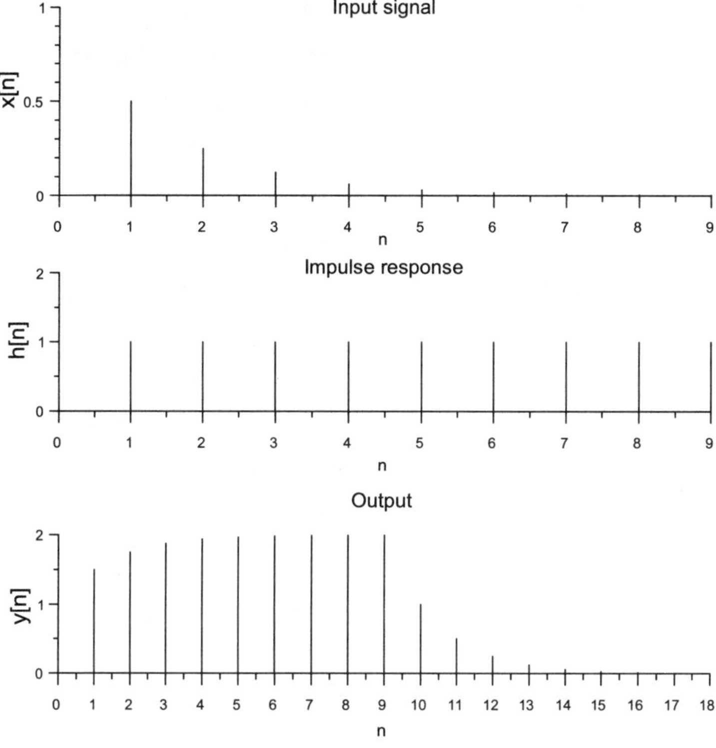

Fig. 6.3.2 Convolution sum for x[n] = 0.5^{n}u[n] and h[n] = u[n]

CONVOLUTION INTEGRAL

1.   x(t) = e^{-at}.u(t) and h(t) = u(t)

```
clear;                                  // clear all variables stored in memory
clc;                                    // clear the display screen
close;                                  // close all graphic windows
max_limit=10;
a=0.5;
t1=0:1:max_limit-1;                     // t1=0 1 2 3 4 5 6 7 8 9
x=exp(-a*t1);                           // defining the input signal
h=ones(1,max_limit);                    // h=1 1 1 1 1 1 1 1 1 1
t2=0:length(h)-1;                       // t2=0 1 2 3 4 5 6 7 8 9
y=convol(x,h);                          // convolution sum
t=0:length(x)+length(h)-2;              // t=0:1:18
subplot(3,1,1);                         // divides graphical window in 3 rows 1 column
```

title('Input signal','fontsize',3);	// label title with font size 3
xlabel('t','fontsize',2);	// label x axis with font size 2
ylabel('x(t)','fontsize',2);	// label y axis with font size 2
plot(t1,x);	// plot input signal
subplot(3,1,2);	// plot in 2nd part of graphical window
plot(t2,h);	// plot impulse response
title('Impulse response','fontsize',3);	// label title with font size 3
xlabel('t','fontsize',2);	// label x axis with font size 2
ylabel('h(t)','fontsize',2);	// label y axis with font size 2
subplot(3,1,3);	// plot in 3rd part of graphical window
plot(t,y);	// plot output
title('Output','fontsize',3);	// label title with font size 3
xlabel('t','fontsize',2);	// label x axis with font size 2
ylabel('y(t)','fontsize',2);	// label y axis with font size 2

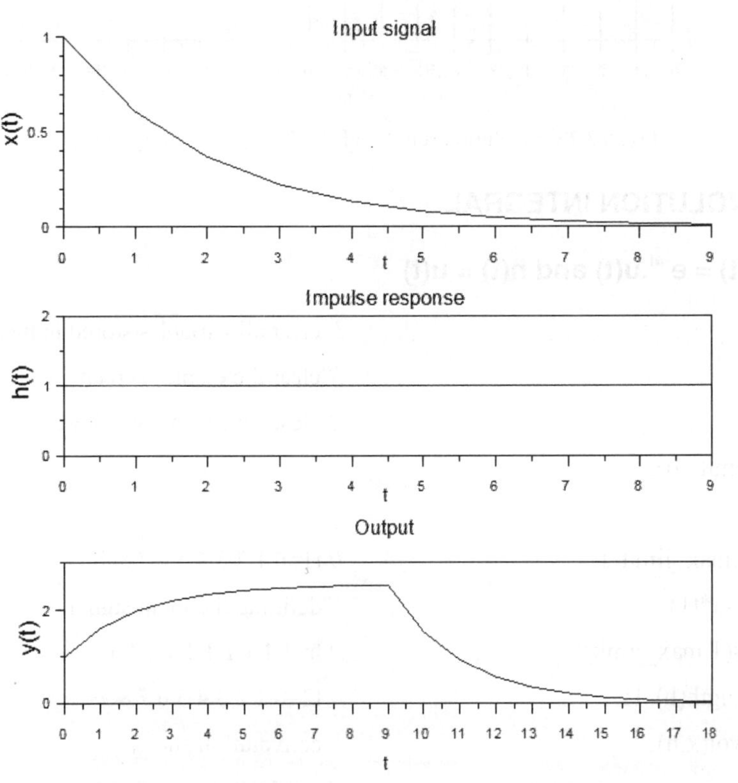

Fig. 6.3.3 Convolution integral for $x(t) = e^{-at}u(t)$ and $h(t) = u(t)$

2.   x(t) = u(t) and h(t) = t

```
clear;                                    // clear all variables stored in memory
clc;                                      // clear the display screen
close;                                    // close all graphic windows
max_limit=10;
x=ones(1,max_limit);                      // input signal x=1 1 1 1 1 1 1 1 1 1
t1=0:length(x)-1;                         // t1=0 1 2 3 4 5 6 7 8 9
t2=0:2*max_limit-1;                       // t2=0:1:19
h=t2;                                     // impulse response
y=convol(x,h);                            // convolution sum
t=0:length(x)+length(h)-2;               // t=0:1:18
subplot(3,1,1);                           // divides graphical window in 3 rows 1
                                          // column

title('Input signal','fontsize',3);      // label title with font size 3
xlabel('t','fontsize',2);                 // label x axis with font size 2
ylabel('x(t)','fontsize',2);             // label y axis with font size 2
plot(t1,x);                               // plot input signal
subplot(3,1,2);                           // plot in second part of graphical
                                          // window

plot(t2,h);                               // plot impulse response
title('Impulse response','fontsize',3);   // label title with font size 3
xlabel('t','fontsize',2);                 // label x axis with font size 2
ylabel('h(t)','fontsize',2);             // label y axis with font size 2
subplot(3,1,3);                           // plot in third part of graphical window
plot(t,y);                                // plot output
title('Output','fontsize',3);             // label title with font size 3
xlabel('t','fontsize',2);                 // label x axis with font size 2
ylabel('y(t)','fontsize',2);             // label y axis with font size 2
```

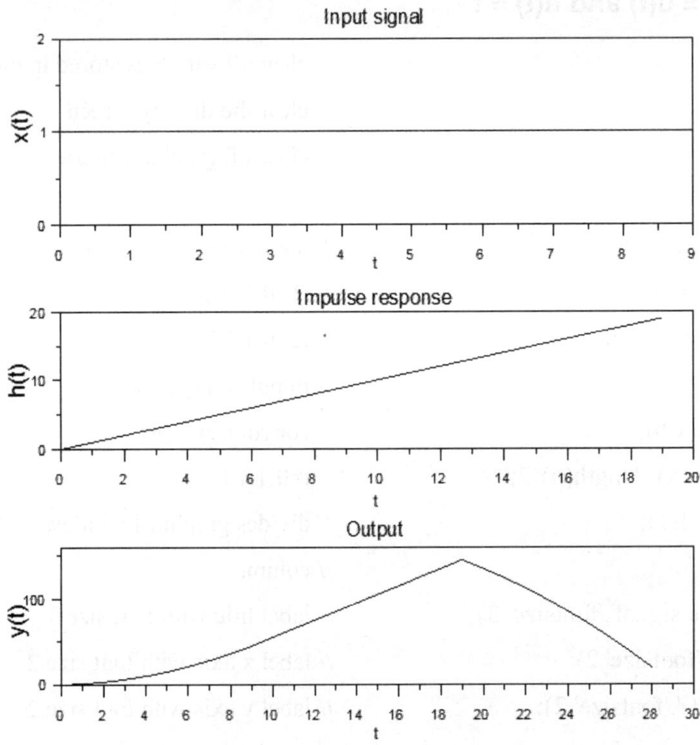

Fig. 6.3.4 Convolution integral for x(t) = u(t) and h(t) = t

Experiment 4

ORDINARY DIFFERENCE EQUATION

AIM

Solution of the Difference Equation dx/dt = sin(2t) with Initial Value x_O = -0.5 at t_O = 0.

PROGRAM

```
clear;                          // clear all variables stored in memory
clc;                            // clear the display screen
close;                          // close all graphic windows

function dx=f(t, x)             // dx: output, f: name of function
                                // t,x: input
   dx = sin(2*t);               // defining ordinary difference equation
endfunction                     // end of function f

t=0:0.1:10;
x0=-0.5;                        // initial value
t0=0;                           // initial value
x=ode(x0,t0,t,f);               // ode is inbuilt function which calls
                                // function f with initial values x0 and t0
plot2d(t,x);                    // plot the solution
```

RESULT

Expected result is

$$x = {}^{-1}/_{2} \cos 2t + \text{constant}$$

Using initial conditions, x_O = -0.5 at t_O = 0, constant comes out to be zero. Therefore,

$$x = {}^{-1}/_{2} \cos 2t$$

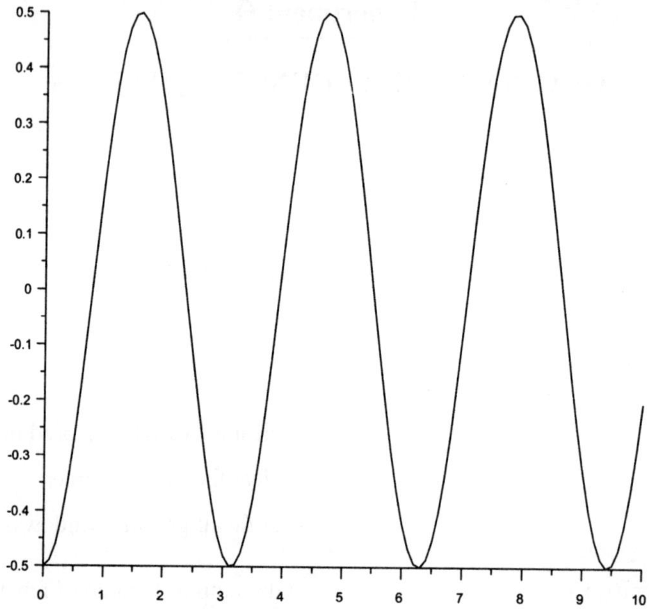

Fig. 6.4.1 Solution of ordinary difference equation

Experiment 5

INTRODUCTION to SIMULINK or XCOS

AIM

Introduction to SIMULINK in MATLAB / XCOS in SCILAB and Calculation of Output of Systems Represented by Block Diagrams.

LOW PASS FILTER

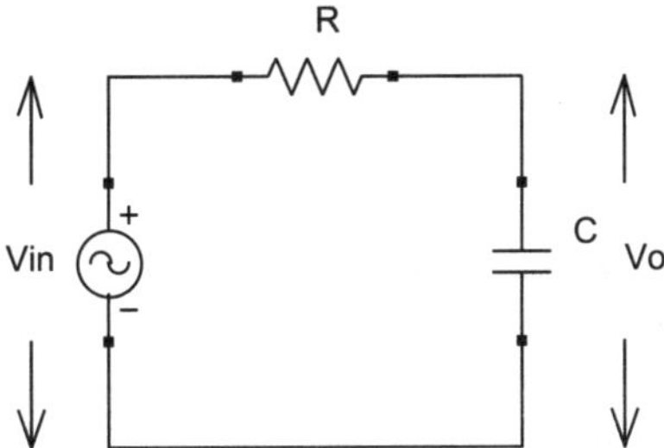

Fig. 6.5.1 Circuit for low pass filter

Transfer Function of LPF

The output voltage across the capacitor is given by

$$V_o = \frac{V_{in} \cdot (-jX_C)}{R - jX_C}$$

$$\Rightarrow \quad V_o = \frac{V_{in}}{1 - \dfrac{R}{jX_C}}$$

$$\Rightarrow \quad V_o = \frac{V_{in}}{1 + j2\pi fRC}$$

$$\Rightarrow \quad \frac{V_o}{V_{in}} = \frac{1}{1 + j2\pi fRC}$$

$$\Rightarrow \quad \frac{V_o}{V_{in}} = \frac{1/RC}{1/RC + j2\pi f}$$

Taking $s = j2\Pi f$

$$\Rightarrow \qquad H(s) = \frac{^1/_{RC}}{(s + ^1/_{RC})}$$

Where, R: resistance, C: capacitance, X_C: capacitive reactance, V_{in}: maximum input voltage, 1/RC: angular cut-off frequency, H(s): transfer function.

Simulation

Let angular cut-off frequency ω_C = 1/RC = 10rad/s. Then, transfer function H(s) for low pass filter becomes

$$H(s) = {^{10}}/_{(10 + s)}$$

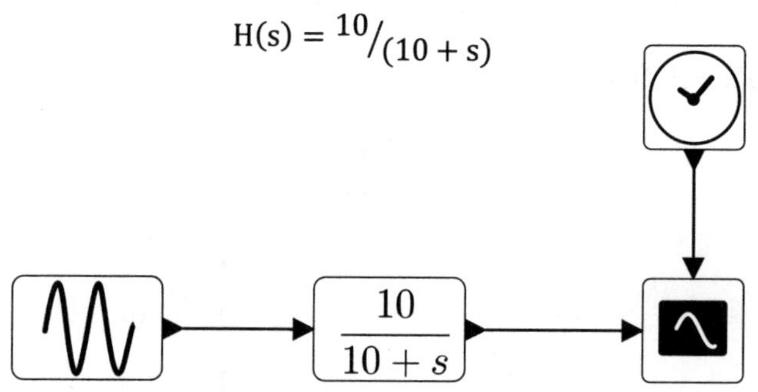

Fig. 6.5.2 Simulation of low pass filter

HIGH PASS FILTER

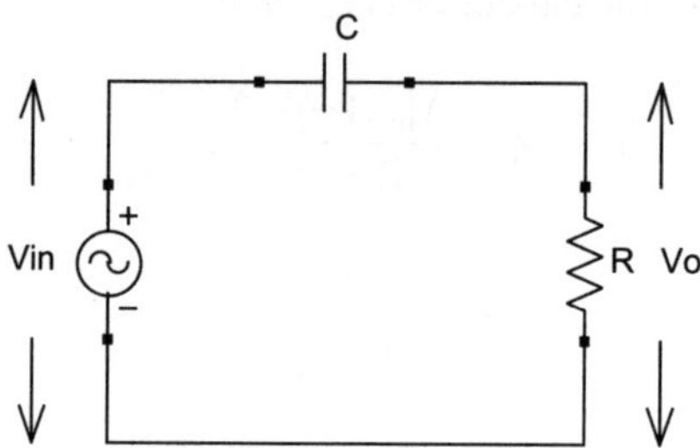

Fig. 6.5.3 Circuit for high pass filter

Transfer Function of HPF

The output voltage across the resistor is given by

$$V_o = \frac{V_{in} \cdot R}{R - jX_C}$$

$$\Rightarrow \qquad V_o = \frac{V_{in}}{1 - \dfrac{jX_C}{R}}$$

$$\Rightarrow \qquad V_o = \frac{V_{in}}{1 + 1/j2\pi fRC}$$

$$\Rightarrow \qquad \frac{V_o}{V_{in}} = \frac{1}{1 + 1/j2\pi fRC}$$

$$\Rightarrow \qquad \frac{V_o}{V_{in}} = \frac{j2\pi fRC}{1 + j2\pi fRC}$$

$$\Rightarrow \qquad \frac{V_o}{V_{in}} = \frac{j2\pi f}{{}^{1}/_{RC} + j2\pi f}$$

Taking $s = j2\Pi f$

$$\Rightarrow \qquad H(s) = \frac{s}{(s + {}^{1}/_{RC})}$$

Where, R: resistance, C: capacitance, X_C: capacitive reactance, V_{in}: maximum input voltage, 1/RC: angular cut-off frequency, H(s): transfer function.

Simulation

Let angular cut-off frequency $\omega_C = 1/RC = 10\,rad/s$. Then, transfer function H(s) for high pass filter becomes

$$H(s) = {}^{s}/_{(10 + s)}$$

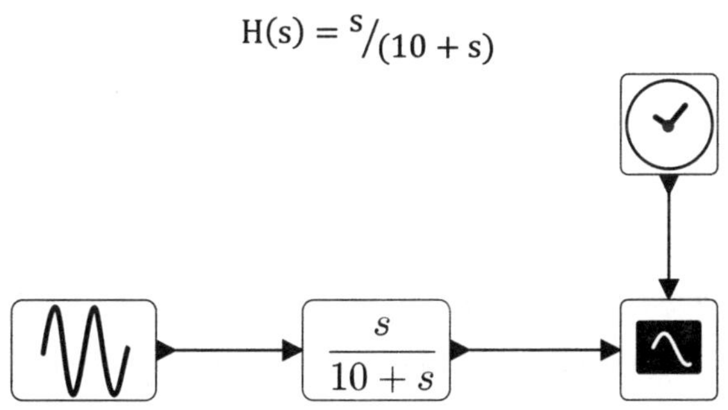

Fig. 6.5.4 Simulation of high pass filter

PROCEDURE

1. In scilab console, go to applications.
2. Click on xcos.
3. Drag blocks of sinusoid generator and clock from sources, CRO from sinks and system from continuous time systems and connect them according to the block diagram shown in Fig. 6.5.2 and Fig. 6.5.4.
4. Double click on continuous time system and set the transfer function as 10/(10+s) for low pass filter and s/(10+s) for high pass filter.

5. Set the magnitude from sinusoid generator to 10V and input frequency to (a) 1 rad/s (below cut off), (b) 10 rad/s (at cut off), (c) 100 rad/s (above cut off).

6. In xcos window, go to simulation and then go to set up.

7. Set the final integration time to 30 (time till which you will get the output).

8. Click play on xcos window.

RESULT

Low Pass Filter

Below cut off, input frequency = 1rad/s

The signals below cut-off frequency are passed without any attenuation and the output voltage is same as the input voltage.

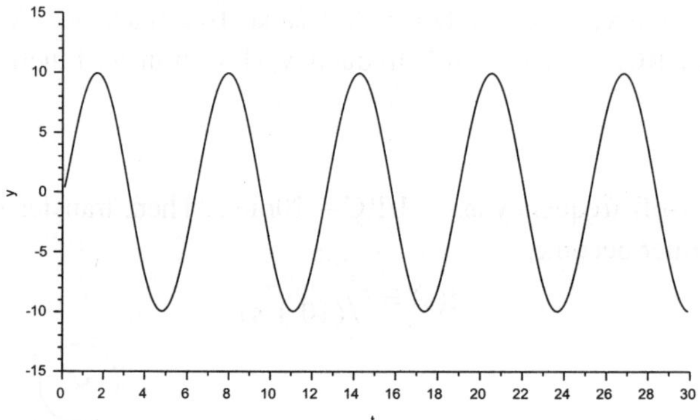

Fig. 6.5.5 Output waveform for below cut off frequency

At cut off, input frequency = 10rad/s

At cut-off frequency, the output voltage is 0.707 times the input voltage.

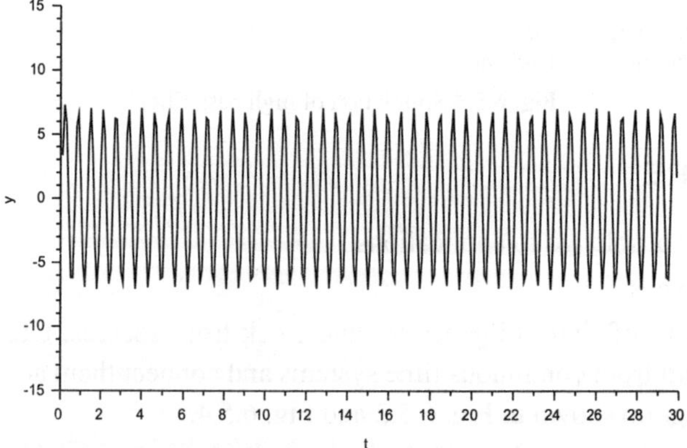

Fig. 6.5.6 Output waveform at cut off frequency

Beyond cut off, input frequency = 100rad/s

The signals beyond cut-off frequency are highly attenuated and the output voltage is very less as compared to the input voltage.

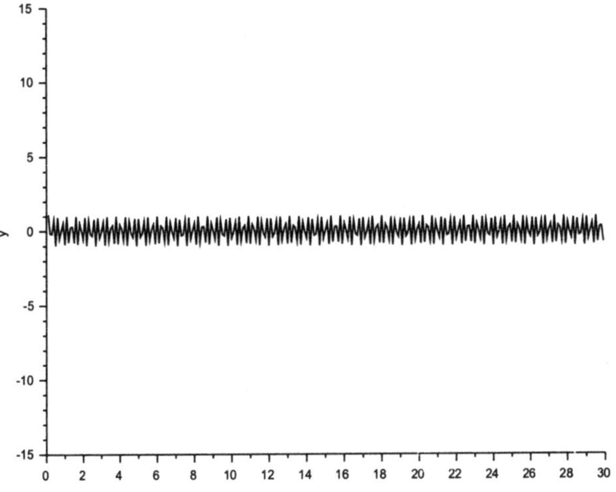

Fig. 6.5.7 Output waveform for above cut off frequency

High Pass Filter

Below cut off, input frequency = 1rad/s

The signals below cut-off frequency are highly attenuated and the output voltage is very less as compared to the input voltage.

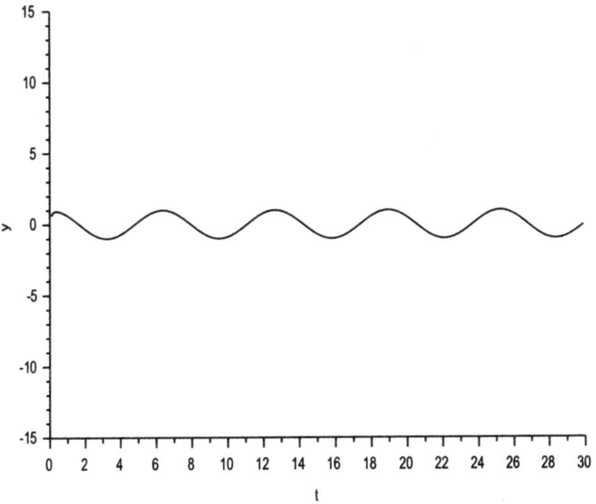

Fig. 6.5.8 Output waveform for below cut off frequency

At cut off, input frequency = 10rad/s

At cut-off frequency, the output voltage is 0.707 times the input voltage.

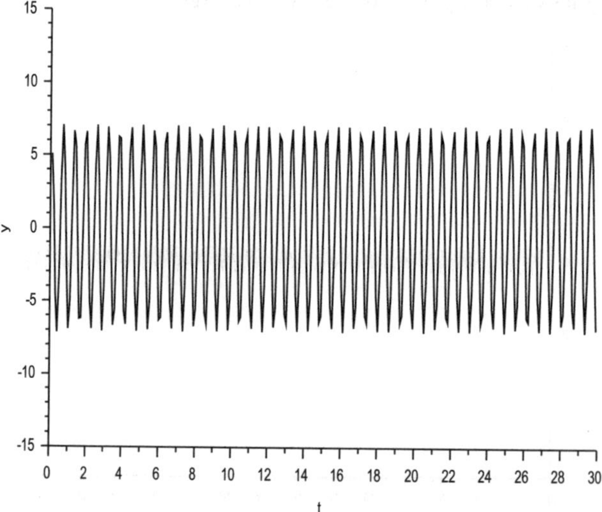

Fig. 6.5.9 Output waveform at cut off frequency

Beyond cut off, input frequency = 50rad/s

The signals beyond cut-off frequency are passed without any attenuation and the output voltage is same as the input voltage.

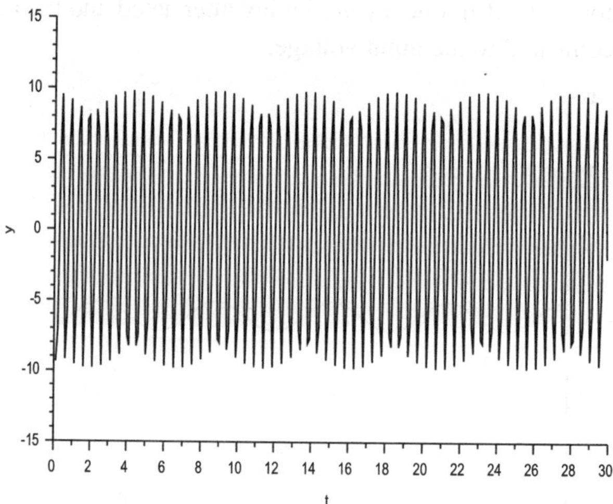

Fig. 6.5.10 Output waveform for above cut off frequency

Experiment 6

CONTINUOUS TIME FOURIER SERIES

AIM

Fourier Series Representation of Continuous Time Signals.

PROGRAM

1. $x(t) = \sin(w_o t)$

```
clear;                            // clear all variables stored in memory
clc;                              // clear the display screen
close;                            // close all graphic windows
t = 0:0.01:1;                     // t = 0 to 1 in step size of 0.01
T = 1;                            // time period
F = 1/T                           // frequency
wo = 2*%pi*F;                     // ω = 2Πf
x = sin(wo*t);                    // given function
for k = 1:5                       // loop to find a0 a1 a2 a3 a4 coefficients
   a(k) = x*(exp(-sqrt(-1)*wo*t.*(k-1)))'/length(t);   // formula for FS
   if (abs(a(k)) <= 0.001)        // if absolute value is small
      a(k) = 0;                   // make coefficient to zero
   end                            // end of if condition
end                               // end of for loop
a=a'                              // transpose
a_conjugate = conj(a);            // taking conjugate
ak = [a_conjugate($:-1:1),a(2:$)];   // a-4 a-3 a-2 a-1 a0 a1 a2 a3 a4
ak_imag = imag(ak);               // imaginary part of coefficients
disp(ak, 'Continuous time fourier series coefficients of the periodic signal
sin(wot) are:')                   // display coefficients
k=-4:4;
a=gca();                          // get the current axes
a.y_location='origin';            // y axis at origin, can see -x and +x axis
```

```
a.x_location='origin';              // x axis at origin, can see -y and +y axis
plot2d3(k,-ak_imag);                // plot imaginary part of coefficients
title('Fourier series coefficients of x(t) = sin(wot)','fontsize',3);
xlabel('k','fontsize',2);           // label x axis with font size 2
ylabel('ak','fontsize',2);          // label y axis with font size 2
```

Expected result

$a_0 = 0$, $a_1 = 1/2j$, $a_{-1} = -1/2j$

Result

Continuous time fourier series coefficients of the periodic signal sin(wot) are:

0   0   0   -6.959D-18 + 0.4950495i   0   -6.959D-18 - 0.4950495i   0   0   0

Note: Real part is too small and hence can be ignored.

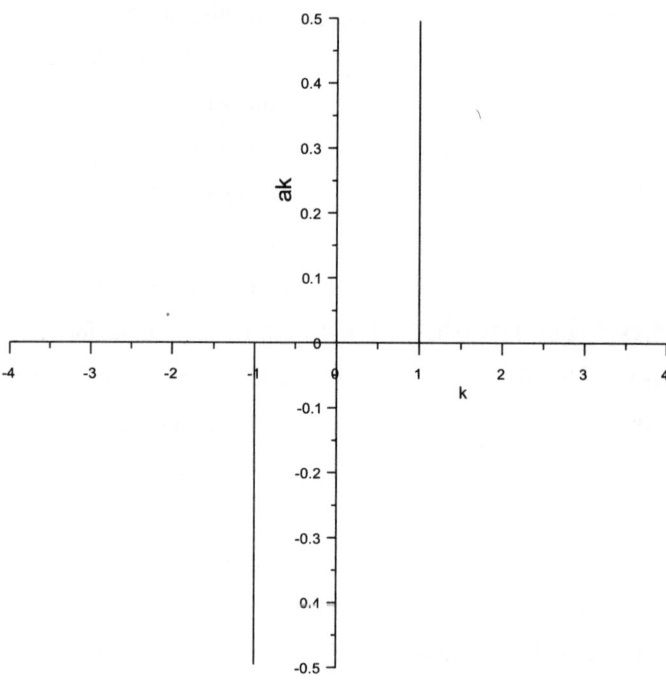

Fourier series coefficients of x(t) = sin(wot)

Fig. 6.6.1 Fourier series coefficients of $\sin(\omega_0 t)$

2. x(t) = cos (w₀t)

```
clear;              // clear all variables stored in memory
clc;                // clear the display screen
```

```
close;                              // close all graphic windows
t = 0:0.01:1;                       // t = 0 to 1 in step size of 0.01
T = 1;                              // time period
F = 1/T                             // frequency
wo = 2*%pi*F;                       // ω = 2Πf
x = cos(wo*t);                      // given function
for k = 1:5                         // loop to find a0 a1 a2 a3 a4 coefficients
   a(k) = x*(exp(-sqrt(-1)*wo*t.*(k-1)))'/length(t);   // formula for FS
   if (abs(a(k)) <= 0.01)           // if absolute value is small
       a(k) = 0;                    // make coefficient to zero
   end                              // end of if condition
end                                 // end of for loop
a=a'                                // transpose
a_conjugate = conj(a);              // taking conjugate
ak = [a_conjugate($:-1:1),a(2:$)];  // a-4 a-3 a-2 a-1 a0 a1 a2 a3 a4
ak_real = real(ak);                 // real part of coefficients
disp(ak, 'Continuous time fourier series coefficients of the periodic signal cos(wot) are:')
k=-4:4;
a=gca();                            // get the current axes
a.y_location='origin';              // y axis in origin, can see -x and +x axis
a.x_location='origin';              // x axis in origin, can see -y and +y axis
plot2d3(k,ak_real);                 // discrete time plot of real part
title('Fourier series coefficients of x(t) = cos(wot)','fontsize',3);
xlabel('k','fontsize',2);           // label x axis with font size 2
ylabel('ak','fontsize',2);          // label y axis with font size 2
```

Expected result

$a_0 = 0$, $a_1 = 1/2$, $a_{-1} = 1/2$

Result

Continuous time fourier series coefficients of the periodic signal cos(wot) are:

0   0   0   0.5049505 + 2.247D-18i   0   0.5049505 - 2.247D-18i   0   0   0

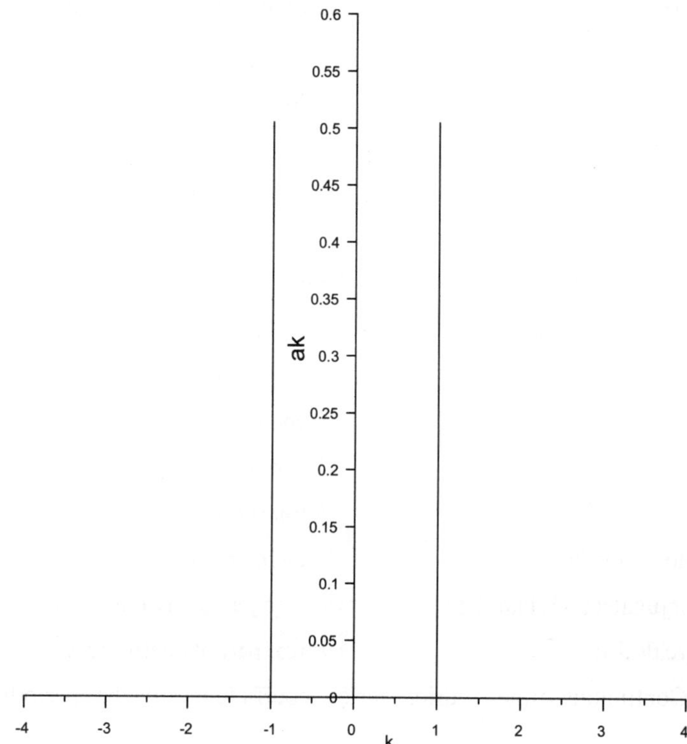

Fig. 6.6.2 Fourier series coefficients of $\cos(\omega_0 t)$

3. x(t) = 1 + 2sin(wot) + 4cos(wot)

```
clear;                                          // clear all variables stored in memory
clc;                                            // clear the display screen
close;                                          // close all graphic windows
t = 0:0.01:1;                                   // t = 0 to 1 in step size of 0.01
T = 1;                                          // time period
F = 1/T                                         // frequency
wo = 2*%pi*F;                                   // ω = 2Πf
x = ones(1,length(t))+2*sin(wo*t)+4*cos(wo*t);              // given function
for k = 1:5                                     // loop to find a0 a1 a2 a3 a4 coefficients
   a(k) = x*(exp(-sqrt(-1)*wo*t.*(k-1)))'/length(t);        // formula for FS
   if (abs(a(k)) <= 0.1)                        // if absolute value is small
      a(k) = 0;                                 // make coefficient to zero
```

```
end                                         // end of if condition
end                                         // end of for loop
a=a';                                       // transpose
a_conjugate = conj(a);                      // taking conjugate
ak = [a_conjugate($:-1:1),a(2: $)];         // a_-4 a_-3 a_-2 a_-1 a_0 a_1 a_2 a_3 a_4
k=-4:4;
subplot(2,1,1);                             // divides graphical window in 2 part
b = get("current_axes");                    // get the current axes
b.y_location = "origin";                    // y axis at origin
b.x_location = "origin";                    // x axis at origin
plot2d3(k,abs(ak));                         // plot magnitude of FS coefficients
b.children(1).children(1).thickness=2 ;     // increases thickness of plotted values
title('Magnitude of fourier series coefficients','fontsize',3);
xlabel('k','fontsize',2);
for i = 1:length(ak)                        // loop to find phase of FS coefficients
    phase(i) = atan(imag(ak(i))/(real(ak(i))+0.00001));
end                                         // add 0.00001 to avoid division by 0
phase = phase'                              // transpose
subplot (2,1,2);                            // plot in 2nd part of graphical window
b = get("current_axes");                    // get the current axes
b.y_location = "origin";                    // y axis at origin
b.x_location = "origin";                    // x axis at origin
plot2d3(k,phase);                           // plot phase of FS coefficients
c = gce();                                  // get the handle of newly creates axis
c.children(1).thickness=2;                  // increases thickness of plotted values
title('Phase of fourier series coefficients','fontsize',3);
xlabel('k','fontsize',2);
```

Magnitude of fourier series coefficients

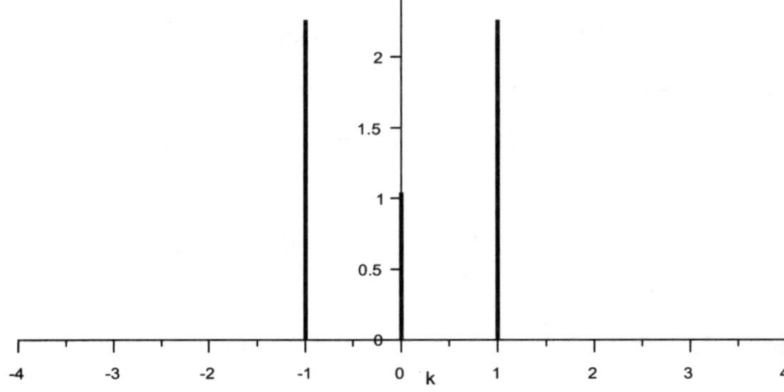

Phase of fourier series coefficients

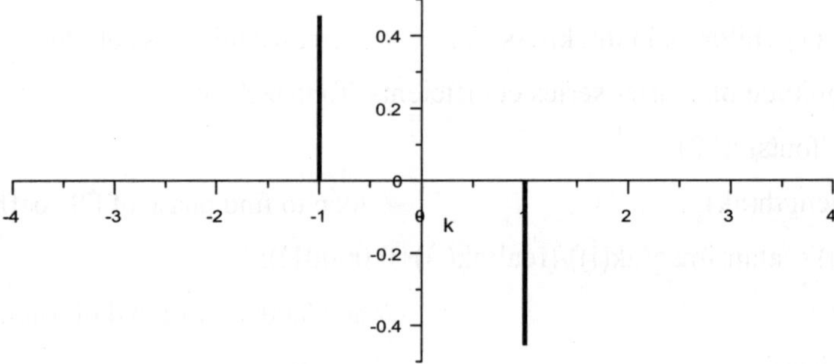

Fig. 6.6.3 Fourier series coefficients of $1 + 2\sin(wot) + 4\cos(wot)$

Experiment 7

DISCRETE TIME FOURIER SERIES

AIM

Discrete Time Fourier Analysis of Periodic Signals.

PROGRAM

1. x[n] = sin(3w$_o$n)

```
clear;                                       // clear all variables stored in memory
clc;                                         // clear the display screen
close;                                       // close all graphic windows
n = 0:0.1:10;                                // n = 0 to 10 in step size of 0.1
N = 10;                                      // period
F = 1/N;                                     // frequency
wo = 2*%pi*F;                                // ω = 2Πf
x = sin(3*wo*n)                              // given function
for k = 1:N-2                                // loop to find a₀ a₁ a₂....a₇ coefficients
   a(k) = x*(exp(-sqrt(-1)*wo*n.*(k-1)))'/length(n);      // formula for FS
      if (abs(a(k)) <= 0.01)                 // if absolute value is small
      a(k) = 0;                              // make coefficient to zero
   end                                       // end of if condition
end                                          // end of for loop
a=a';                                        // transpose
a_conjugate = conj(a);                       // taking conjugate
ak = [a_conjugate($:-1:1),a(2:$)];           // ak = a₋₇ a₋₆....a₅ a₆ a₇
ak_imag = -imag(ak);                         // imaginary part of ak
k=-(N-3):(N-3);                              // k = -7 to 7
k1=N+k;                                       // k1 = 3 to 17
k2=-(N+k);                                    // k2 = -3 to -17
b=gca();                                      // get the current axes
b.y_location='origin';                        // y axis at origin, can see -x and +x axis
```

```
b.x_location='origin';                      // x axis at origin, can see -y and +y axis
plot2d3(k,ak_imag);                         // plot a₋₇ a₋₆ a₋₅...a₄ a₅ a₆ a₇
b.children(1).children(1).thickness=2;      // increases thickness of plotted values
plot2d3(k1,ak_imag);                        // In DTFS, aₖ = aₙ₊ₖ
b.children(1).children(1).thickness=2;      // increases thickness of plotted values
plot2d3(k2,ak_imag($:-1:1));                // In DTFS, aₖ = aₙ₊ₖ
b.children(1).children(1).thickness=2;      // increases thickness of plotted values
title('DTFS coefficients of periodic signal sin(3won)','fontsize',3);
xlabel('k','fontsize',2);                   // label x axis with font size 2
ylabel('ak','fontsize',2);                  // label y axis with font size 2
```

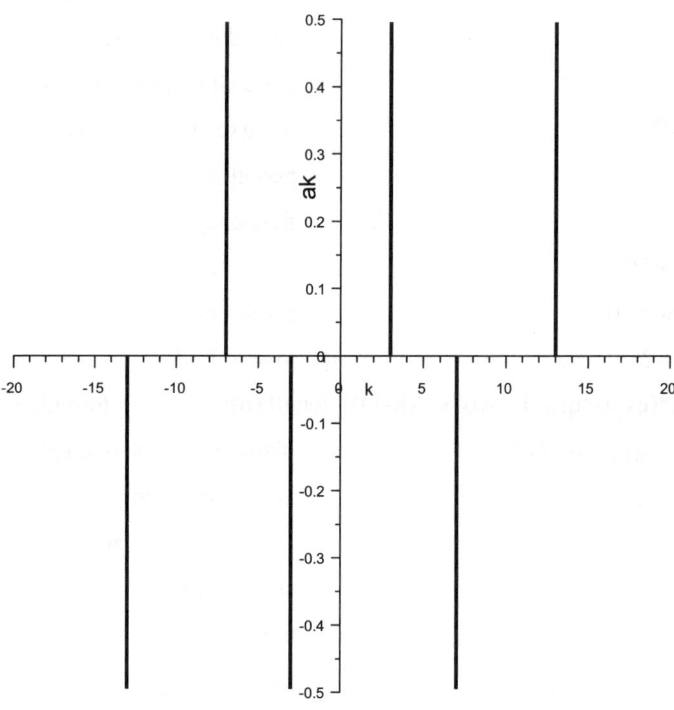

Fig. 6.7.1 Discrete time fourier series coefficients of $\sin(3\omega_0 n)$

2. x[n] = cos(wₒn)

```
clear;                                       // clear all variables stored in memory
clc;                                         // clear the display screen
close;                                       // close all graphic windows
```

```
n = 0:0.1:10;                                        // n = 0 to 10 in step size of 0.1
N = 10;                                              // period
F = 1/N;                                             // frequency
wo = 2*%pi*F;                                        // ω = 2Πf
x = cos(wo*n)                                         // given function
for k = 1:N-2                                         // loop to find a0 a1 a2....a7 coefficients
    a(k) = x*(exp(-sqrt(-1)*wo*n.*(k-1)))'/length(n);        // formula for FS
        if (abs(a(k)) <= 0.01)                       // if absolute value is small
        a(k) = 0;                                    // make coefficient to zero
    end                                              // end of if condition
end                                                  // end of for loop
a=a';                                                // transpose
a_conjugate = conj(a);                               // taking conjugate
ak = [a_conjugate($:-1:1),a(2:$)];                   // ak = a-7 a-6....a5 a6 a7
ak_real = real(ak);                                  // real part of ak
k=-(N-3):(N-3);                                      // k = -7 to 7
k1=N+k;                                              // k1 = 3 to 17
k2=-(N+k);                                           // k2 = -3 to -17
b=gca();                                             // get the current axes
b.y_location='origin';                               // y axis at origin, can see -x and +x axis
b.x_location='origin';                               // x axis at origin, can see -y and +y axis
plot2d3(k,ak_real);                                  // plot a-7 a-6 a-5...a4 a5 a6 a7
b.children(1).children(1).thickness=2;               // increases thickness of plotted values
plot2d3(k1,ak_real);                                 // In DTFS, ak = aN+k
b.children(1).children(1).thickness=2;               // increases thickness of plotted values
plot2d3(k2,ak_real($:-1:1));                          // In DTFS, ak = aN+k
b.children(1).children(1).thickness=2;               // increases thickness of plotted values
title('DTFS coefficients of periodic signal cos(won)','fontsize',3);
xlabel('k','fontsize',2);                            // label x axis with font size 2
ylabel('ak','fontsize',2);                           // label y axis with font size 2
```

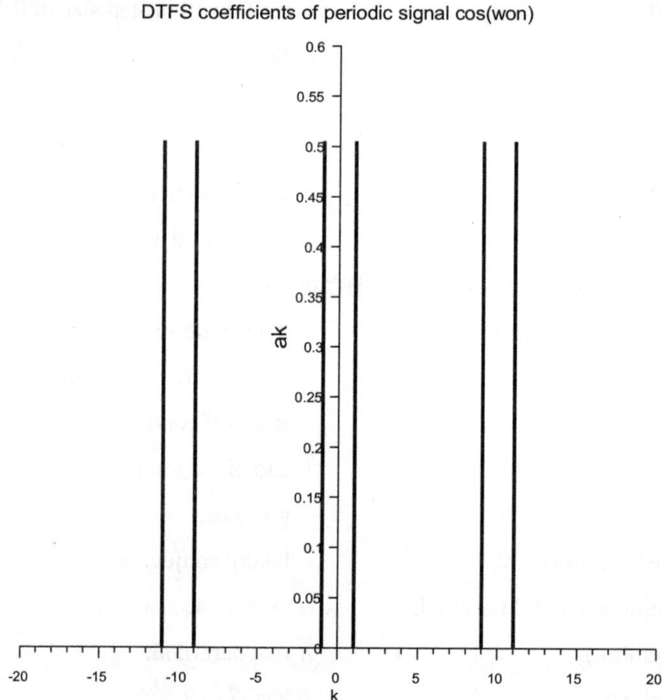

Fig. 6.7.2 Discrete time fourier series coefficients of $\cos(\omega_o n)$

Experiment **8**

FOURIER TRANSFORM

AIM

Fourier Transform of Continuous Time Signals.

PROGRAM

1. Square Waveform x(t) = 4 from -T to T

```
clear;                              // clear all variables stored in memory
clc;                                // clear the display screen
close;                              // close all graphic windows
amp=4;                              // amplitude of square wave
T=10;
dt=0.001;
t=-T:dt:T;
for i=1:length(t)
   x(i)=amp;                        // defining amplitude of signal at all pts.
end
x=x';                               // transpose
subplot(2,1,1);                     // divides graphical window into 2 parts
a=get("current_axes");              // get the current axes
a.y_location='origin';              // y axis at origin, can see -x and +x axis
plot(t,x);                          // plot the input signal
xlabel('time','fontsize',2);        // label x axis with font size 2
ylabel('x(t)','fontsize',2);        // label y axis with font size 2
title('Continuous time signal x(t)','fontsize',3);
frequency=1;
w_max=2*%pi*frequency;              // w=2Πf
w=0:w_max/100:w_max;
Xw=x*exp(-sqrt(-1)*t'*w)*dt;        // formula for Fourier transform
Xw_Mag=real(Xw);                    // taking the real part
```

```
w=[-mtlb_fliplr(w),w(2:$)];                // w going from –wmax to wmax
Xw_Mag=[mtlb_fliplr(Xw_Mag),Xw_Mag(2:$)];
subplot(2,1,2);                            // plot in 2nd part of graphical window
a=get("current_axes");                     // get the current axes
a.y_location='origin';                     // y axis at origin
a.x_location='origin';                     // x axis at origin
plot(w,Xw_Mag);                            // plot of Fourier transform
xlabel('frequency','fontsize',2);          // label x axis with font size 2
title('Continuous time fourier transform','fontsize',3);
```

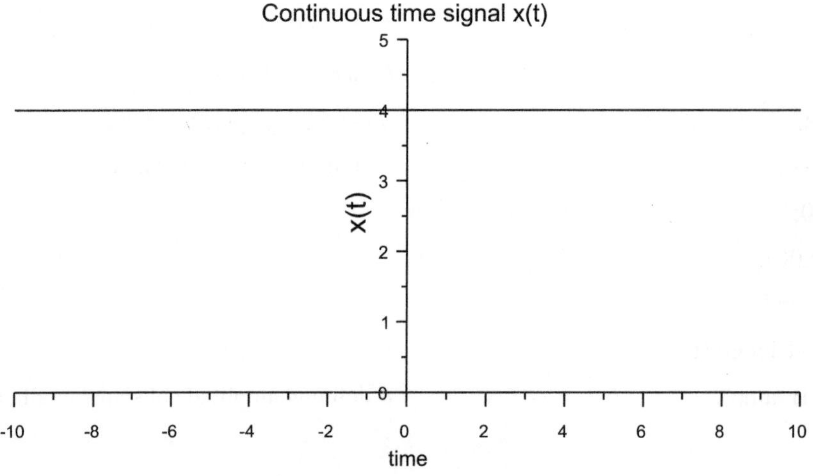

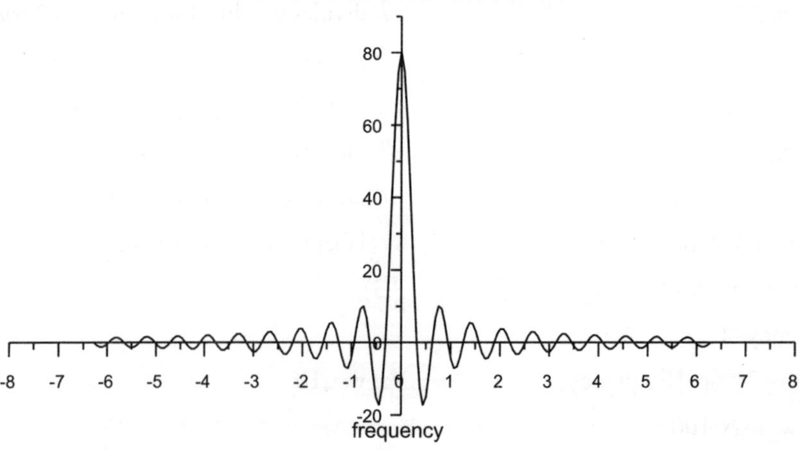

Fig. 6.8.1 Continuous time fourier transform

2. $x(t) = e^{-at}\, u(t)$

```
clear;                                  // clear all variables stored in memory
clc;                                    // clear the display screen
close;                                  // close all graphic windows
a=2;
T=10;
dt=0.001;
t=0:dt:T;
x=exp(-a*t);                            // defining the input signal
subplot(2,1,1);                         // divides graphical window into 2 parts
b=gca();                                // get the current axes
b.y_location='origin';                  // y axis at origin, can see -x and +x axis
plot(t,x);                              // plot the input signal
xlabel('time','fontsize',2);            // label x axis with font size 2
ylabel('x(t)','fontsize',2);            // label y axis with font size 2
title('Continuous time signal','fontsize',3); // label title
frequency=1;
w_max=2*%pi*frequency;                  // w=2Πf
w=0:w_max/100:w_max;
Xw=x*exp(-sqrt(-1)*t'*w)*dt;            // formula for Fourier transform
Xw_Mag=abs(Xw);
w=[-mtlb_fliplr(w),w(2:$)];             // w going from –wmax to wmax
Xw_Mag=[mtlb_fliplr(Xw_Mag),Xw_Mag(2:$)];
subplot(2,1,2);                         // plot in 2nd part of graphical window
b=gca();                                // get the current axes
b.y_location='origin';                  // y axis at origin, can see -x and +x axis
b.x_location='origin';                  // x axis at origin, can see -y and +y axis
plot(w,Xw_Mag);                         // plot of Fourier transform
xlabel('frequency','fontsize',2);       // label x axis with font size 2
title('Continuous time fourier transform','fontsize',3);
```

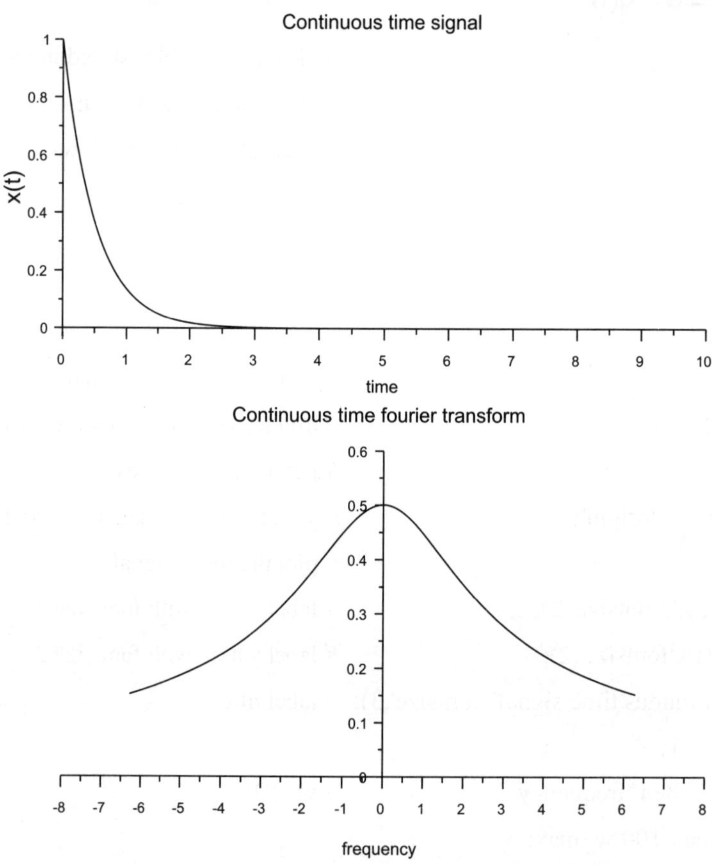

Fig. 6.8.2 Continuous time fourier transform of x(t)=e^{-at}u(t)

<table><tr><td>**7**</td><td># Experiments on Microprocessor 8086</td></tr></table>

LIST OF EXPERIMENTS

1. To Write Assembly Language Program to Add Two – 8 Bit, 16 Bit and 32 Bit Hexadecimal Numbers.

2. To Write Assembly Language Program to Subtract Two – 8 Bit, 16 Bit Hexadecimal Numbers.

3. To Write Assembly Language Program to Transfer a Block of Data.

4. To Write Assembly Language Program to Multiply Two – 8 Bit, 16 Bit Hexadecimal Numbers.

5. To Write Assembly Language Program to Convert a 16 Bit Hexadecimal Number to Decimal Number.

6. To Write Assembly Language Program to Generate Fibonacci Series.

7. To Write Assembly Language Program to Sort Hexadecimal Numbers in Ascending/Descending Order.

8. To Write Assembly Language Program to Find the Square Root of an Integer.

9. To Write Assembly Language Program to Generate Digital Clock.

10. To Write Assembly Language Program to Find Factorial of a Given Number.

8086 Microprocessor

INTRODUCTION

The 8086 is a 16-bit microprocessor chip designed by the Intel Corporation which brought a great revolution in high speed data processing. It is capable of addressing one megabyte of memory. It is fabricated by using high-performance metal-oxide semiconductor (HMOS) technology.

PIN DIAGRAM DESCRIPTION

The 8086 chip can operate in three different clock frequencies, *i.e.* 5, 8 and 10 MHz, housed in a 40 pin Dual-inline-package (DIP). The pins of 8086 are as follows.

AD_{15}-AD_0

These are time-multiplexed address and data lines. Address is available during T_1 state while the data is available during T_2, T_3, T_W and T_4.

A_{19}/S_6, A_{18}/S_5, A_{17}/S_4, A_{16}/S_3

The four most significant address lines A_{19}-A_{16} are time-multiplexed with status lines S_6-S_3. The address line is available during T_1 state while status information is available during T_2, T_3, T_W and T_4.

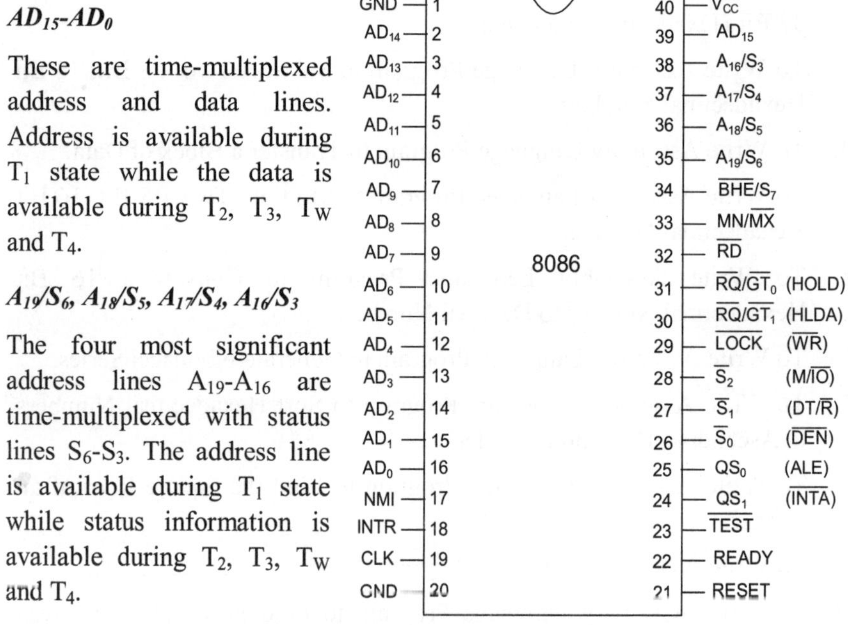

Fig. 7a Pin diagram of 8086

$\overline{BHE}/S_7$ (Bus high enable/ Status)

BHE is an active low signal which indicates the transfer of data over higher order data bus. The signal goes low during T_1 state for read, write and interrupt acknowledge cycles while status information is available during T_2, T_3 and T_4.

$\overline{RD}$ *(Read)*

It is an active low signal used to indicate the peripheral devices that the processor is doing a memory or I/O read operation.

READY

It is an active high signal and is used as an acknowledgement signal from the slow devices or memory regarding completion of the data transfer.

INTR (Interrupt Request)

It is a level triggered input signal and is used for interrupting the processor. In case of any pending interrupt request, processor goes into the interrupt acknowledge cycle.

$\overline{TEST}$

It is an active low signal and is examined by wait instruction. If the signal is low, execution continues otherwise, the processor goes into an idle state.

NMI (Non-maskable interrupt)

It is an edge triggered non-maskable interrupt.

RESET

It is an active high signal which forces the processor to terminate its present ongoing activity immediately and start the execution from FFF0H.

CLK (Clock input)

This input is an asymmetric square wave with 33% duty cycle and provides the basic timing for the processor and bus controller.

V_{CC}

+5V input power supply.

GND

It provides ground for internal circuit.

$MN/\overline{MX}$ *(Minimum/Maximum)*

This signal indicates mode of operation of the processor whether minimum (single processor) or maximum (multi-processor) mode.

The pins used in minimum mode are as follows.

$M/\overline{I/O}$ *(Memory/Input output)*

When the signal is high, processor is doing memory operation and when the signal is low, processor is doing I/O operations.

$\overline{INTA}$ *(Interrupt acknowledge)*

It is an active low signal which indicates that the processor has accepted the interrupt.

ALE (Address latch enable)

The signal indicates the availability of the valid address on the address/data lines.

$DT/\overline{R}$ *(Data transmit/ receive)*

This signal is used to control the direction of data flow through the transceiver.

$\overline{WR}$ *(Write)*

It is an active low signal used to indicate that a write memory or write I/O cycle is being performed by the processor.

$\overline{DEN}$ *(Data enable)*

It is an active low signal used to indicate the availability of valid data on the address and data lines.

HOLD/ HLDA (Hold/ Hold acknowledge)

HOLD signal indicates that another master requests for the bus access. Once the processor receives the HOLD request, it sends an acknowledgement signal on HLDA pin.

The pins used in maximum mode are as follows:

$\overline{S_2}$, $\overline{S_1}$, $\overline{S_0}$ *(Status lines)*

These lines indicate the type of operation carried out by the processor.

$\overline{LOCK}$

When the signal is low, the other system bus masters will not be able to gain control of the system bus.

QS_1, QS_0 *(Queue status)*

These lines indicate the status of code-prefetch queue.

$\overline{RQ/GT_0}$, $\overline{RQ/GT_1}$ *(Request/Grant)*

These pins are used by other local bus masters to force the processor to release the local bus.

ARCHITECTURE

The robust internal architecture of 8086 microprocessor is shown in Fig. 7b.

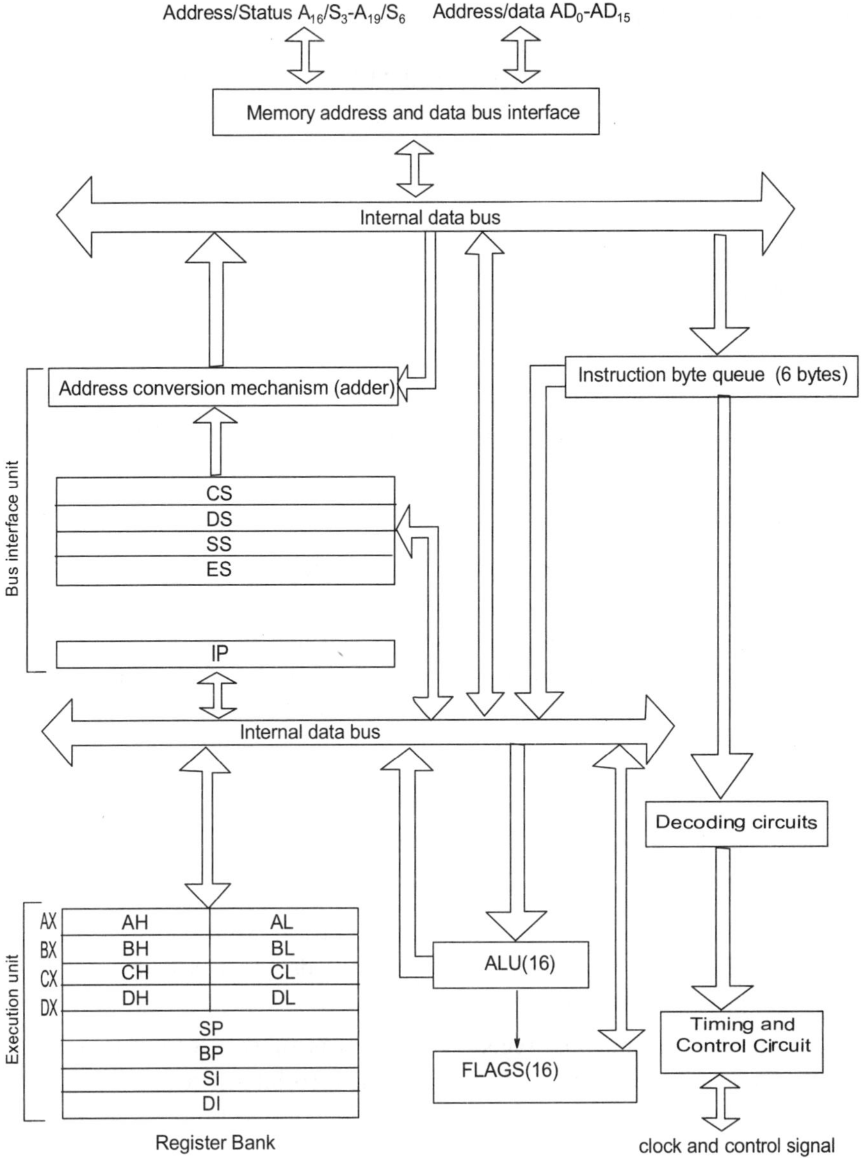

Fig. 7b Architecture of 8086

It is divided broadly into two parts: Bus interface unit (BIU) and Execution unit (EU).

Bus Interface Unit (BIU)

It consists of the bus interface logic, segment registers, memory addressing logic and six bytes long instruction queue. This unit controls the overall data transfer and addresses on the bus for the execution unit.

Execution Unit (EU)

It consists of control circuitry, a decoder and ALU. The 16- bit ALU can perform various mathematical operations. The timing and control unit derives the necessary control signals to execute the instructions received from the queue.

Pipeline

While the execution of the fetched instruction is going on, the external bus fetches the machine code of the next instruction and stores it in a 6 byte long queue called as pre-decoded instruction byte queue. It is a first in first out queue. Whenever the EU is ready to fetch the next instruction, it directly reads from the queue itself. This enhances the processing speed to a great extent. The process of fetching the next instruction while the execution of current instruction is taking place is called as pipelining.

Physical address generation

For generating the complete physical address which is of 20 bits, the 8086 processor uses segment and offset registers each of 16 bit size. The content of a segment register is called as segment address and that of an offset register is called as offset address. In order to generate physical address, the segment address is shifted left bitwise four times and then added to the offset address.

Registers in 8086

 The 8086 microprocessor has a set of 16 bit registers containing general purpose and special purpose registers.

1. *General purpose registers*

    The general purpose registers can be used either as 8-bit or 16-bit register. These registers are responsible for storing the data, offset address and for other purpose such as a counter.

a. AX register (Accumulator register): It consists of two 8-bit AH and AL registers or it can be combined to be used as a 16-bit AX register. It is used for I/O operations, rotate and string manipulation instructions.

b. BX register (Base register): It is used as offset storage for generating physical addresses. It generally holds the initial base location of a memory.

c. CX register (Count Register): It is used in case of string and loop instructions.

d. DX register (Data register): It is used as an implicit operand or destination for few instructions.

2. *Special purpose registers*

    The special purpose registers consist of segment registers, pointers, index registers or offset storage registers.

a. Segment registers

Code segment (CS): It is used to address a memory location in the code segment of the memory. It is used for the instructions that are referenced by instruction pointer (IP) register.

Stack segment (SS): It is used to address a memory location in the stack segment of the memory. All the data referenced by stack pointer (SP) and base pointer (BP) registers are stored in the stack segment.

Data segment (DS): It is used to address a memory location in the data segment of the memory. All the data referenced by general registers (AX, BX, CX and DX) and index registers (SI, DI) are stored in the data segment.

Extra segment (ES): It is used to address a memory location in the extra segment of the memory.

b. Pointers

These registers are used to contain the offset within a particular segment.

Index pointer (IP): It contains the offset within the code segment.

Stack pointer (SP): It contains offset within the stack segment.

Base pointer (BP): It contains the offset within the data segment.

c. Index registers

These registers are used for general purpose as well as offset storage. They are also used in string manipulations.

Source index (SI): It is used to store the source data offset in data segment of the memory.

Destination index (DI): It is used to store the destination data offset in data segment of the memory.

3. *Flag registers*

The 8086 processor has nine flags which determine the current status of the processor. The flags are broadly classified into two categories.

a. Conditional flags

Carry flag (CY): Whenever there is carry out or borrow of MSB in case of addition or subtraction respectively, the flag is set.

Auxiliary carry flag (AC): Whenever a carry or borrow is generated by third bit during addition or subtraction respectively, the flag is set.

Parity flag (PF): If the lower 8 bits of the result consists of even number of 1s, the flag is set.

Zero flag (ZF): Whenever the result of any arithmetic or logical operation is zero, the flag is set.

Sign flag (SF): If result of any operation is negative, the flag is set.

Overflow flag (OF): Whenever the result of a signed operation is large enough to be stored in destination register, an overflow occurs and the flag is set.

b.   Control Flags

Trap flag (TF): When the flag is set, the processor executes the program in a single step mode *i.e.* it executes only one instruction at a time.

Interrupt flag (IF): When the flag is set, the processor recognizes the maskable interrupt.

Direction flag (DF): When the flag is set, string bytes are accessed from higher to lower memory address and when the flag is reset, string bytes are accessed from lower to higher memory address.

Experiment 1

ADDITION of 8 BIT, 16 BIT and 32 BIT NUMBERS

AIM

To Write Assembly Language Program to Add Two - 8 Bit, 16 Bit and 32 Bit Hexadecimal Numbers.

ADDITION OF TWO 8 BIT NUMBERS

Register Addressing Mode

Table 7.1.1 Addition of two 8 bit numbers using register addressing mode

Label	Mnemonic	Comments
	MOV CL,00H	Clear CL to save carry
	MOV AL,11H	AL = 11H (first number)
	MOV BL,22H	BL = 22H (second number)
	ADD AL,BL	AL = AL + BL
	JNC *End*	Jump to end if no carry
	INC CL	Increment CL in case of carry
End	INT 3	Interrupt

Result

Table 7.1.2

Input	Output
AL = 11	AL = 33
BL = 22	BL = 22
CL = 00	CL = 00

Description

The program adds two numbers present in registers AL and BL and shows the result with carry in AL and CL registers respectively.

Immediate Addressing Mode

Table 7.1.3 Addition of two 8 bit numbers using immediate addressing mode

Label	Mnemonic	Comments
	MOV CL,00H	Clear CL to save carry
	MOV AL,22H	AL = 22H (first number)

	ADD AL,33H	AL = AL + 33H
	JNC *End*	Jump to end if no carry
	INC CL	Increment CL in case of carry
End	INT 3	Interrupt

Result

Table 7.1.4

Input	Output
AL = 22	AL = 55
BL = 33	BL = 33
CL = 00	CL = 00

Description

The program adds the number present in register AL with 33H and shows the result with carry in AL and CL registers respectively.

Direct Data Addressing Mode

Table 7.1.5 Addition of two 8 bit numbers using direct data addressing mode

Label	Mnemonic	Comments
	MOV CL,00H	Clear CL to save carry
	MOV AL,[1234]	AL = data present in memory location 1234 (first number)
	MOV BL,[0011]	BL = data present in memory location 0011 (second number)
	ADD AL,BL	AL = AL + BL
	JNC *End*	Jump to end if no carry
	INC CL	Increment CL in case of carry
End	INT 3	Interrupt

Result

Table 7.1.6

Input	Output
AL = xx	AL = 00
BL = xx	BL = 01
CL = 00	CL = 01
[1234] = FF	[1234] = FF
[0011] = 01	[0011] = 01

Experiment **2**

SUBTRACTION of 8 BIT and 16 BIT NUMBERS

AIM

To Write Assembly Language Program to Subtract Two – 8 Bit, 16 Bit Hexadecimal Numbers.

SUBTRACTION OF TWO 8 BIT NUMBERS

Table 7.2.1 Subtraction of two 8 bit numbers

Label	Mnemonic	Comments
	MOV CL,00H	Clear CL to save borrow
	MOV AL,0FH	AL = 0FH (first number)
	MOV BL,01H	BL = 01H (second number)
	SUB AL,BL	AL = AL – BL
	JNC *End*	Jump to end if no borrow
	INC CL	Increment CL in case of borrow
End	INT 3	Interrupt

Result

Table 7.2.2

Input	Output
AL = 0F	AL = 0E
BL = 01	BL = 01
CL = 00	CL = 00

Description

The program subtracts two numbers present in registers AL and BL and shows the result with borrow in AL and CL registers respectively.

SUBTRACTION OF TWO 16 BIT NUMBERS

Table 7.2.3 Subtraction of two 16 bit numbers

Label	Mnemonic	Comments
	MOV CX,0000H	Clear CX to save borrow
	MOV AX,000AH	AX = 000AH (first 16 bit number)

	JNC *End*	Jump to end if no carry
	INC CL	Increment CL in case of carry
End	INT 3	Interrupt

Result

Table 7.1.10

Input	Output
AL = 15	AL = 16
BX = 0100	BX = 0100
CL = 00	CL = 00
SI = 0200	SI = 0200
[0300] = 01	[0300] = 01

Description

The program adds two numbers present in register AL and memory location specified in registers BX and SI and shows the result with carry in AL and CL registers respectively.

Register Relative Addressing Mode

Table 7.1.11 Addition of two 8 bit numbers using register relative addressing mode

Label	Mnemonic	Comments
	MOV CL,00H	Clear CL to save carry
	MOV AL,0FH	AL = 0FH (first number)
	MOV BX,0100H	Load BX with memory address 0100H
	ADD AL,[BX+0200]	AL = AL + [0300] *i.e.* AL = AL + data present at memory location 0300H
	JNC *End*	Jump to end if no carry
	INC CL	Increment CL in case of carry
End	INT 3	Interrupt

Result

Table 7.1.12

Input	Output
AL = 0F	AL = 11
BX = 0100	BX = 0100
CL = 00	CL = 00
[0300] = 02	[0300] = 02

Description

The program adds the number present in register AL with the number present in memory location specified in register BX plus 0200H and stores the result with carry in registers AL and CL respectively.

ADDITION OF TWO 16 BIT NUMBERS

Table 7.1.13 Addition of two 16 bit numbers

Label	Mnemonic	Comments
	MOV CX,0000H	Clear CX to save carry
	MOV AX,00FFH	AX = 00FFH (first 16 bit number)
	MOV BX,FF01H	BX = FF01H (second 16 bit number)
	ADD AX,BX	AX = AX + BX
	JNC *End*	Jump to end if no carry
	INC CX	Increment CX in case of carry
End	INT 3	Interrupt

Result

Table 7.1.14

Input	Output
AX = 00FF	AX = 0000
BX = FF01	BX = FF01
CX = 0000	CX = 0001

Description

The program adds two 16 bit numbers present in registers AX and BX and shows the result with carry in AX and CX registers respectively.

ADDITION OF TWO 32 BIT NUMBERS

Table 7.1.15 Addition of two 32 bit numbers

Mnemonic	Comments
MOV SI,0500H	SI = 0500H (location of first 32 bit number)
MOV DI,0600H	DI = 0600H (location of second 32 bit number)
MOV AX,[SI]	AX = contents of 0500 and 0501 memory location *i.e.* AX = lower 16 bit of first number
MOV BX,[DI]	BX = contents of 0600 and 0601 memory location *i.e.* BX = lower 16 bit of second number
ADD AX, BX	AX = AX + BX

0508 : 09	0608 : xx	0508 : 09	0608 : 09
0509 : 0A	0609 : xx	0509 : 0A	0609 : 0A
050A : 0B	060A : xx	050A : 0B	060A : 0B
050B : 0C	060B : xx	050B : 0C	060B : 0C
050C : 0D	060C : xx	050C : 0D	060C : 0D
050D : 0E	060D : xx	050D : 0E	060D : 0E
050E : 0F	060E : xx	050E : 0F	060E : 0F
050F : 10	060F : xx	050F : 10	060F : 10

Description

The program transfers the data stored at memory location 0500H – 050FH to memory location 0600H – 060FH. The starting address 0500H and 0600H are stored in registers SI and DI respectively. Register CL is used to store the total number of values to be shifted.

Description

The program adds two numbers present in memory location with offset 1234H and 0011H and shows the result with carry in AL and CL registers respectively.

Register Indirect Addressing Mode

Table 7.1.7 Addition of two 8 bit numbers using register indirect addressing mode

Label	Mnemonic	Comments
	MOV CL,00H	Clear CL to save carry
	MOV SI,0500H	Load SI with memory address where first number is stored
	MOV AL,11H	AL = 11H (second number)
	ADD AL,[SI]	AL = AL + [0500] *i.e.* AL = AL + data present in memory location 0500H
	JNC *End*	Jump to end if no carry
	INC CL	Increment CL in case of carry
End	INT 3	Interrupt

Result

Table 7.1.8

Input	Output
AL = 11	AL = 33
CL = 00	CL = 00
SI = 0500	SI = 0500
[0500] = 22	[0500] = 22

Description

The program adds two numbers present in register AL and memory location specified in register SI and shows the result with carry in AL and CL registers respectively.

Base Plus Index Addressing Mode

Table 7.1.9 Addition of two 8 bit numbers using base plus index addressing mode

Label	Mnemonic	Comments
	MOV CL,00H	Clear CL to save carry
	MOV AL,15H	AL = 15H (first number)
	MOV BX,0100H	Load BX with memory address 0100H
	MOV SI,0200H	Load SI with memory address 0200H
	ADD AL,[BX+SI]	AL = AL + [0300] *i.e.* AL = AL + data present at memory location 0300H

		MOV BX,000BH	BX = 000BH (second 16 bit number)
		SUB AX,BX	AX = AX − BX
		JNC *End*	Jump to end if no borrow
		INC CX	Increment CX in case of borrow
	End	INT 3	Interrupt

Result

Table 7.2.4

Input	Output
AX = 000A	AX = FFFF
BX = 000B	BX = 000B
CX = 0000	CX = 0001

Description

The program subtracts two 16 bit numbers present in registers AX and BX and shows the result with borrow in AX and CX registers respectively.

Experiment 3

TRANSFER a BLOCK of DATA

AIM

To Write Assembly Language Program to Transfer a Block of Data.

PROGRAM

Table 7.3.1 Transfer a block of data

Label	Mnemonic	Comments
	MOV CL,10H	CL = 10H (total number of data to be transferred)
	MOV SI,0500H	SI = 0500 (memory location where data is present)
	MOV DI, 0600H	DI = 0600 (memory location where data is to be transferred)
Loop	MOV AL,[SI]	AL = data present at memory location SI
	MOV [DI],AL	Move AL to memory location specified by DI
	INC SI	SI = SI + 1
	INC DI	DI = DI + 1
	DEC CL	CL = CL − 1
	JNZ *Loop*	If CL != 0 *i.e.* if complete block of data is not shifted, then shift next byte
	INT 3	Interrupt

Result

Table 7.3.2

Input		Output	
0500 : 01	0600 : xx	0500 : 01	0600 : 01
0501 : 02	0601 : xx	0501 : 02	0601 : 02
0502 : 03	0602 : xx	0502 : 03	0602 : 03
0503 : 04	0603 : xx	0503 : 04	0603 : 04
0504 : 05	0604 : xx	0504 : 05	0604 : 05
0505 : 06	0605 : xx	0505 : 06	0605 : 06
0506 : 07	0606 : xx	0506 : 07	0606 : 07
0507 : 08	0607 : xx	0507 : 08	0607 : 08

MOV [SI], AX	Move result to 0500 and 0501 memory location
MOV AX, [SI+2]	AX = contents of 0502 and 0503 memory location *i.e.* AX = higher 16 bit of first number
MOV BX, [DI+2]	BX = contents of 0602 and 0603 memory location *i.e.* BX = higher 16 bit of second number
ADC AX, BX	AX = AX + BX with previous carry
MOV [SI+2], AX	Move result to 0502 and 0503 memory location
INT 3	Interrupt

Result

Table 7.1.16

Input				Output	
First number		Second number		Memory Location	Data
Memory Location	Data	Memory Location	Data		
0500	01	0600	01	0500	02
0501	02	0601	01	0501	03
0502	03	0602	01	0502	04
0503	04	0603	01	0503	05

Description

The program adds two 32 bit numbers stored in memory location with starting offset as 0500H and 0600H respectively. The result is stored in memory location 0500H to 0503H.

Experiment 6

FIBONACCI SERIES

AIM

To Write Assembly Language Program to Generate Fibonacci Series.

PROGRAM

Table 7.6.1 Fibonacci series

Label	Mnemonic	Comments
	MOV AX,0000H	AX = 0000 (first term)
	MOV BX,0001H	BX = 0001 (second term)
	MOV SI,0500H	SI = 0500 (Starting memory location where Fibonacci series is to be stored)
	MOV [SI], AX	Copy first term in memory
	MOV CX,000AH	CX = 000A, loop for 10 terms
Loop	MOV DX,AX	DX = present term
	ADD AX,BX	Add previous and present terms to generate next term
	INC SI	SI = SI + 1
	INC SI	SI = SI + 1
	MOV [SI], AX	Copy the generated term in memory
	MOV BX, DX	BX = previous term
	DEX CX	CX = CX – 1
	JNZ Loop	If CX != 0, generate next term
	INT 3	Interrupt

Result

Table 7.6.2

Memory Location	Data	Memory Location	Data
0500	00	050B	05
0501	00	050C	00
0502	00	050D	08
0503	01	050E	00
0504	00	050F	0D
0505	01	0510	00

Result

Table 7.4.4

Input	Output
AX = 0002	AX = 0006
BX = 0003	BX = 0003
DX = 0000	DX = 0000

Description

The program multiplies two 16 bit numbers present in registers AX and BX and stores the 32 bit answer in DX and AX registers.

Experiment 5

CONVERSION of 16 BIT HEXADECIMAL to DECIMAL NUMBER

AIM

To Write Assembly Language Program to Convert a 16-Bit Hexadecimal Number to Decimal Number.

PROGRAM

Table 7.5.1 Conversion of 16 bit hexadecimal to decimal number

Mnemonic	Comments
MOV SI,0500H	Load SI with address where result is to be stored
MOV AX,FFFFH	AX = FFFFH (hexadecimal number which is to be converted to decimal number)
MOV [0300],AX	Save number to memory location 0300H
MOV BX,2710H	BX = 2710_{16} or 10000_{10}
CALL *Division*	Call subroutine
MOV BX,03E8H	BX = $03E8_{16}$ or 1000_{10}
CALL *Division*	Call subroutine
MOV BX,0064H	BX = 0064_{16} or 100_{10}
CALL *Division*	Call subroutine
MOV BX,000AH	BX = $000A_{16}$ or 10_{10}
CALL *Division*	Call subroutine
MOV [SI],DX	Move remainder to source destination
INT 3	Interrupt

Table 7.5.2 Program for division subroutine

Label	Mnemonic	Comments
Division	MOV AX,[0300]	AX = number saved at memory location 0300H
	MOV DX,0000H	DX = 0000H
	DIV BX	DX AX = DX AX / BX (divides number by BX)
	MOV [SI],AX	Move quotient to memory location

Experiment 7

SORTING in ASCENDING/DESCENDING ORDER

AIM

To Write Assembly Language Program to Sort Hexadecimal Numbers in Ascending/Descending Order.

PROGRAM

Table 7.7.1 Sorting in Ascending/Descending order

Label	Mnemonic	Comments
	MOV SI,0500H	SI = starting memory location where data is stored
	MOV DI,SI	DI = starting memory location
	MOV DX, 0004H	DX = number of data to be sorted
Outer loop	MOV CX, DX	CX = DX CX is used for inner loop
Inner loop	CMP CX, 0001H	Compare CX with 0001H
	JE *Next*	If CX = 0001, jump to decrement outer loop counter
	DEC CX	CX = CX − 1
	MOV BX, [SI]	Copy the content at location addressed by SI to BX
	INC SI	SI = SI + 1
	INC SI	SI = SI + 1
	MOV AX, [SI]	Copy the content at location addressed by SI to AX
	CMP BX, AX	Compare BX with AX *i.e.* compare two terms
	JAE / JBE *Inner loop*	JAE – descending *i.e.* jump if BX is above or equal to AX JBE –ascending *i.e.* jump if BX is below or equal to AX
	XCHG AX, BX	Interchange the two terms
	MOV [SI], AX	Copy AX to location addressed by SI

Experiment 4

MULTIPLICATION of 8 BIT and 16 BIT NUMBERS

AIM

To Write Assembly Language Program to Multiply Two - 8 Bit, 16 Bit Hexadecimal Numbers.

MULTIPLICATION OF TWO 8 BIT NUMBERS

Table 7.4.1 Multiplication of two 8 bit numbers

Mnemonic	Comments
MOV AL,02H	AL = 02 (first number)
MOV BH,03H	BH = 03 (second number)
MUL BH	AX = AL * BH
INT 3	Interrupt

Result

Table 7.4.2

Input	Output	Input	Output
AL = 02	AX = 0006	AL = 11	AX = 0242
BH = 03		BH = 22	

Description

The program multiplies two 8 bit numbers present in registers AL and BH and stores the 16 bit answer in AX register.

MULTIPLICATION OF TWO 16 BIT NUMBERS

Table 7.4.3 Multiplication of two 16 bit numbers

Mnemonic	Comments
MOV DX,0000H	DX = 0000
MOV AX,0002H	AX = 0002 (first number)
MOV BX,0003H	BX = 0003 (second number)
MUL BX	DX AX = AX * BX
INT 3	Interrupt

0506	00	0511	15
0507	02	0512	00
0508	00	0513	22
0509	03	0514	00
050A	00	0515	37

Description

First two terms of Fibonacci series are initially stored in registers AX and BX. Then the other terms are generated by adding the current term and the previously stored term. The generated terms are stored in memory location specified in register SI. CX contains the number of terms which are to be generated.

Table 7.6.3 Description for Fibonacci series

CX	AX	BX	DX	SI	Data stored in memory = AX
0A	0000	0001	....	0500	00
					00
			AX 0000		
	AX + BX 0001			0502	00
					01
09		DX 0000			
			AX 0001		
	AX + BX 0001			0504	00
					01
08		DX 0001			
			AX 0001		
	AX + BX 0002			0506	00
					02
07		DX 0001			
			AX 0002		
	AX + BX 0003			0508	00
					03

06		DX 0002			
			AX 0003		
	AX + BX 0005			050A	00
					05
05		DX 0003			
			AX 0005		
	AX + BX 0008			050C	00
					08
04		DX 0005			
			AX 0008		
	AX + BX 000D			050E	00
					0D
03		DX 0008			
			AX 000D		
	AX + BX 0015			0510	00
					15
02		DX 000D			
			AX 0015		
	AX + BX 0022			0512	00
					22
01		DX 0015			
			AX 0022		
	AX + BX 0037			0514	00
					37
00		DX 0022			

		specified by SI
	MOV [0300],DX	Move remainder to memory location 0300H
	INC SI	Increment source index pointer
	INC SI	Increment source index pointer
	RET	Return to main program

Result

Table 7.5.3

Input	Output
AX = FFFFH	0500 : 00
	0501 : 06
	0502 : 00
	0503 : 05
	0504 : 00
	0505 : 05
	0506 : 00
	0507 : 03
	0508 : 00
	0509 : 05

Description

Table 7.5.4 Description for Hexadecimal number to decimal number

Number	Dividend	Remainder	Quotient	Data stored in memory
FFFFH or 65535_{10}	10000	5535	06 $\longrightarrow$	06
5535_{10}	1000	535	05 $\longrightarrow$	05
535_{10}	100	35	05 $\longrightarrow$	05
35_{10}	10	05	03 $\longrightarrow$	03
				05

	DEC SI	SI = SI − 1
	DEC SI	SI = SI − 1
	MOV [SI], BX	Copy BX to location addressed by SI
	INC SI	SI = SI + 1
	INC SI	SI = SI + 1
	JMP *Inner loop*	Jump back to inner loop to compare next terms
Next	DEC DX	Decrement the counter for outer loop
	MOV SI, DI	SI = starting memory location
	CMP DX, 0001H	Compare DX with 0001
	JNE *Outer loop*	If DX != 0001, jump back to the outer loop
	INT 3	Interrupt

Result

Table 7.7.2

Before execution		After execution	
Memory Location	Input	Descending	Ascending
0500	00	00	00
0501	03	04	01
0502	00	00	00
0503	04	03	02
0504	00	00	00
0505	02	02	03
0506	00	00	00
0507	01	01	04

Description

Register SI stores the memory location where data is stored and DX stores the number of data to be sorted. The program consists of two loops – inner loop and outer loop. If the number of data is four, then the outer loop is executed three times and for each outer loop, inner loop is also executed three times. For the first outer loop, first data is compared with the second data in first inner loop. Then, second data is compared with the third data in second inner loop and third data is compared with the fourth data in third inner loop. The values are exchanged whenever required. Then the outer loop is decremented and the inner loop is executed again three times. The cycle repeats till the outer counter becomes one.

Table 7.7.3 Description for sorting in ascending order

| | | First outer loop DX = 0004 | | | Second outer loop DX = 0003 | | | Third outer loop DX = 0002 | | |
| | | Inner loop number | | | Inner loop number | | | Inner loop number | | |
Memory location	Data	1st	2nd	3rd	1st	2nd	3rd	1st	2nd	3rd
0500	00	00	00	00	00	00	00	00	00	00
0501	03	03	03	03	02	02	02	01	01	01
0502	00	00	00	00	00	00	00	00	00	00
0503	04	04	02	02	03	01	01	02	02	02
0504	00	00	00	00	00	00	00	00	00	00
0505	02	02	04	01	01	03	03	03	03	03
0506	00	00	00	00	00	00	00	00	00	00
0507	01	01	01	04	04	04	04	04	04	04

Table 7.7.4 Description for sorting in descending order

| | | First outer loop DX = 0004 | | | Second outer loop DX = 0003 | | | Third outer loop DX = 0002 | | |
| | | Inner loop number | | | Inner loop number | | | Inner loop number | | |
Memory location	Data	1st	2nd	3rd	1st	2nd	3rd	1st	2nd	3rd
0500	00	00	00	00	00	00	00	00	00	00
0501	01	02	02	02	03	03	03	04	04	04
0502	00	00	00	00	00	00	00	00	00	00
0503	02	01	03	03	02	04	04	03	03	03
0504	00	00	00	00	00	00	00	00	00	00
0505	03	03	01	04	04	02	02	02	02	02
0506	00	00	00	00	00	00	00	00	00	00
0507	04	04	04	01	01	01	01	01	01	01

Experiment **8**

SQUARE ROOT

AIM

To Write Assembly Language Program to Find the Square Root of an Integer.

PROGRAM

Table 7.8.1 Square root of an integer

Label	Mnemonic	Comments
	MOV SI, 0500H	SI = 0500
	MOV BX, 0019H	BX = 19_{16} or 25_{10} (number)
	MOV CX, 0000H	Start counter with 0000
Loop	MOV AX, CX	AX = CX
	MUL AX	DX AX = AX * AX
	MOV [SI], CX	Copy CX to location specified by SI
	INC CX	CX = CX + 1
	CMP AX, BX	Compare multiplied number with the given number
	JB *Loop*	If multiplied number < given number, jump back to the loop
	JZ *End*	If both are equal, jump to end and square root of given number is displayed in the memory location
	MOV [SI], BX	If number is not a perfect square, number is displayed in the memory location
End	INT 3	Interrupt

Result

Table 7.8.2

Input	Output	Input	Output
BX = 0019	BX = 0019	BX = 0003	BX = 0003
0500 = xx	0500 = 00	0500 = xx	0500 = 00
0501 = xx	0501 = 05	0501 = xx	0501 = 03

Description

Table 7.8.3 Description for finding square root of number

BX = Number	CX	AX = CX	DX AX = AX *AX	AX < BX	AX = BX	Memory
25	0000	0000	0000 0000	Yes		
	0001	0001	0000 0001	Yes		
	0002	0002	0000 0004	Yes		
	0003	0003	0000 0009	Yes		
	0004	0004	0000 0016	Yes		
	0005	0005	0000 0025	No	Yes	0005

Note

The program works when the multiplied result is 16 bit only *i.e.* DX = 0000.

Experiment 9

DIGITAL CLOCK

AIM

To Write an Assembly Language Program to Generate Digital Clock.

PROGRAM

Table 7.9.1 Digital clock

Label	Mnemonic	Comments
	MOV AX,0000H	Clear AX to store time. AL stores seconds and AH stores minutes
Display	CALL F000:F07C	Clear display
	MOV BL,80H	BL is input parameter of subprogram. BL=80 represents the position of first line
	CALL F000:F078	Clears first line
	CALL F000:F094	Display AX on BL indicated location of LCD (first line)
	MOV BX,0003H	Outer counter
Inner loop	MOV CX,FFFFH	Inner counter
***	LOOP ***	Decrement CX and if CX!=0 execute the same line
	DEC BX	Decrement outer counter
	JNZ *Inner loop*	If BX!=0, start the inner loop again
	ADD AL,01	AL = AL + 01
	DAA	Decimal adjust after addition
	CMP AL,60	Compare AL with 60 seconds
	JNZ *Display*	If AL!= 60, jump to display time on LCD screen
	MOV AL,00	If AL = = 60, clear AL
	INC AH	Increment minute register
	CMP AH,03	Compare AH with 3 minutes
	JNZ *Display*	If AH !=3, jump to display time on screen
	INT 3	If AH = = 3, stop timer and Interrupt

Result

The program displays clock on the LCD screen. It starts with 1 second and goes till 3 minutes.

Description

AX register is used to store the time. AL stores the seconds and AH stores the minutes. With every second, AL register is incremented. As the second register *i.e.* AL reaches to 60 seconds, it is cleared and the minute register *i.e.* AH is incremented. The program runs for 3 minutes.

BX and CX registers are used to create a time delay of 1 second. Only a single loop cannot produce a time delay of 1 second, hence nested loops are used. Inbuilt subroutines are used to clear the LCD screen and display the contents of AX register on it.

Experiment **10**

FACTORIAL

AIM

To Write Assembly Language Program to Find the Factorial of a Given Number.

PROGRAM

Table 7.10.1 Factorial of a given number

Label	Mnemonic	Comments
	MOV DX, 0000H	Clear DX
	MOV AX, 0001H	AX = 0001
	MOV BX, 0005H	BX = 0005H (number)
	CMP BX,0000H	Compare number with 0000
	JE *End*	If number = 0000, then factorial is 01 which is stored in AX. Jump to end
	MOV AX, BX	AX = BX = number
Loop	DEC BX	BX = BX - 1
	MUL BX	DX AX = AX * BX
	CMP BX, 0001H	Compare BX with 0001
	JNZ *Loop*	If BX != 0001, then go back to loop for next multiplication
End	INT 3	Interrupt

Result

Table 7.10.2

Input	Output	Input	Output
BX = 0005	AX = 0078 DX = 0000	BX = 0000	AX = 0001 DX = 0000

Description

If the number is equal to zero, then the factorial is 1 which is stored in AX register otherwise the program computes the factorial of the number and stores it in AX and DX registers.

<table><tr><td>**8**</td><td></td></tr></table>

Experiments on Communication Electronics

LIST OF EXPERIMENTS

1. Study of Amplitude Modulation and Demodulation.
2. Study of Frequency Modulation and Demodulation.
3. Study of Pulse Amplitude Modulation.
4. Study of Pulse Width Modulation.
5. Study of Pulse Position Modulation.
6. Study of Pulse Code Modulation.
7. Study of Delta Modulation.
8. Study of Time Division Multiplexing.
9. Study of Amplitude Shift Keying.
10. Study of Phase Shift Keying.
11. Study of Frequency Shift Keying.
12. Study of Single Side Band Modulation and Demodulation.

Experiment 1

AMPLITUDE MODULATION and DEMODULATION

AIM

Study of Amplitude Modulation and Demodulation

APPARATUS REQUIRED

Bread board, transistor - BC547, diode, cathode ray oscilloscope, function generators, probes, connecting wires, capacitors, resistors, dc power supply.

THEORY

Modulation

It is a process in which one of the parameters of a high frequency carrier signal is varied in accordance to the instantaneous value of the message signal. The message signal is also called as modulating signal and the signal resulting from the process of modulation is called as modulated signal.

Analog or Continuous Wave Modulation

When a sine wave is used as a high frequency carrier signal to modulate the message signal and message signal is also continuous in nature then the modulation process is known as analog modulation. Since, the carrier wave used is continuous in nature; therefore, it is also known as continuous wave modulation.

Amplitude Modulation (AM)

It is one of the continuous wave modulation techniques in which amplitude of the high frequency carrier signal is varied according to the instantaneous value of the message signal. Hence, amplitude of the carrier signal contains the information about the message signal.

Assume a message signal with frequency f_m and a carrier signal with frequency f_C such that,

$$f_C \gg f_m$$

Let the modulating voltage $v_m(t)$ and carrier voltage $v_c(t)$ are represented by

$$v_m(t) = V_m \sin\omega_m t$$

$$v_c(t) = V_c \sin\omega_c t$$

Since, the amplitude of carrier signal is varied according to the message signal, expression for amplitude modulated wave can be written as

$$v(t) = (V_c + V_m \sin\omega_m t) \sin\omega_c t$$

$$\Rightarrow \qquad v(t) = V_c(1 + {}^{V_m}\!/_{V_c} \sin\omega_m t) \sin\omega_c t$$

$$\Rightarrow \qquad v(t) = V_c(1 + m \sin\omega_m t) \sin\omega_c t$$

$$\Rightarrow \qquad v(t) = V_c \sin\omega_c t + \frac{mV_c}{2}\cos(\omega_c - \omega_m)t - \frac{mV_c}{2}\cos(\omega_c + \omega_m)t \quad - \text{eqn 1}$$

Where,

m: modulation index and is given as

$$m = {}^{V_m}\!/_{V_c}$$

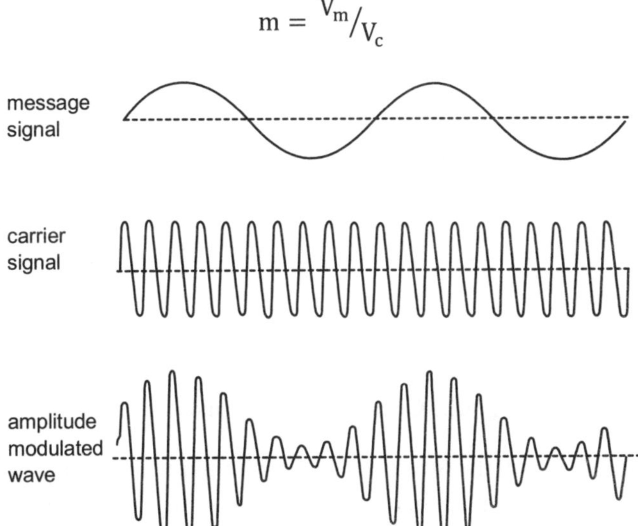

Fig. 8.1.1 Amplitude modulation

Frequency spectrum of AM wave

Equation 1 gives the expression for amplitude modulated wave and it shows the presence of three components *i.e.* carrier signal with frequency ω_c, lower side band with frequency $\omega_c - \omega_m$ and upper side band with frequency $\omega_c + \omega_m$ as shown in Fig. 8.1.2.

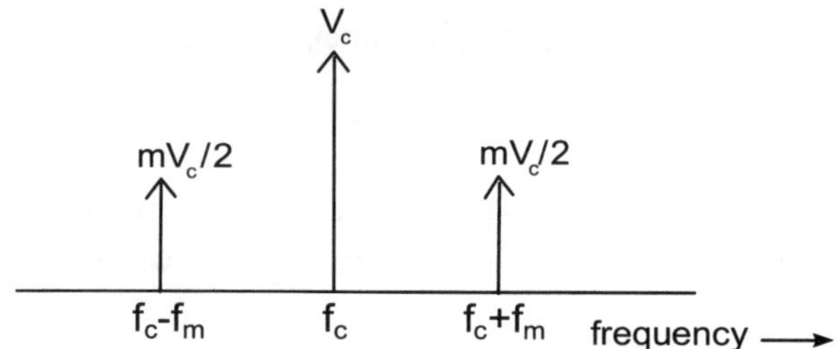

Fig. 8.1.2 Frequency spectrum of AM wave

Bandwidth

It is defined as the difference between upper and lower frequencies or it can be defined as the range of frequency over which an information signal is transmitted. For AM, bandwidth required is given as

$$BW = (f_c + f_m) - (f_c - f_m) = 2f_m$$

Modulation index

It tells the quality of amplitude modulated wave and is defined as

$$m = \frac{(V_{max} - V_{min})}{(V_{max} + V_{min})}$$

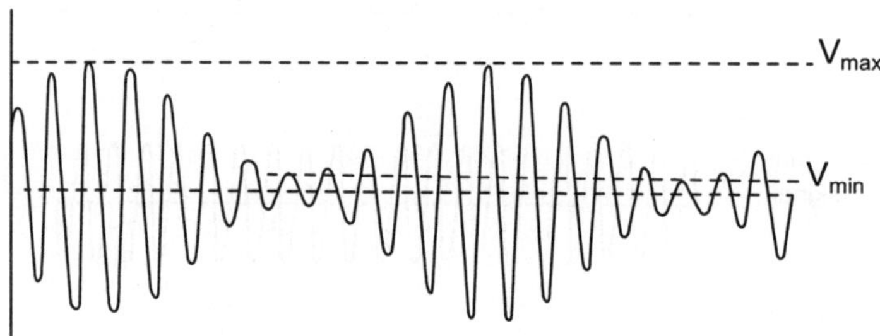

Fig. 8.1.3 Calculation of modulation index

V_{max} and V_{min} are the maximum and minimum amplitude of amplitude modulated wave. Modulation index should be less than 1 otherwise the carrier wave becomes over modulated that leads to distortion.

Evaluation of modulation index using trapezoidal display

Modulation index can be calculated by using trapezoidal display on CRO screen. The amplitude modulated wave is connected to the vertical deflection plates and modulating signal to the horizontal deflection plates of CRO as shown in Fig. 8.1.4.

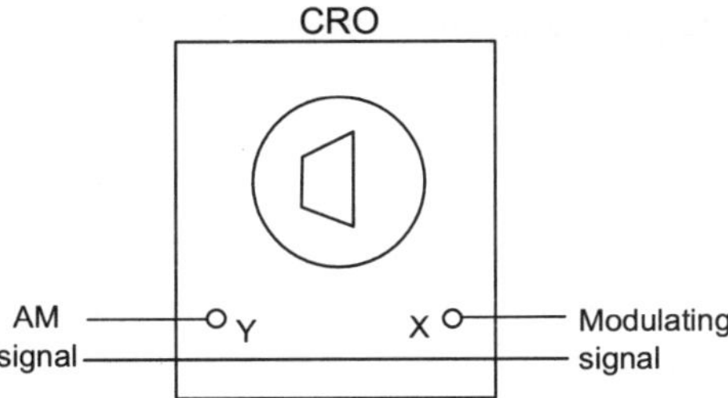

Fig. 8.1.4 Connection on CRO for calculation of modulation index

The trapezoidal display on the CRO screen is as shown in Fig. 8.1.5.

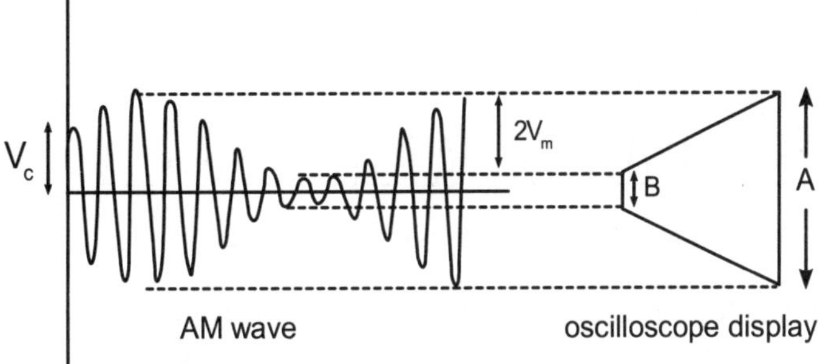

Fig. 8.1.5 Trapezoidal display of AM wave

Modulation index is given as

$$m = \frac{V_m}{V_c} = \frac{A - B}{A + B}$$

Amplitude Demodulation

Amplitude demodulation is a process of recovering the message signal back from the modulated signal at the receiver end.

Envelope detector

Envelope detector consists of diode, resistor and capacitor as shown in Fig. 8.1.6.

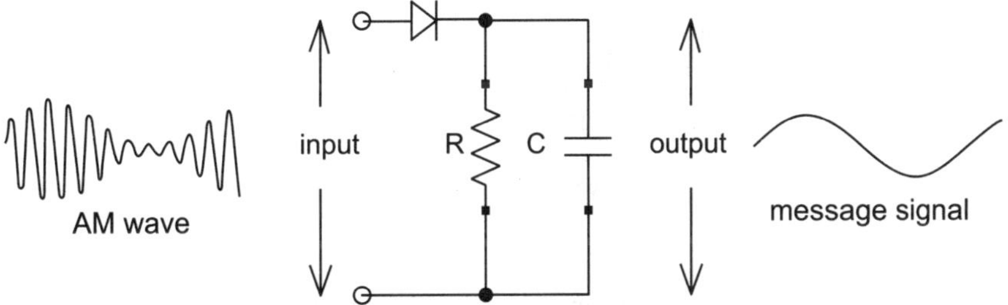

Fig. 8.1.6 Envelope detector for amplitude demodulation

It produces an output signal that follows the shape of the envelope of the amplitude modulated signal.

Working

Assume that the capacitor is not connected in the circuit shown in Fig. 8.1.6. During the positive half cycle of the amplitude modulated wave, diode is forward biased and conducts. Therefore, it acts as a closed switch and complete input voltage appears across resistor R. During the negative half cycle of the input wave, diode is reversed biased and does not conduct. Therefore, it acts as an open switch and no current flows through the circuit. Hence, voltage drop across resistor is zero. Therefore, output voltage across the resistor will be a half rectified carrier wave as shown in Fig. 8.1.7.

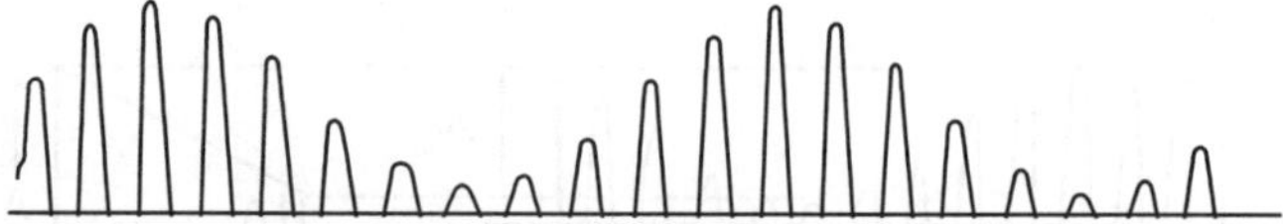

Fig. 8.1.7 Output across resistor without capacitor

Now, consider the case when capacitor is connected in the circuit. During the positive half cycle of the input wave, diode is forward biased and conducts. Therefore, capacitor charges to the peak value of carrier voltage through resistance R + r_d, where r_d is the resistance offered by the forward biased diode. During the negative half cycle, diode is reversed biased and it does not conduct. The input voltage gets disconnected from the RC circuit and hence capacitor starts discharging through resistance R. Time constant RC is chosen such that the capacitor discharges slowly. For the next cycle of the input wave, capacitor again charges to the new peak value of the input voltage during the positive half and tries to retain that value during the negative half. This process continues and therefore, the voltage across the capacitor is same as the envelope of the amplitude modulated wave as shown in Fig. 8.1.8. This voltage is nothing but modulating signal.

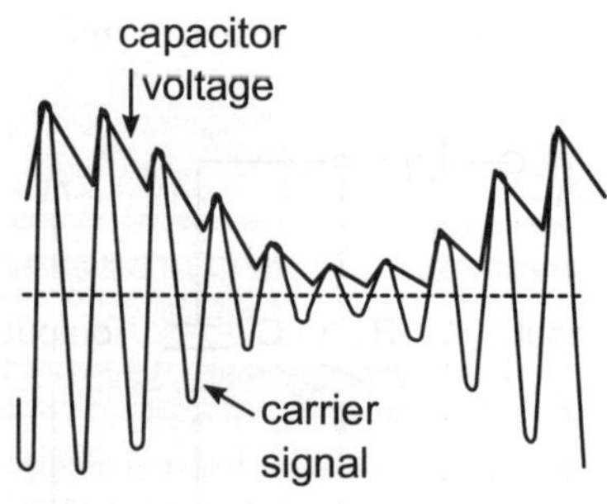

Fig. 8.1.8 Output voltage across capacitor

Condition

RC time constant should be properly chosen. RC time constant should be greater than time period of carrier signal but less than time period of modulating signal to avoid error due to diagonal peak clipping.

CIRCUIT DIAGRAM

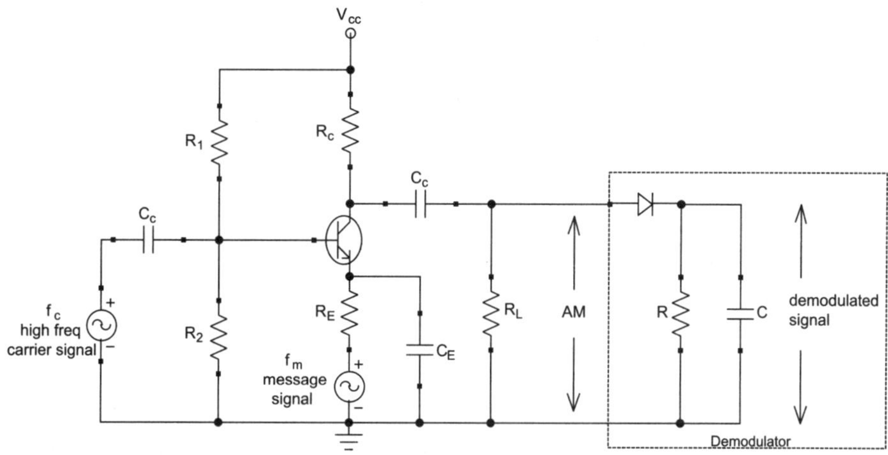

Fig. 8.1.9 Circuit diagram for amplitude modulation and demodulation

Working

It is a common emitter amplifier with the message signal and carrier signal connected at the emitter and base terminals of the transistor respectively. Let the message signal $v_m(t)$ and carrier signal $v_c(t)$ are represented by

$$v_m(t) = V_m \sin\omega_m t$$

$$v_c(t) = V_c \sin\omega_c t$$

As the modulating signal is applied at the emitter terminal, the instantaneous emitter current i_E changes according to the message signal and is given as

$$i_E = I_E + K_1 v_m(t)$$

Where,

i_E: instantaneous current, I_E: quiescent emitter current, $v_m(t)$: message signal, K_1: constant.

Rewriting the above equation, we get

$$i_E = I_E + K_1 V_m \sin\omega_m t$$

Since, voltage amplification factor A_V of transistor depends on the emitter current, hence, we can write

$$A_v = K_2 i_E$$

$\Rightarrow$
$$A_v = K_2(I_E + K_1 V_m \sin\omega_m t)$$

Where, K_2 is a constant. The output voltage v_o of amplifier is given as

$$v_o = A_v v_i$$

Where, v_i is input voltage. Since, input voltage is the carrier signal which is applied at base terminal, therefore output voltage can be written as

$$v_o = A_v V_c \sin\omega_c t$$

$\Rightarrow$
$$v_o = K_2(I_E + K_1 V_m \sin\omega_m t)V_c \sin\omega_c t$$

$\Rightarrow$
$$v_o = K_2 I_E \left(1 + \frac{K_1 V_m}{I_E} \sin\omega_m t\right) V_c \sin\omega_c t$$

$\Rightarrow$
$$v_o = K_2 I_E \left(V_c \sin\omega_c t + \frac{K_1 V_m V_c}{I_E} \sin\omega_m t \sin\omega_c t\right)$$

$\Rightarrow$
$$v_o = K_2 I_E \left(V_c \sin\omega_c t + \frac{mV_c}{2} (\cos(\omega_c - \omega_m)t - \cos(\omega_c + \omega_m)t)\right)$$

Where, m is modulation index and is given as

$$m = \frac{K_1 V_m}{I_E}$$

The expression for output voltage shows that it is an amplitude modulated wave.

CALCULATIONS

Demodulation

Carrier signal frequency $f_c = \ldots$
Time period of carrier signal $= 1/f_c = \ldots$
Message signal frequency $f_m = \ldots$
Time period of message signal $= 1/f_m = \ldots$
Let $C = 0.1\,\mu F$. Choose R such that,

Time period of carrier signal $<$ RC $<$ time period of message signal.

PROCEDURE

1. Design the CE amplifier according to the method as mentioned in experiment 6 of chapter 3.

2. Connect the circuit as shown in Fig. 8.1.9 on breadboard.

3. Connect one function generator to the base of transistor to give the high frequency carrier signal. Set it to sine wave of frequency around 30-50 kHz.

4. Connect another function generator at the emitter junction of the transistor to give the low frequency message signal. Set it to sine wave of frequency around 500-800 Hz.

5. Connect channel 2 of CRO across the output terminal *i.e.* R_L and observe the waveform.

6. Adjust the amplitude of message and carrier signal to get AM wave at the output and take the trace.

7. Connect channel 1 of CRO across the carrier signal and take the trace.

8. Reconnect channel 1 of CRO across the message signal. Measure its frequency and take the trace.

9. Switch on the x-y mode on CRO and observe the trapezoidal display on the screen as shown in Fig. 8.1.5.

10. Measure A and B and calculate the value of modulation index.

11. For demodulator, calculate the value of R and C such that RC is greater than time period of carrier signal and less than time period of modulating signal.

12. Connect channel 2 of CRO across the capacitor and trace the demodulated output waveform.

13. Measure the frequency of demodulated signal and compare it with the actual message signal.

14. Change the frequency of message signal and repeat the above steps.

Note

Amplitude of message signal should be less than amplitude of carrier signal to avoid distortion due to over modulation.

OBSERVATION TABLE

Table 8.1.1 Observation table for amplitude modulation and demodulation

S.No.	V_m (V)	f_m (Hz)	Modulation index (trapezoidal display)			Demodulated signal frequency (Hz)
			A	B	m	
1.						
2.						
3.						

OBSERVATIONS

Attach the traces for amplitude modulated and demodulated wave.

RESULT

Amplitude modulation and demodulation circuits have been designed successfully. The demodulated and the modulating message signal frequency comes out to be almost same.

DISCUSSION

Advantages

1. AM transmitters and receivers are simple to design.

2. AM signals can be transmitted over longer distances.

Disadvantages

1. Poor noise immunity. Noise is an undesired signal that changes the amplitude of the desired signal in a random manner. Since, envelope of AM wave contains the information about the message signal; once it gets distorted by noise it becomes difficult to recover the original signal back at the receiver.

2. Presence of any undesired harmonics in amplitude modulated wave can lead to distortion at the receiver end.

3. Bandwidth required for AM is twice the modulating signal frequency. This makes the system highly inefficient.

Applications

1. It is used in radio systems. AM radio stations, also called as medium wave stations, was the first technique to broadcast radio signals to the public. AM radio is in a band of 550 KHz – 1700 KHz.

2. AM is used for transmitting video signals in TV broadcasting. Video signal contains very low as well as very high frequency components and hence, bandwidth requirement is very high for transmitting such a signal. If FM is used instead of AM, the circuitry becomes very complex therefore, AM is preferred over FM.

Experiment 2

FREQUENCY MODULATION and DEMODULATION

AIM

Study of Frequency Modulation and Demodulation

APPARATUS REQUIRED

FM kit, Cathode ray oscilloscope, connecting wires, probes.

THEORY

Modulation

It is a process in which one of the parameters of a high frequency carrier signal is varied in accordance to the instantaneous value of the message signal. The message signal is also called as modulating signal and the signal resulting from the process of modulation is called as modulated signal.

Analog or Continuous Wave Modulation

When a sine wave is used as a high frequency carrier signal to modulate the message signal and message signal is also continuous in nature then the modulation process is known as analog modulation. Since, the carrier wave used is continuous in nature; therefore, it is also known as continuous wave modulation.

Frequency Modulation (FM)

It is one of the continuous wave modulation techniques in which frequency of the carrier signal is varied according to the instantaneous value of the message signal. Therefore, frequency of the carrier signal contains the information about the message signal keeping the amplitude and phase of the carrier as constant.

Assume a message signal with frequency f_m and a carrier signal with frequency f_c such that,

$$f_c \gg f_m$$

Let the modulating voltage $v_m(t)$ and carrier voltage $v_c(t)$ are represented by

$$v_m(t) = V_m \cos \omega_m t$$

$$v_c(t) \;=\; V_c \cos \omega_c t$$

The frequency modulated wave is given as

$$v(t) = V_c \cos(\omega_c t + m_f \sin\omega_m t)$$

Where, m_f is the modulation index for FM and is given as

$$m_f = \frac{\Delta f}{f_m}$$

Where, Δf is the frequency deviation.

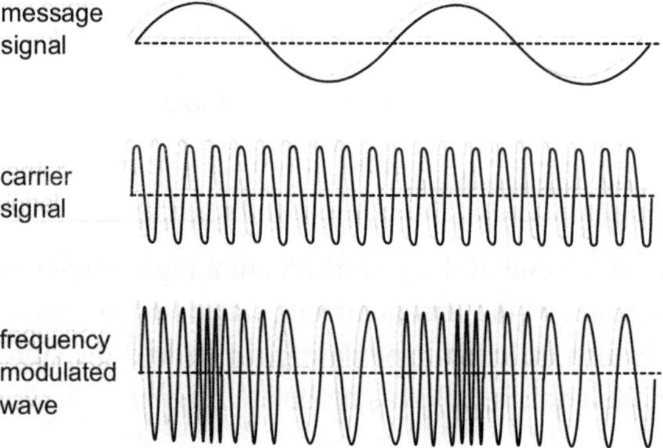

Fig. 8.2.1 Frequency modulation

Frequency modulation using VCO

A very simple method of generating an FM signal is by the use of a voltage controlled oscillator. A VCO is a circuit that generates oscillations with frequency proportional to the applied input voltage. It uses a varactor diode that converts the instantaneous value of the message signal to the change in frequency.

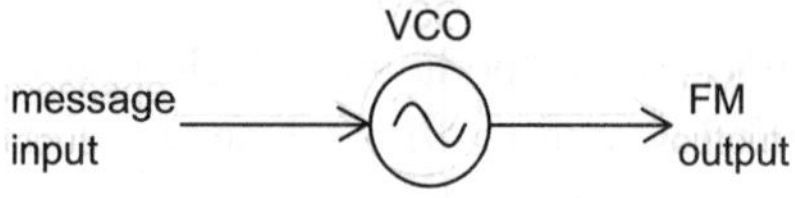

Fig 8.2.2 Frequency modulation by using VCO

Working

Initially, in the absence of any input signal, VCO operates at its center frequency which is set at the carrier frequency. The expression for the center frequency is given as

$$F_C = \frac{1}{2\Pi\sqrt{LC}} \qquad\qquad - \text{eqn 1}$$

Where,

F_c: center frequency, L: inductance, C: capacitance of varactor diode.

The modulating signal is applied at the control voltage terminal of the VCO. As the modulating signal voltage changes, the reverse bias of varactor diode also changes. This changes the capacitance of varactor diode and hence VCO output frequency also changes (refer eqn1). Therefore, as the input voltage increases or decreases, VCO output frequency also increases or decreases respectively. Hence, frequency modulated wave is produced at the output of VCO.

The advantage of using VCO for frequency modulation is that the variation of frequency with respect to control voltage is linear over a large range of frequency. But the disadvantage is that it is not a very stable source of frequency hence, is not preferred.

Frequency Demodulation

Frequency demodulation is a process of recovering the message signal back from the modulated signal at the receiver end.

Phase locked loop (PLL)

It is a frequency or phase sensitive circuit having three basic elements known as phase detector, VCO and a low pass filter as shown in Fig. 8.2.3.

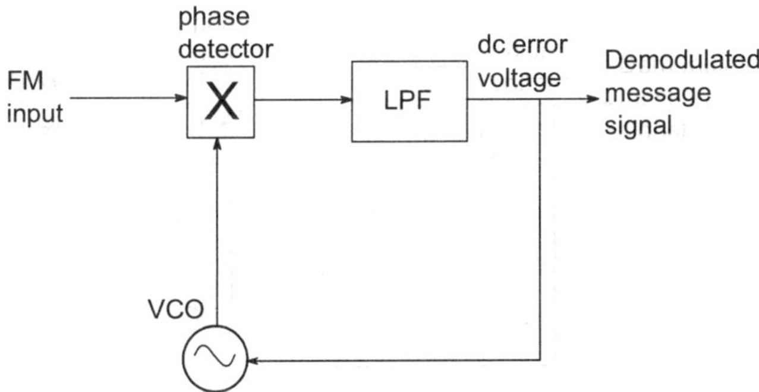

Fig. 8.2.3 PLL demodulator

Phase Detector: It compares the frequency or phase of two input signals *i.e.* frequency modulated signal and signal coming from VCO, and generates a dc output voltage signal.

Low Pass Filter: Low pass filter is used to pass only the dc voltage signal to VCO.

VCO: It generates oscillations with frequency proportional to the applied input voltage.

Working

Initially, when the input signal is not applied to PLL, phase detector circuit produces zero output voltage which is fed to VCO. With zero voltage input, VCO works at its free running frequency that is set at carrier frequency f_C. As the frequency modulated signal is applied as input signal to the phase detector, it compares the two frequencies *i.e.* FM input frequency and VCO frequency and generates a dc output voltage proportional to the difference of the two frequencies. If the input frequency is higher than f_C, phase detector produces a positive dc error voltage which is fed to VCO through low pass filter. Role of low pass filter is to pass only the dc signal to VCO and rejects all other harmonics produced by phase detector. This positive dc error voltage forces VCO to produce oscillations with higher frequency such that this dc error voltage reduces. After some cycles, VCO output frequency becomes same as input frequency and PLL is said to be in locked condition. At this point, although the frequencies at the input and output of VCO are equal but due to the phase difference between the two signals, phase detector still produces the dc error voltage which causes VCO to produce the same frequency as that of the input signal. If the input signal frequency changes to some other value, phase detector will produce a new dc error voltage that will force VCO to change its output frequency according to the applied signal. Hence, if frequency modulated signal is applied as input to the phase detector circuit, dc error voltage will keep on varying according to the change in the input signal frequency. This varying dc error voltage is nothing but the modulating signal.

PROCEDURE

1. Make connections on FM kit according to block diagram shown in Fig. 8.2.2.
2. Connect channel 1 of CRO to the output of VCO.
3. Take the trace of the carrier signal.
4. Give low frequency message signal using function generator to the input of VCO.
5. Now, connect channel 1 of CRO to the input of VCO and take the trace of message signal.
6. Connect channel 2 of CRO to the output of VCO and take the trace of FM wave.
7. For demodulation, connect FM wave as input to the PLL circuit as shown in Fig. 8.2.3.
8. Connect channel 2 of CRO at the output of PLL and trace the demodulated waveform.

9. Measure the frequency of message signal and demodulated signal and compare their values.

10. Change the frequency of message signal and observe the changes in FM wave and demodulated wave and take the traces.

OBSERVATION TABLE

Table 8.2.1 Observation table for frequency demodulation

S.No.	V_m (V)	f_m (Hz)	Demodulated signal frequency (Hz)
1.			
2.			
3.			

OBSERVATIONS

Attach the traces for frequency modulation and demodulation.

RESULT

Frequency modulation and demodulation circuits have been studied successfully. The demodulated and the modulating signal frequency comes out to be almost same.

DISCUSSION

Advantages

1. FM is less prone to noise as compared to AM.

2. It has better sound quality than AM.

3. Power requirement is less as compared to AM.

4. Guard bands are provided between two FM channels and therefore, adjacent channel interference is less as compared to AM.

5. Since, FM is more immune to noise, therefore, signal to noise ratio is higher as compared to AM.

Disadvantages

1. Transmitters and Receivers are complex to design as compared to AM.

2. Limited reception area because of line of sight transmission.

3. Bandwidth requirement is higher as compared to AM.

Applications

1. FM is used in radio broadcasting (88 MHz – 108 MHz).

2. FM is used in satellite communication because of less power requirement.

3. FM is used in police and hospital communications; emergency channels; TV sound; wireless telephone systems; radio amateur bands above 30MHz *etc*.

Experiment 3

PULSE AMPLITUDE MODULATION

AIM

Study of Pulse Amplitude Modulation.

APPARATUS REQUIRED

Breadboard, transistor BC547, resistors – 220Ω, 10kΩ, capacitor, connecting wires, cathode ray oscilloscope, probes, function generators.

THEORY

Modulation

It is a process in which one of the parameters of a high frequency carrier signal is varied in accordance to the instantaneous value of the message signal. The message signal is also called as modulating signal and the signal resulting from the process of modulation is called as modulated signal.

Pulse Modulation

When a pulse train is used as a high frequency carrier signal as shown in Fig. 8.3.1 to modulate the message signal, then the modulation process is known as pulse modulation. One of the parameters of the pulse train *i.e.* amplitude, width or position is varied in accordance to the instantaneous value of the message signal.

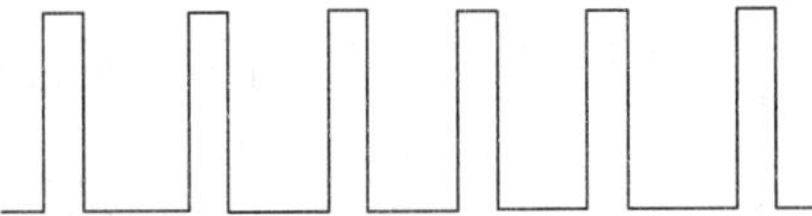

Fig. 8.3.1 Pulse train as carrier signal

The advantage of using pulse modulation is that power is transmitted in short bursts rather than continuously as in the case of continuous wave modulation and time between two pulses can be used to send samples of other message signals. But the disadvantage is the large bandwidth requirement.

Pulse Amplitude Modulation (PAM)

It is the process of varying the amplitude of the high frequency pulse train in proportion to the instantaneous value of the message signal. It is the simplest form of pulse modulation scheme.

PAM using natural sampling

In this, top of the high frequency pulse train follows the shape of the message signal and is obtained by multiplying message signal with the periodic train of pulses with unit amplitude and width dt.

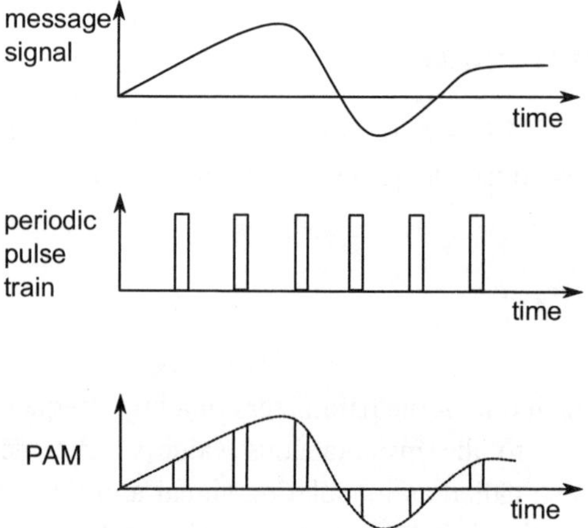

Fig. 8.3.2 PAM using natural sampling

Generation of PAM

Generation of PAM is shown in Fig. 8.3.3.

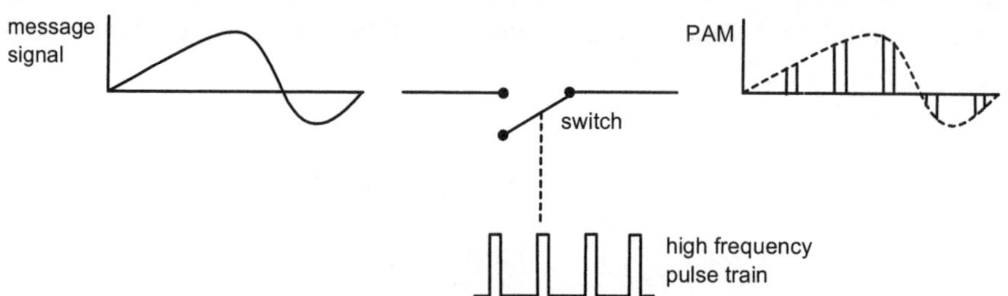

Fig. 8.3.3 Generation of PAM

The message signal is applied to a switch which is controlled by a high frequency pulse train. When the pulse is present, switch is closed and the message signal appears at the output and when the pulse is absent, the switch is open and there is no output. In this way, pulse amplitude modulated wave is obtained at the output.

Pulse Amplitude Demodulation

It is a process of recovering the message signal back from the modulated signal at the receiver end.

CIRCUIT DIAGRAM

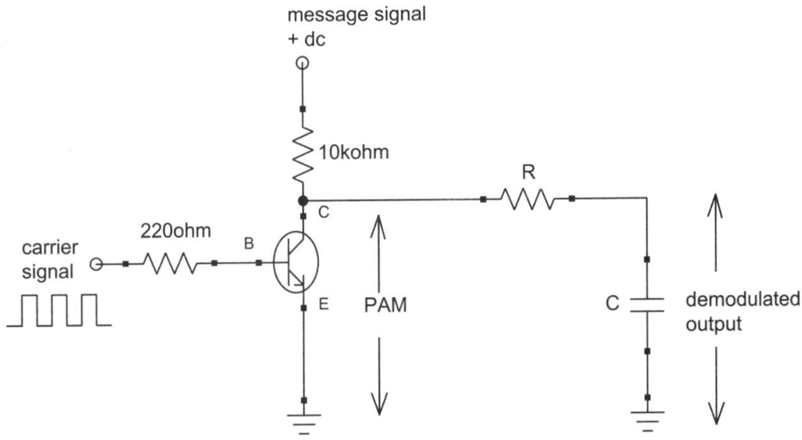

Fig. 8.3.4 Circuit for pulse amplitude modulation and demodulation

Working

Circuit shown in Fig. 8.3.4 consists of a BJT working as a switch in common emitter configuration. The high frequency carrier signal is applied at the base of transistor and low frequency message signal with dc offset voltage is applied at the collector of transistor. Dc offset is used to provide biasing voltage to the transistor. When the pulse at base is high, transistor goes into saturation region and voltage at output terminal becomes zero. When the pulse is low, transistor goes into cut off region and voltage at output terminal becomes same as input message signal plus dc offset. Hence, PAM is produced at the collector terminal.

For demodulation, PAM is fed to a simple low pass filter which is designed to pass low frequency message signal and reject high frequency carrier signal. Therefore, demodulated message signal is produced across the capacitor.

CALCULATIONS

Demodulation

Carrier signal frequency f_c = ...

Message signal frequency f_m = ...

For LPF, set the cut off frequency equal to the message signal frequency *i.e.*

$$f_m = 1/(2\pi RC)$$

Choose C = 0.01uF

Find R using the relation

$$R = 1/(2\pi f_m C) = \ldots$$

PROCEDURE

1. Connect the circuit shown in Fig. 8.3.4 on breadboard.
2. Connect function generator at the base of transistor to give carrier signal.
3. Set it for square wave with a high frequency of 5 kHz.
4. Connect another function generator at the collector of transistor to give message signal.
5. Set it for sine wave with a low frequency of 500Hz and set the dc offset.
6. Connect channel 2 of CRO at the collector terminal and observe the output waveform.
7. Adjust the amplitude and frequency of message and carrier signal to get PAM wave at the output and take the trace.
8. Connect channel 1 of CRO across the carrier signal and take the trace.
9. Reconnect channel 1 of CRO across the message signal. Measure its frequency and take the trace.
10. Design the low pass filter and find the values of R and C.
11. Give PAM as input to low pass filter.
12. Reconnect channel 1 of CRO at output of low pass filter and take the trace of demodulated signal and measure the output frequency.
13. Compare the frequencies of message signal and demodulated signal.
14. Change the frequency of message and carrier signal and repeat the above steps.

OBSERVATION TABLE

Table 8.3.1 Observation table for pulse amplitude demodulation

S.No.	f_m (Hz)	f_c (kHz)	Demodulated signal frequency
1.			
2.			
3.			

OBSERVATIONS

Attach the traces for pulse amplitude modulation and demodulation.

RESULT

Pulse amplitude modulation and demodulation circuits have been designed successfully. The demodulated and the modulating signal frequency comes out to be almost same.

DISCUSSION

Advantages

1. Simple circuitry for PAM generation and detection.

Disadvantages

1. Poor noise performance.

2. Power depends on the amplitude and the width of pulses. All the pulses in a PAM wave differs in amplitude, therefore power will also differ. Hence, transmitter must be able to handle the power required to transmit pulse having maximum amplitude.

Experiment 4

PULSE WIDTH MODULATION

AIM

Study of Pulse Width Modulation.

APPARATUS REQUIRED

Breadboard, connecting wires, digital storage oscilloscope, probes, function generators, IC-741, ±12V dual power supply, resistor, capacitor.

THEORY

Modulation

It is a process in which one of the parameters of a high frequency carrier signal is varied in accordance to the instantaneous value of the message signal. The message signal is also called as modulating signal and the signal resulting from the process of modulation is called as modulated signal.

Pulse Modulation

When a pulse train is used as a high frequency carrier signal as shown in Fig. 8.4.1 to modulate the message signal, then the modulation process is known as pulse modulation. One of the parameters of the pulse train *i.e.* amplitude, width or position is varied in accordance to the instantaneous value of the message signal.

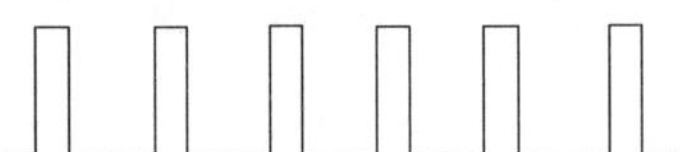

Fig 8.4.1 Pulse train as carrier signal

The advantage of using pulse modulation is that power is transmitted in short bursts rather than continuously as in the case of continuous wave modulation and time between two pulses can be used to send samples of other message signals. But the disadvantage is the large bandwidth requirement.

Pulse Width Modulation (PWM)

It is a process of varying the width of the high frequency pulse train in proportion to the instantaneous value of the message signal. It is also known as Pulse Duration Modulation (PDM) or Pulse Length Modulation (PLM).

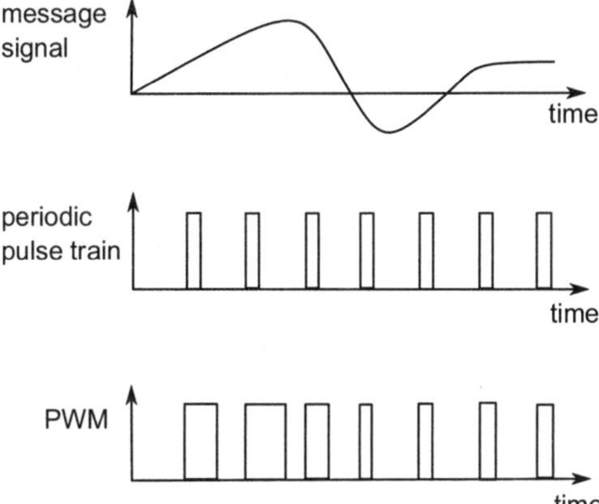

Fig 8.4.2 Pulse width modulation

Fig. 8.4.2 shows that the width of the pulse train contains the information about the message signal. As the amplitude of message signal increases or decreases, the pulse width increases or decreases respectively.

Generation of PWM using op-amp

Method 1

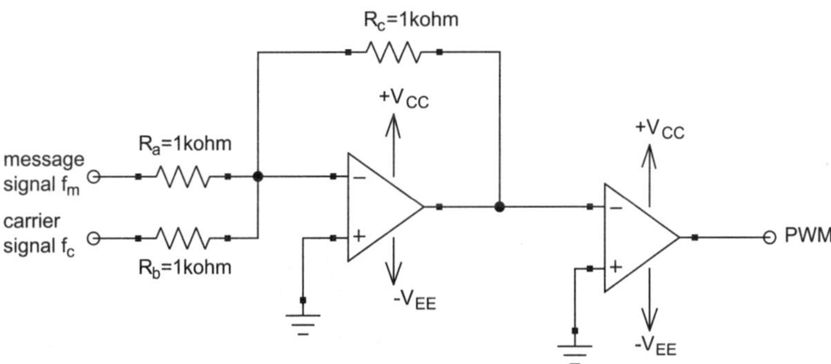

Fig. 8.4.3 Method 1: Circuit for PWM generation

Circuit shown in Fig. 8.4.3 consists of two op-amps. First op-amp works as an adder circuit which adds two inputs *i.e.* message signal which is a sine wave of low frequency and carrier signal which is a triangular wave of high frequency. The added signal is then fed to the zero level comparator which produces PWM at the output.

Method 2

Circuit shown in Fig. 8.4.4 consists of a single op-amp working as a comparator. It compares low frequency message signal with the high frequency carrier signal. The carrier signal can be a triangular or a saw-tooth waveform.

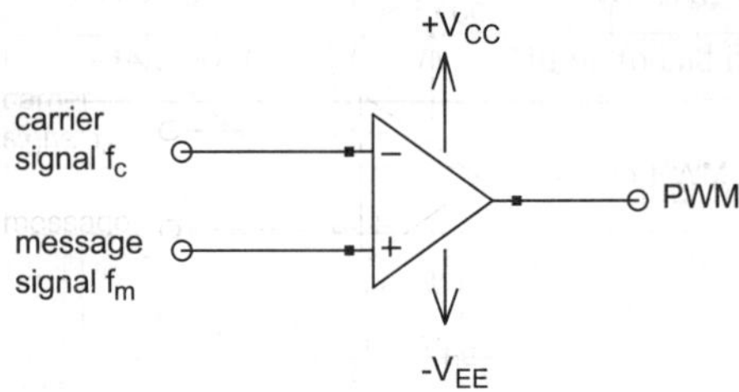

Fig. 8.4.4 Method 2: Circuit for PWM generation

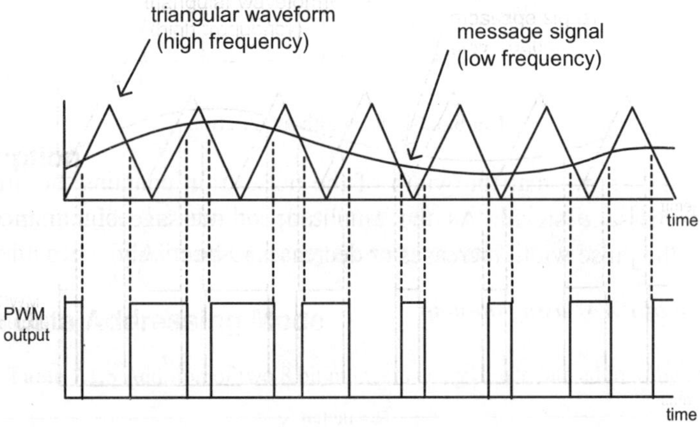

Fig. 8.4.5 Waveforms for PWM generation

Working

High frequency carrier signal *i.e.* triangular wave is applied at the inverting terminal of op-amp and low frequency message signal *i.e.* sine wave is applied at the non-inverting terminal of op-amp. Op-amp that is working as a comparator compares the two input signals. When the amplitude of triangular wave is greater than sine wave, the input voltage at inverting terminal becomes greater than non-inverting terminal. Since, op-amp is used in open loop configuration (very high gain amplifier), op-amp goes into saturation and output becomes $-$Vsat *i.e.* -12V. As the amplitude of triangular wave decreases and becomes less than sine wave, then input voltage at non-inverting terminal becomes greater than inverting terminal. Op-amp goes into saturation and output becomes +Vsat *i.e.* +12V. Hence, magnitude of input message signal decides the duration for which comparator output is high or low which as a result decides the width of the pulses. Therefore, as the amplitude of message signal changes, the width of the pulse also changes.

Condition

Peak voltage of carrier signal should be greater than message signal to obtain PWM.

Pulse Width Demodulation

It is a process of recovering the message signal back from the modulated signal at the receiver end.

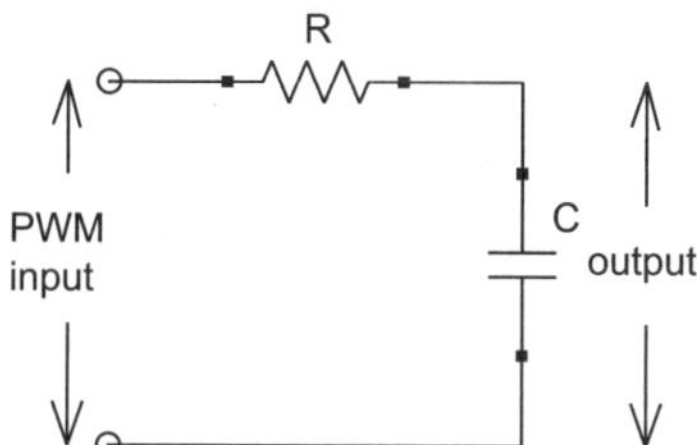

Fig. 8.4.6 Circuit for pulse width demodulation

Circuit shown in Fig. 8.4.6 is a low pass filter which is designed to pass low frequency message signal and reject high frequency carrier signal.

CALCULATIONS

Demodulation

Carrier signal frequency f_c = ...

Message signal frequency f_m = ...

For LPF, set the cut off frequency equal to the message signal frequency *i.e.*

$$f_m = 1/(2\pi RC)$$

Choose C = 0.01uF.

Find R using the relation

$$R = 1/(2\pi f_m C) = ...$$

PROCEDURE

1. Mount 741 IC on breadboard and check it in inverting mode whether it is working properly or not before making connections.

2. Connect the circuit shown in Fig. 8.4.4 on breadboard.

3. Connect one function generator to the inverting terminal of op-amp and set it to triangular wave with high frequency of around 3 kHz.

4. Connect another function generator to the non-inverting terminal of op-amp and set it to sine wave with low frequency of around 1 kHz.

5. Connect channel 2 of CRO at the output terminal and observe the output waveform.

6. Adjust the amplitude of both the input signals such that amplitude of carrier signal is more than that of message signal.

7. Trace the PWM output.

8. Connect channel 1 of CRO across the carrier signal and take the trace.

9. Reconnect channel 1 of CRO across the message signal. Measure its frequency and take the trace.

10. Connect the circuit shown in Fig. 8.4.6 on breadboard.

11. Connect the PWM output to the input of demodulator circuit.

12. Reconnect channel 2 of CRO at the output of demodulator circuit and take the trace of demodulated signal and measure the output frequency.

13. Compare the frequency of message signal and demodulated signal.

14. Change the frequency of message and carrier signals and repeat the above steps.

OBSERVATION TABLE

Table 8.4.1 Observation table for pulse width demodulation

S.No.	f_m (Hz)	f_c (kHz)	Demodulated signal frequency
1.			
2.			
3.			

OBSERVATIONS

Attach the traces for pulse width modulation and demodulation.

RESULT

Pulse width modulation and demodulation circuit have been designed successfully. The demodulated and the modulating signal frequency comes out to be almost same.

DISCUSSION

Advantages

1. Noise is less as compared to PAM since PWM contains information in its width and not in the amplitude.

2. Amplitude limiter circuits are used to remove the portion of signal affected by noise and thus provide good noise immunity.

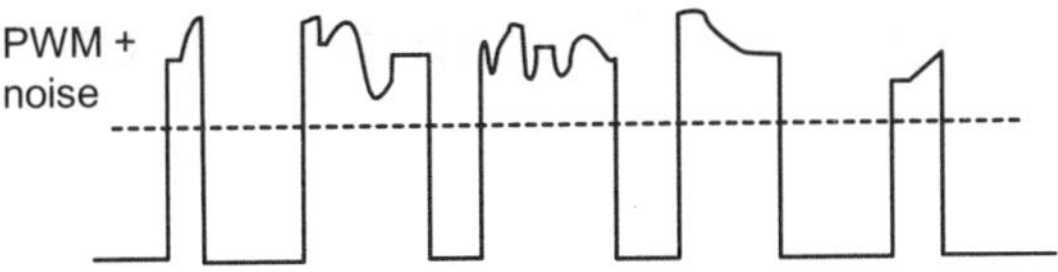

Fig. 8.4.7 Noise in PWM

Disadvantages

1. Power depends on the amplitude and width of the pulse. Since, all the pulses in PWM wave differs in width, therefore power will also differ. Transmitter must be able to handle the power required to transmit the pulse with maximum width.

2. Time division multiplexing is difficult to achieve because of the varying pulse width. The pulses from different sample may overlap with each other.

3. Bandwidth requirement is more as compared to PAM.

Experiment 5

PULSE POSITION MODULATION

AIM

Study of Pulse Position Modulation.

APPARATUS REQUIRED

Breadboard, connecting wires, DSO, probes, function generators, IC-741, 555 timer, ±12V and 5V power supply, resistors, capacitors, diode.

THEORY

Modulation

It is a process in which one of the parameters of a high frequency carrier signal is varied in accordance to the instantaneous value of the message signal. The message signal is also called as modulating signal and the signal resulting from the process of modulation is called as modulated signal.

Pulse Modulation

When a pulse train is used as a high frequency carrier signal as shown in Fig. 8.5.1 to modulate the message signal, then the modulation process is known as pulse modulation. One of the parameters of the pulse train *i.e.* amplitude, width or position is varied in accordance to the instantaneous value of the message signal.

Fig. 8.5.1 Pulse train as carrier signal

The advantage of using pulse modulation is that power is transmitted in short bursts rather than continuously as in the case of continuous wave modulation and time between two pulses can be used to send samples of other message signals. But the disadvantage is the large bandwidth requirement.

Pulse Position Modulation (PPM)

It is a process of varying the position of the high frequency pulse train in proportion to the instantaneous value of the message signal while keeping the amplitude and width of pulse constant.

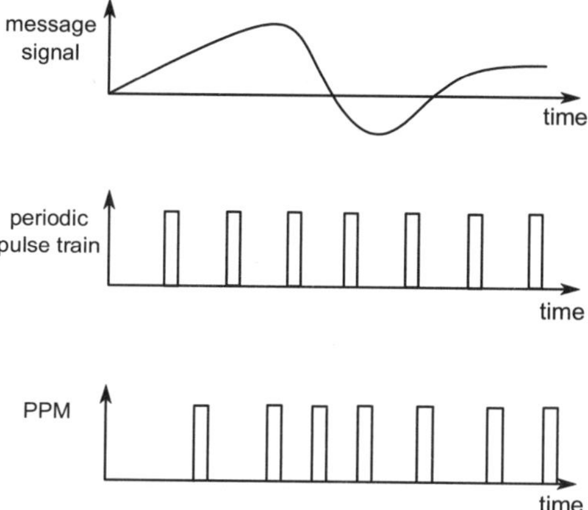

Fig. 8.5.2 Pulse Position Modulation

Fig. 8.5.2 shows that position of the pulse train contains the information about the message signal.

Generation of PPM

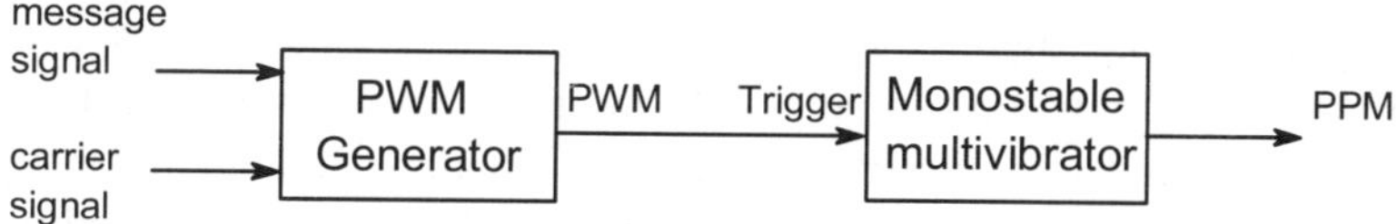

Fig. 8.5.3 Block diagram of PPM generation

Fig. 8.5.3 shows the block diagram of PPM generation. The message signal and carrier signal are fed to PWM generator which produces PWM wave. It triggers the mono-stable multivibrator which produces pulses of constant width on trailing edges of PWM pulses.

CIRCUIT DIAGRAM

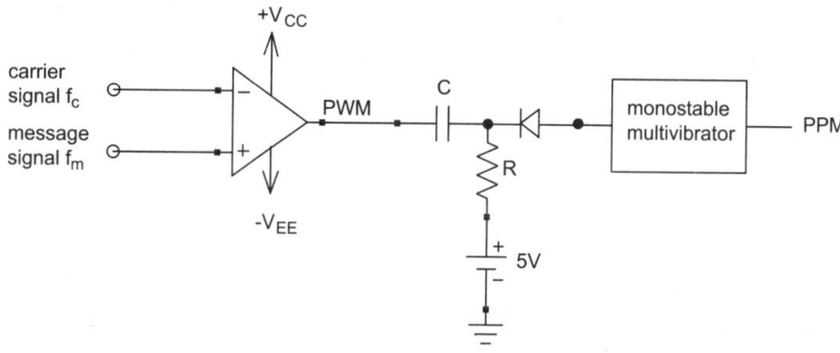

Fig. 8.5.4 Circuit for PPM generation

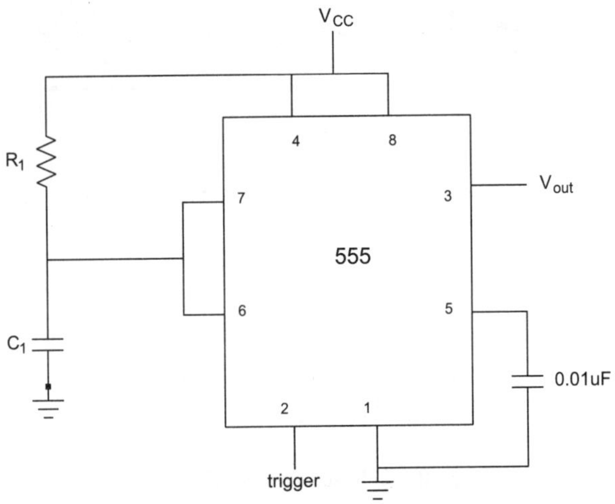

Fig. 8.5.5 Circuit for mono-stable multivibrator

Working

Circuit shown in Fig. 8.5.4 consists of a comparator circuit followed by a mono-stable multivibrator. High frequency carrier signal *i.e.* triangular wave is applied at the inverting terminal of op-amp and low frequency message signal *i.e.* sine wave is applied at the non-inverting terminal of op-amp. Op-amp that is working as a comparator compares the two input signals. When the amplitude of triangular wave is greater than sine wave, the input voltage at inverting terminal becomes greater than non-inverting terminal. Since, op-amp is used in open loop configuration (very high gain amplifier); op-amp goes into saturation and output becomes −Vsat *i.e.* -12V. As the amplitude of triangular wave decreases and becomes less than sine wave, then input voltage at non-inverting terminal becomes greater than inverting terminal. Op-amp goes into saturation and output becomes +Vsat *i.e.* +12V. Hence, magnitude of input message signal decides the duration for which comparator output is high or low which as a result decides the width of the pulses. Therefore, as the amplitude of message signal changes, width of the pulse also changes. Hence, PWM is obtained at the output of the comparator circuit. Comparator is followed by a differentiator circuit which is used to convert pulses into spikes. If the amplitude of spike is ±12V, it can damage the 555 timer. In order to avoid that, 5V dc voltage is added to produce the dc shift in the differentiated output such that amplitude of negative spikes reduces. Then the diode is used to eliminate positive spikes and pass the negative spikes which are applied at the trigger terminal (second pin) of 555 timer. The 555 timer as shown in Fig. 8.5.5 is working as a mono-stable multivibrator which produces pulses of fixed width and fixed amplitude when it gets a triggering negative pulse.

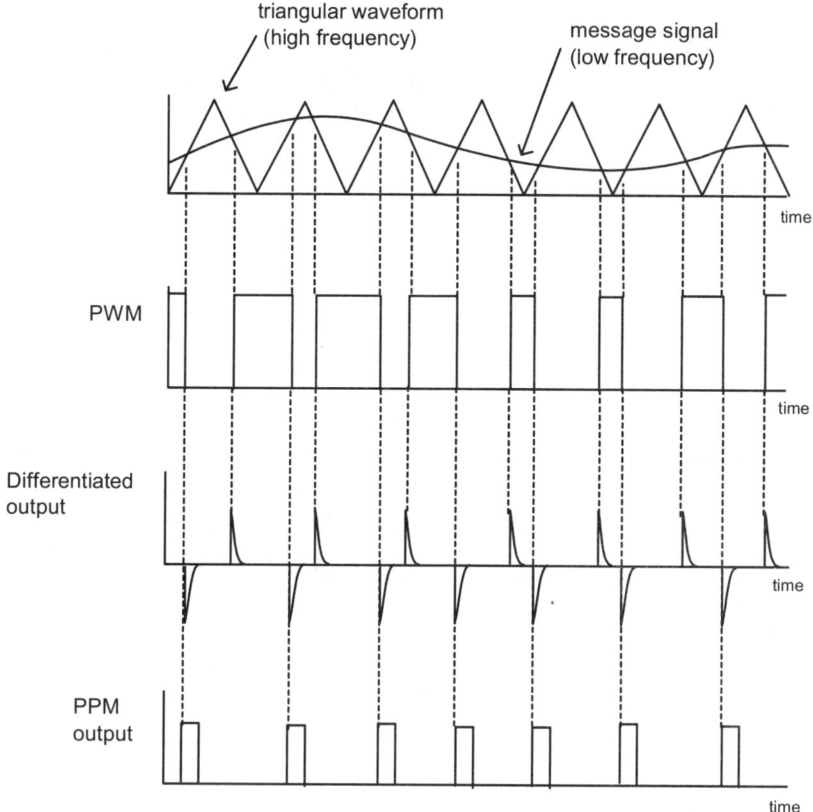

Fig. 8.5.6 Waveforms for PPM generation

Figure 8.5.6 shows the waveforms for PPM generation. Each trailing edge of PWM pulse is a starting point of the pulse in PPM.

CALCULATIONS

Carrier signal frequency f_c = ...

Message signal frequency f_m = ...

Differentiator

Let C = 0.01uF.

Choose R such that RC is much less than the time period of message signal *i.e.*

$$T \gg RC$$

Mono-stable Multivibrator

The time during which the timer output remains high is given as

$$t = 1.1 \times R_1 \times C_1$$

Choose t such that it is five times of pulse width t' of the differentiated negative pulse.

$$t = 5 . t' = \dots$$

Let $C_1 = 0.01uF$, calculate R_1 by using the relation

$$R_1 = {}^{t}\!/_{1.1C_1}$$

PROCEDURE

1. Mount 741 IC on breadboard and check it in inverting mode whether it is working properly or not before making connections.

2. Connect the circuit as shown in Fig. 8.5.4 on breadboard.

3. Connect one function generator to the inverting terminal of op-amp and set it to triangular wave with high frequency of around 3 kHz.

4. Connect another function generator to the non-inverting terminal of op-amp and set it to sine wave with low frequency of around 1 kHz.

5. Connect channel 2 of CRO at the output terminal of op-amp and observe the output waveform.

6. Adjust the amplitude of both the input signals such that amplitude of carrier signal is more than that of message signal.

7. Trace the PWM output.

8. Connect channel 1 of CRO across the carrier signal and take the trace.

9. Reconnect channel 1 of CRO across the message signal and take the trace.

10. Design the differentiator circuit according to the method as mentioned above.

11. Give PWM output to the differentiator and take the trace of the differentiated output waveform by connecting channel 2 of CRO across resistor.

12. Connect diode such that it passes only negative spikes.

13. Design the mono-stable multivibrator according to the method as mentioned above.

14. Connect the circuit for mono-stable multivibrator as shown in Fig. 8.5.5 on breadboard.

15. Give negative spikes to the trigger input of 555 timer.

16. Trace the PPM output waveform at the 3rd pin of 555 timer.

OBSERVATIONS

Attach the traces for PPM wave.

RESULT

Pulse position modulation circuit has been designed and traces have been taken successfully.

DISCUSSION

Advantages

1. Noise is less as compared to PAM since PPM contains information in position of pulses and not in the amplitude.

2. Amplitude limiter circuits are used to remove the portion of signal affected by noise and thus provide good noise immunity.

3. In PPM, amplitude and width of the pulses are constant. Therefore, power required for transmitting PPM wave is also constant.

4. Depending on the available bandwidth, the pulse width can be decided. Narrower is the pulse width, more is the bandwidth required to transmit such a pulse. Hence, if the bandwidth available is narrow, wide pulses are used and if bandwidth available is large, narrow pulses are used. This is called as bandwidth optimization.

Disadvantages

1. Time division multiplexing is difficult to achieve because of the varying pulse positions. The pulses from different sample may overlap with each other.

2. Large bandwidth as compared to PAM.

3. Pulses are varying in position so chances of crosstalk between different samples are more in time division multiplexing.

Experiment 6

PULSE CODE MODULATION

AIM

Study of Pulse Code Modulation.

APPARATUS REQUIRED

PCM kit, connecting wires, probes, digital storage oscilloscope.

THEORY

Digital Communication

When the message signal which is in form of codes 0 and 1, is transmitted from one place to another, either directly or by some form of modulation technique, then such kind of communication is called as digital communication.

Pulse Code Modulation

It is a form of digital modulation scheme which converts an analog message signal into digital form. It basically represents signal in form of codes which are transmitted to the receiver through the communication channel.

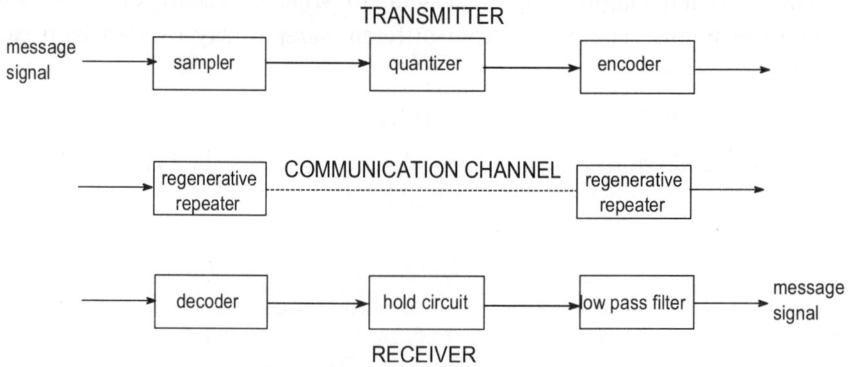

Fig. 8.6.1 Block diagram of PCM system

Figure 8.6.1 shows the block diagram of PCM system. It consists of three main parts *i.e.* transmitter, communication channel and receiver.

Steps involved in PCM

Sampling

It is the first step of digital communication which converts a continuous time signal into a discrete time signal.

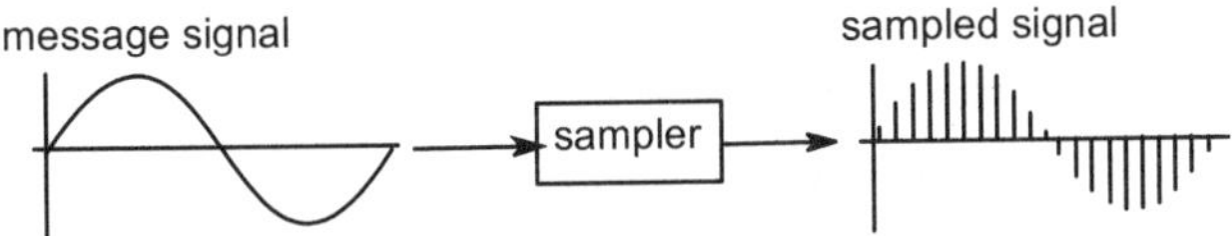

Fig. 8.6.2 Sampling of signal

Figure 8.6.2 shows the sampled signal which is transmitted to the receiver. In order to reconstruct the message signal completely from the samples at the receiver, it is important to transmit sufficient number of samples of the signal. Number of samples to be taken depends on the maximum frequency present in the message signal and is given by the sampling theorem.

Sampling theorem states that a continuous time signal can be completely represented by the samples and recovered back if the samples are taken with the frequency

$$f_s \geq 2f_m$$

Where, f_s is sampling frequency and f_m is message signal frequency. The above relation is called as the Nyquist Criterion.

Quantizer

The sampled signal is discrete in time but continuous in amplitude *i.e.* amplitude can have any of the infinite values in a finite range. Role of quantizer is to convert this discrete time signal into a digital signal.

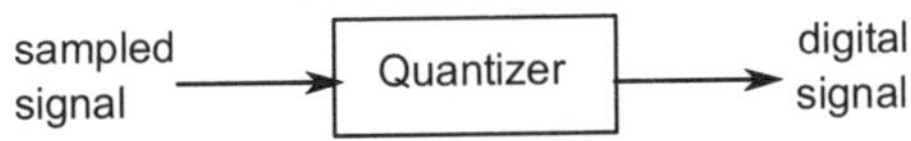

Fig. 8.6.3 Quantizer

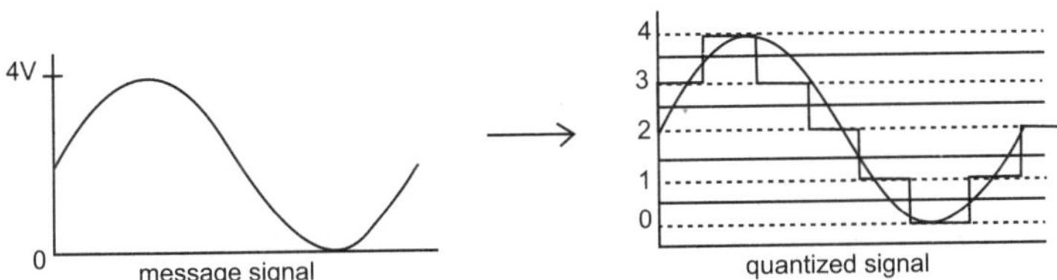

Fig. 8.6.4 Quantized signal

Assume a message signal whose voltage varies from 0 to 4V. This voltage range is divided into 5 levels *i.e.* 0, 1, 2, 3 and 4V as shown in Fig. 8.6.4. These levels are known as quantization levels. If the signal is between -0.5 to 0.5V, it is estimated by 0V and if the signal is between 0.5 to 1.5V, it is estimated by 1V and so on *i.e.* the discrete time signal becomes discrete in amplitude as well and the resulting signal is called as the quantized signal.

Encoder

The quantized signal which is discrete in time as well as amplitude cannot be transmitted directly. The number of voltage levels which are to be transmitted is equal to the number of quantization levels which is quite large. As shown in Fig. 8.6.4, number of quantization levels are 5 and hence circuit required to transmit these five voltage levels will be complex. Therefore, encoder is used to encode these quantization levels.

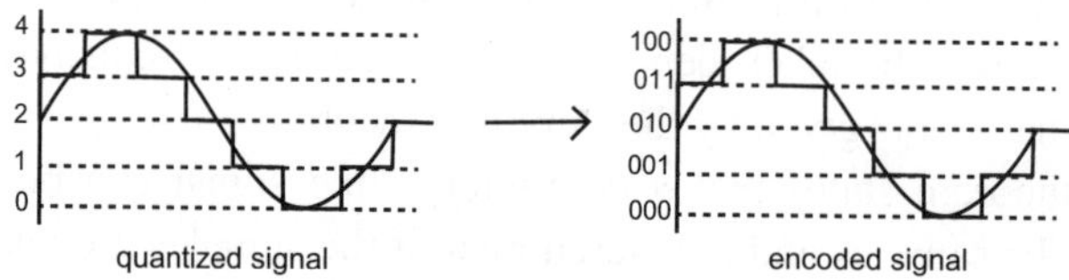

Fig. 8.6.5 Encoded signal

In Fig. 8.6.5, instead of sending 0, 1, 2, 3 and 4, data will be transmitted as 000, 001, 010, 011 and 100. Hence, instead of five voltage levels only two voltage levels *i.e.* '0' and '1' are used which can be transmitted easily. '1' is transmitted by the presence of pulse and '0' is transmitted by the absence of pulse. The most popular code used is the binary code.

Table 8.6.1 Binary code for each quantized level

Quantization level	Binary Code
0	000
1	001
2	010
3	011
4	100

Regenerative repeater

When the signal is transmitted over communication channel, it gets affected by noise due to which the signal cannot be recovered completely at the receiving end. Therefore, repeaters are used in communication path to eliminate the effect of noise on the signal. It separates signal from the noise, amplify the signal and then retransmit it to the receiver. They are placed close to each other in the communication path.

Decoder

Role of decoder is opposite to that of encoder. It regroups the incoming bits and decodes it into a quantized signal.

Hold circuit and low pass filter

They are used to reconstruct the original message signal.

Summary

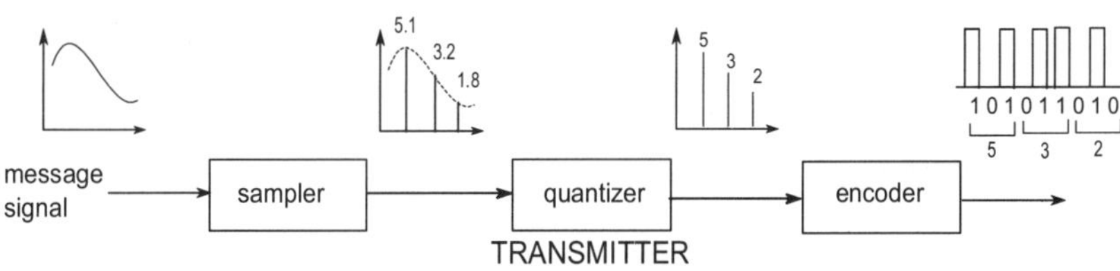

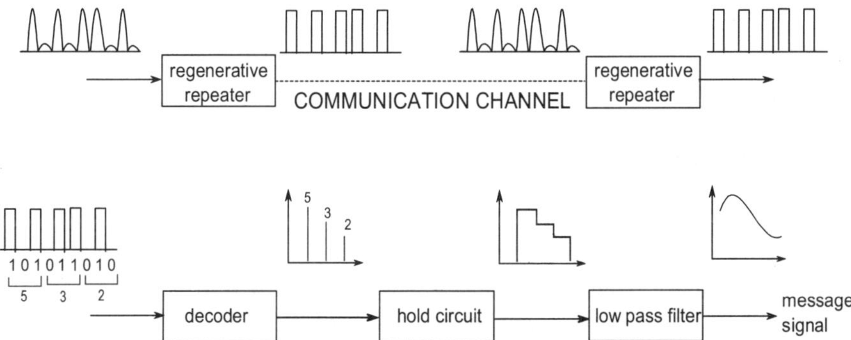

Fig. 8.6.6 Block diagram of PCM system

PROCEDURE

1. Make the connections on the kit according to the block diagram shown in Fig. 8.6.1.

2. Trace the message signal, output of sampler and encoder using DSO.

3. Measure the frequency of message signal f_m.

4. Connect the PCM output to the input of demodulator circuit.

5. Trace the output of decoder and low pass filter using DSO.

6. Measure the frequency of demodulated signal.

7. Change the input signal amplitude and frequency and repeat the above steps.

OBSERVATION TABLE

Table 8.6.2 Observation table for PCM

S.No.	fm (Hz)	Demodulated signal frequency (Hz)
1.		
2.		
3.		

OBSERVATIONS

Attach the traces for pulse code modulation.

RESULT

Pulse code modulation and demodulation have been studied successfully. The demodulated and the modulating signal frequency comes out to be almost same.

DISCUSSION

Advantages of PCM

1. PCM system is more immune to noise as compared to analog system. In PCM, '1' is marked by presence of pulse and '0' is marked by the absence of pulse. Hence, when the digital signal is transmitted, it is not important to evaluate the amplitude, frequency or phase of the pulse at the receiver. The information lies in the absence or presence of pulse. Even if the pulse is distorted by noise present in communication channel, a simple technique is used at the receiver. If the pulse is above reference level, it is taken as 1 and if the pulse is below reference level, it is taken as 0 as shown in Fig 8.6.7. Therefore, the effect of noise is greatly reduced in digital systems and thus PCM system is suitable for long distance transmission.

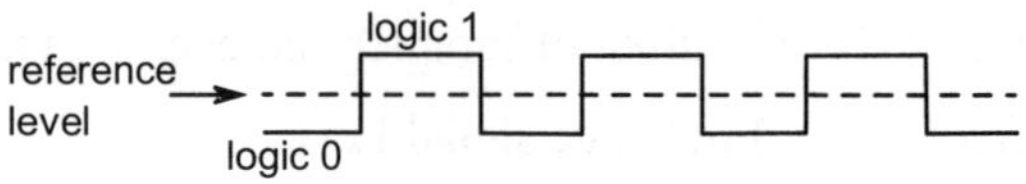

Fig. 8.6.7

2. Digital circuits are easy to handle.

3. By using different techniques, redundant information can be easily removed.

4. Multiplexing techniques can be implemented easily in digital system.

5. Repeaters can be used to eliminate the effect of noise. It separates signal from the noise, amplify the signal and then retransmit it to the receiver.

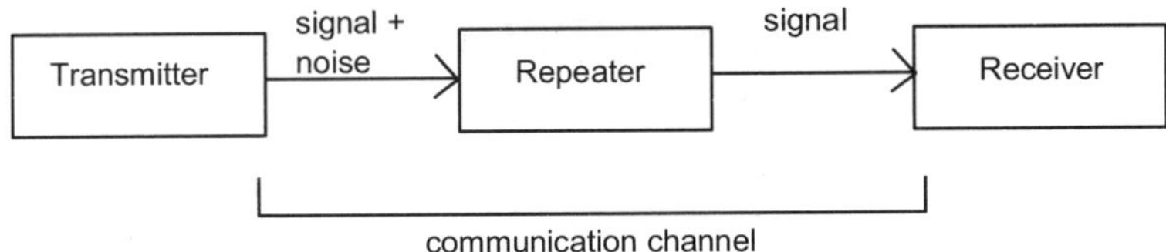

Fig. 8.6.8 Role of repeater

6. In analog modulation, the required signal to noise ratio *i.e.* SNR at the receiver is 40-60 dB for proper detection of the message signal. While in case of digital modulation, the required SNR is 10-12 dB.

Disadvantages of PCM

1. Bandwidth requirement is more as compared to analog modulation technique.

2. The system is more complex.

3. Precise time synchronization is required between transmitter and receiver.

Applications

PCM system is widely used in telephone system to transmit speech signals. They are also used to transmit video data over long distances. Most of the space ships have video camera which collects the information of the space. The video signals are converted into digital format through PCM system and then transmitted to the earth.

Experiment 7

DELTA MODULATION

AIM

Study of Delta Modulation.

APPARATUS REQUIRED

DM kit, connecting wires, probes, digital storage oscilloscope.

THEORY

Redundant Information in Pulse Code Modulation

When the signal is changing slowly, then the difference between the present sample and the next sample of the message signal is very small. When such a signal is sampled, quantized and encoded using PCM scheme, there is a large amount of redundant information. As shown in Fig. 8.7.1, third and fourth, fifth and sixth, seventh and eighth samples carry the same information. This shows that the resulting encoded signal carries the same redundant information.

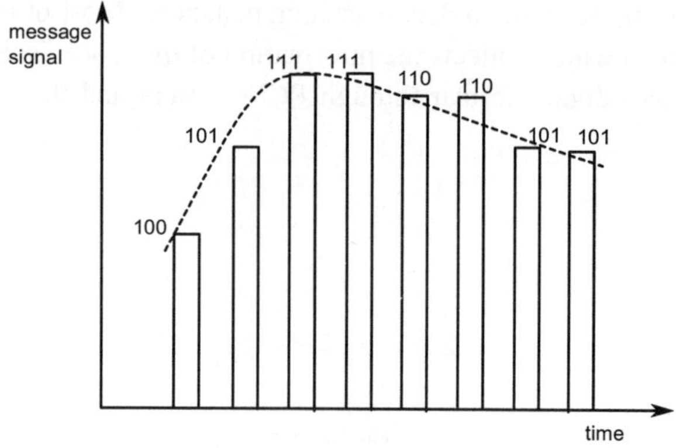

Fig. 8.7.1 Redundant information in PCM

In order to avoid sending redundant information, different modulation schemes such as Differential Pulse Code Modulation, Delta Modulation are used.

Delta Modulation

It is a special form of pulse code modulation technique which uses only two quantization levels and a single bit to transmit the analog signals using digital modulation technique. In PCM, multiple bits are required to represent samples of the signal whereas in case of DM, only a single bit is used to the represent the samples. The basic principle of DM is that it compares the present sample with the past sample and transmits the difference between the two adjacent samples. If the difference between present and past sample is positive, '1' is transmitted and if the difference is negative, '0' is transmitted.

Delta modulator

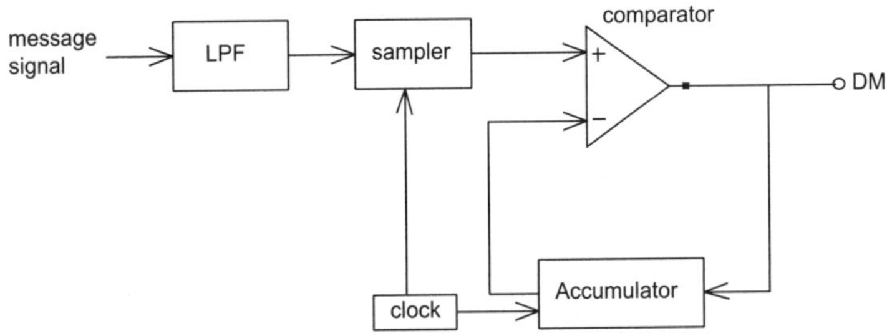

Fig. 8.7.2 Delta modulator

Delta modulator is shown in Fig. 8.7.2. The operation of DM is as follows.

1. The message signal is first passed through the anti-alias low pass filter to remove the unwanted frequencies and band limit the message signal.

2. Filtered message signal is then sampled and converted into a discrete time signal.

3. Comparator compares the present sample with the past sample and produces positive or negative voltage depending on the difference between the two.

4. If the difference is positive, '1' is transmitted and if the difference is negative, '0' is transmitted.

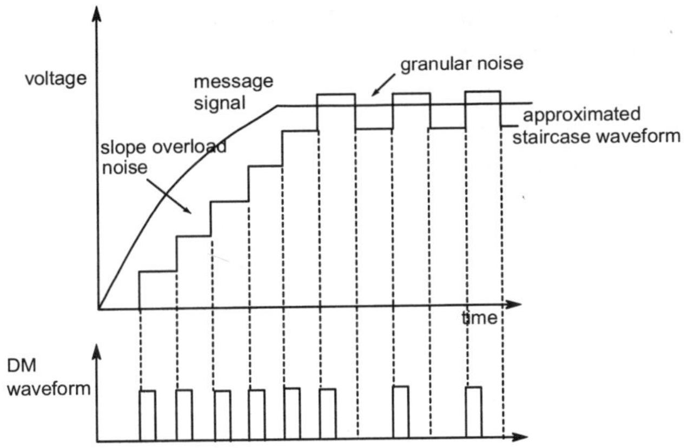

Fig. 8.7.3 DM waveform

Figure 8.7.3 shows that the message signal is approximated by a stair case waveform with fixed step size. If the difference between two adjacent samples is positive, stair case waveform is increased by one step and if the difference is negative, waveform is decreased by one step. DM waveform represents a stream of '1' and '0' which is transmitted to the receiver.

Delta demodulator

Delta demodulator is shown in Fig. 8.7.4. The DM waveform is fed into integrator which converts a stream of '1' and '0' to a staircase waveform. Low pass filter then produces message signal back from the approximated staircase signal.

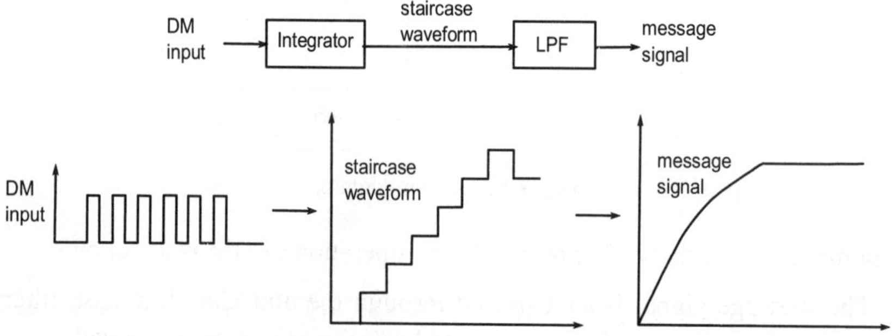

Fig. 8.7.4 Delta demodulator

Disadvantages of DM

Main drawbacks of delta modulation are slope overload distortion and granular noise as shown in Fig. 8.7.3.

Slope overload noise occurs when the message signal has a steep rise and the staircase waveform is not able to match the input signal. In this case, step size should be large so that the approximated signal is able to match the steep slope of the input. Granular noise occurs when the variation in message signal is very small. The staircase waveform keeps on oscillating around the message signal. To reduce this noise, step size should be made small. Therefore, another technique known as adaptive delta modulation is used in which size of step changes depending on the message signal as shown in Fig. 8.7.5.

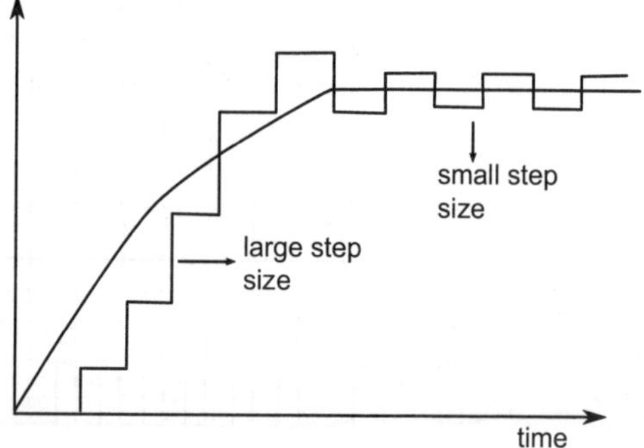

Fig. 8.7.5 Varying step size in adaptive delta modulation

PROCEDURE

1. Make the connections on the kit according to the block diagram as shown in Fig. 8.7.2.
2. Trace the message signal, output of sampler and comparator using DSO.
3. Measure the frequency of message signal f_m.
4. Connect the DM output to the input of demodulator circuit according to the block diagram shown in Fig. 8.7.4.
5. Trace the output of integrator and low pass filter.
6. Measure the frequency of demodulated signal.
7. Change the input signal amplitude and frequency and repeat the above steps.

OBSERVATION TABLE

Table 8.7.1 Observation table for delta modulation

S.No.	f_m (Hz)	demodulated signal frequency (Hz)
1.		
2.		
3.		
4.		

OBSERVATIONS

Attach the traces of delta modulation.

RESULT

Delta modulation and demodulation have been studied successfully. The demodulated and modulating signal frequency comes out to be almost same.

DISCUSSION

Delta modulation uses only two quantization levels which are represented by a single bit whereas PCM uses multiple quantization levels which are represented by multiple bits. Therefore, in case of DM, only a single bit is transmitted per sample whereas in case of PCM, multiple bits are transmitted per sample. Hence, DM requires less transmission channel bandwidth as compared to PCM. Moreover, DM does not require ADC and DAC for encoding and decoding and therefore, cost of DM system is less as compared to PCM and the circuitry is also simple.

Experiment 8

TIME DIVISION MULTIPLEXING

AIM

Study of Time Division Multiplexing.

APPARATUS REQUIRED

TDM kit, connecting wires, probes, digital storage oscilloscope.

THEORY

Multiplexing

Multiplexing is the process of sending more than one signal through the same communication channel.

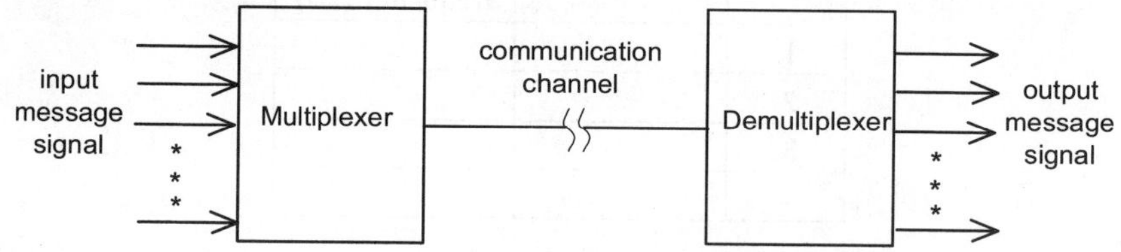

Fig. 8.8.1 Role of multiplexer in communication system

Fig. 8.8.1 shows that multiplexer combines all the input message signals and transmits it over the same communication channel. At the receiver end, de-multiplexer separates different message signal from the composite signal.

Two types of multiplexing

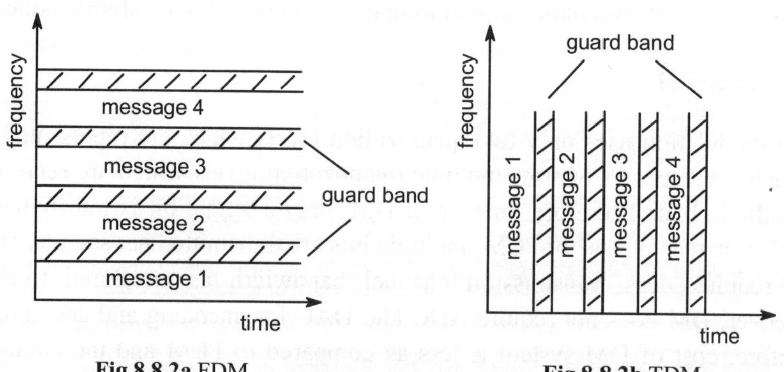

Fig 8.8.2a FDM   **Fig 8.8.2b** TDM

1. Frequency Division Multiplexing: In FDM, different message signals occupying different frequency band are combined together and transmitted at the same time as shown in Fig. 8.8.2a.

2. Time Division Multiplexing: In TDM, different message signals occupying the same frequency band are transmitted at different time slots as shown in Fig. 8.8.2b.

Time Division Multiplexing

When the analog signal is sampled and converted to a discrete time signal, there is a finite time slot between two samples as shown in Fig. 8.8.3a. This time slot can be used to send samples of other message signals as shown in Fig. 8.8.3b. This is known as time division multiplexing.

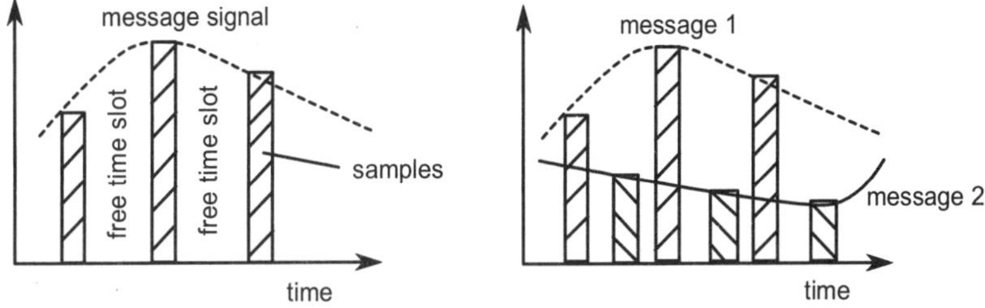

Fig 8.8.3a Free time slots in a sampled signal   **Fig 8.8.3b** Time division multiplexing

Block diagram for TDM

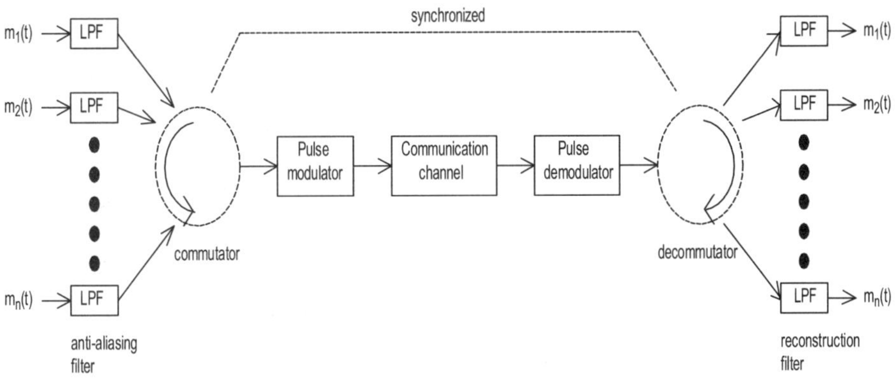

Fig. 8.8.4 Block diagram for TDM

Let $m_1(t)$, $m_2(t)$.. $m_n(t)$ represents the message signal which are to be transmitted over the communication channel. They are first passed through the low pass filter, also known as anti aliasing filter, which is used to remove any unwanted frequency components and band limit the message signal. This is done to ensure that there is no aliasing effect when the signal is sampled according to the Nyquist rate. The output of LPF is then applied to a commutator, an electronic switch, which take a narrow sample of each of the message signal at a rate of Nyquist sampling frequency fs and sequentially interleave them inside the sampling time interval Ts. This time division multiplexed signal is then fed to the pulse modulator which converts the multiplexed signal into the form suitable for transmission over communication channel. At the receiver, pulse demodulator does the opposite of it. It demodulates the signal which is then fed to the decommutator which distributes the samples of message signal to its corresponding low pass filter. Decommutator works in synchronism with the commutator such that samples reach at the right LPF.

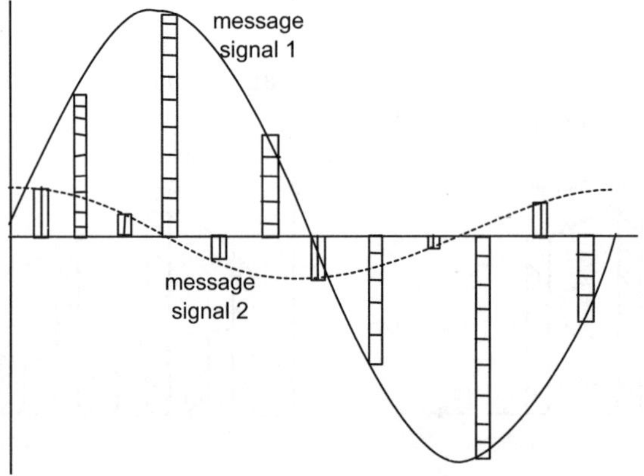

Fig. 8.8.5 TDM of two signals

Concept of guard band

Whenever a pulse travels from transmitter to the receiver through the communication channel, the pulse spreads as shown in Fig. 8.8.6. Hence, two pulses can overlap with each other. In order to avoid that, proper guard band should be present in between two pulses.

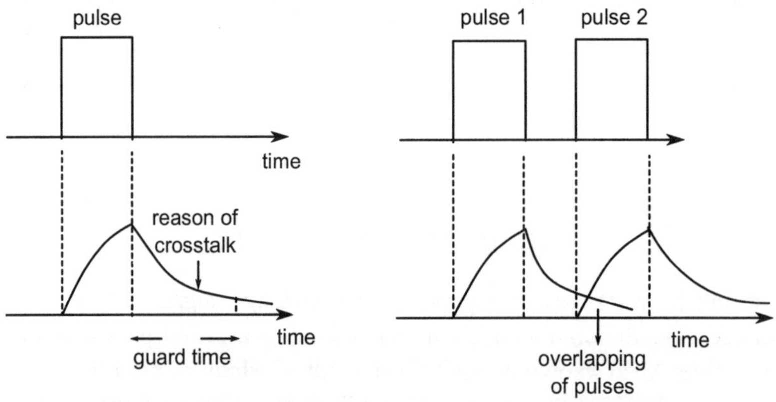

Fig. 8.8.6 Pulse spreading and overlapping of two pulses

PROCEDURE

1. Make the connections on the kit according to the block diagram shown in Fig. 8.8.4.
2. Connect DSO using probe across the input message signals and measure the frequency. Take the traces.
3. Take the traces of output of commutator, pulse modulator and pulse demodulator using DSO.
4. Connect DSO across the demodulated signals and measure the frequency. Take the traces.

OBSERVATION TABLE

Table 8.8.1 Observation table for TDM

Message signal number	f_m (Hz)	Demodulated signal frequency (Hz)
1		
2		
3		
4		

OBSERVATIONS

Attach the traces for time division multiplexing.

RESULT

Time division multiplexing has been studied successfully. The demodulated and modulating signal frequency comes out to be almost same.

DISCUSSION

TDM systems are better than FDM because of the following reasons.

- Simple circuitry.

- More immune to interference.

- When large number of message signal is to be transmitted, FDM system requires additional circuitry with each additional message signal. But in case of TDM, no extra circuitry is required.

But the disadvantage of using TDM is that the perfect synchronization is required between transmitter and receiver which is not the case in FDM.

TDM technique is widely used in multiplexing different digital signals for *e.g.* computer output, digitized voice signal, digitized facsimile, TV signals, *etc.*

Experiment 9

AMPLITUDE SHIFT KEYING

AIM

Study of Amplitude Shift Keying.

APPARATUS REQUIRED

ASK kit, connecting wires, probes, digital storage oscilloscope.

THEORY

Digital Modulation

It is a process in which high frequency carrier signal is used to modulate the digital message signal. The digitally modulated analog carrier signal is then transmitted from one point to another in a communication system.

Amplitude Shift Keying (ASK) or On-Off Keying (OOK)

It is the simplest digital modulation technique. It is the process in which amplitude of the high frequency carrier signal is varied according to the binary information signal *i.e.* amplitude of carrier signal represents the digital data. It is similar to amplitude modulation except that amplitude of the output signal can have only two values.

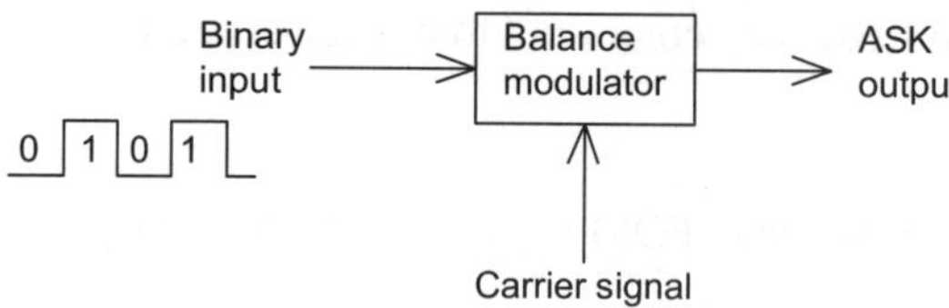

Fig. 8.9.1 Block diagram for ASK generation

Binary input and the carrier signal are applied as two inputs of balance modulator as shown in Fig. 8.9.1 to produce ASK at the output. Let $v_{ask}(t)$ be the amplitude shift keying voltage which is given as

$$v_{ask}(t) = [1 + v_m(t)] [A/2 \cos(\omega_c t)]$$

Where,

$v_m(t)$: Binary message signal, A: amplitude of un-modulated carrier signal, ω_c: frequency of analog carrier signal (radians/sec).

When input is at logic 1 *i.e.* $v_m(t) = +1V$, expression for ASK voltage becomes,

$$v_{ask}(t) = [1 + 1]\,[A/2\,\cos(\omega_c t)]$$

$\Rightarrow$
$$v_{ask}(t) = A\,\cos(\omega_c t)$$

i.e. when input is high, amplitude of ASK waveform is same as amplitude of high frequency carrier signal.

When input is at logic 0 *i.e.* $v_m(t) = -1V$, expression for ASK voltage becomes,

$$v_{ask}(t) = [1 - 1]\,[A/2\,\cos(\omega_c t)]$$

$\Rightarrow$
$$v_{ask}(t) = 0$$

i.e. when input is low, amplitude of ASK waveform becomes zero.

Hence, modulated signal is either $A\cos(\omega_c t)$ or 0V. Therefore, when the input is high the carrier signal is transmitted and when the input is low, no signal is transmitted. Thus, it is also known as On Off keying.

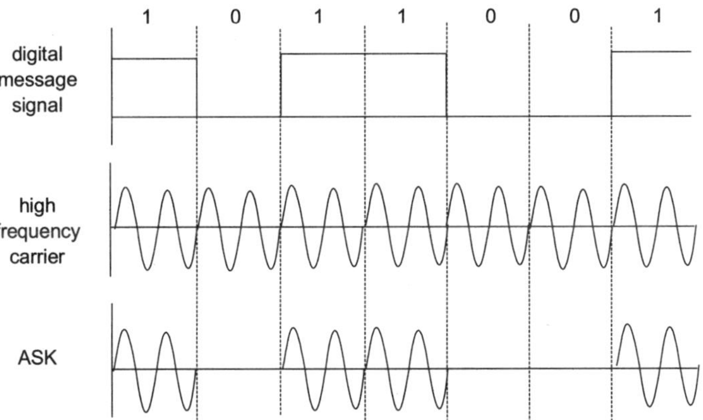

Fig. 8.9.2 Amplitude shift keying waveform

PROCEDURE

1. Make the connections on ASK kit according to the block diagram shown in Fig. 8.9.1.
2. Connect DSO using probes across the message signal and carrier signal and take the traces.
3. Connect DSO across the ASK output and take the trace.
4. Change the input message signal and observe the output waveform.

OBSERVATIONS

Attach the traces for ASK waveform.

RESULT

Amplitude shift keying has been studied successfully.

DISCUSSION

PCM system is used to convert analog message signal into digital form *i.e.* string of 1's and 0's. Such a message signal can be easily transmitted as appropriate voltage levels from one point to another over copper wires. But due to the low frequency of the message signal, a large height antenna is required to transmit it over free space. Therefore, some kind of modulation is required in which digital message signal is modulated by a high frequency analog carrier signal. An example of such modulation scheme is ASK.

ASK is similar to AM and is sensitive to noise. ASK generation and detection circuits are simple in design and relatively inexpensive. It is widely used to transmit digital signal over optical fiber in form of light. When input is 1, a short pulse of light is transmitted over optical fiber using LED and when input is 0, no light is transmitted.

Experiment 10

PHASE SHIFT KEYING

AIM

Study of Phase Shift Keying.

APPARATUS REQUIRED

PSK kit, connecting wires, probes, digital storage oscilloscope.

THEORY

Digital Modulation

It is a process in which high frequency carrier signal is used to modulate the digital message signal. The digitally modulated analog carrier signal is then transmitted from one point to another in a communication system.

Phase Shift Keying

It is the process in which phase of the high frequency carrier signal is varied according to the binary information signal *i.e.* phase of carrier signal represents the digital data. It is similar to phase modulation.

Binary phase shift keying transmitter

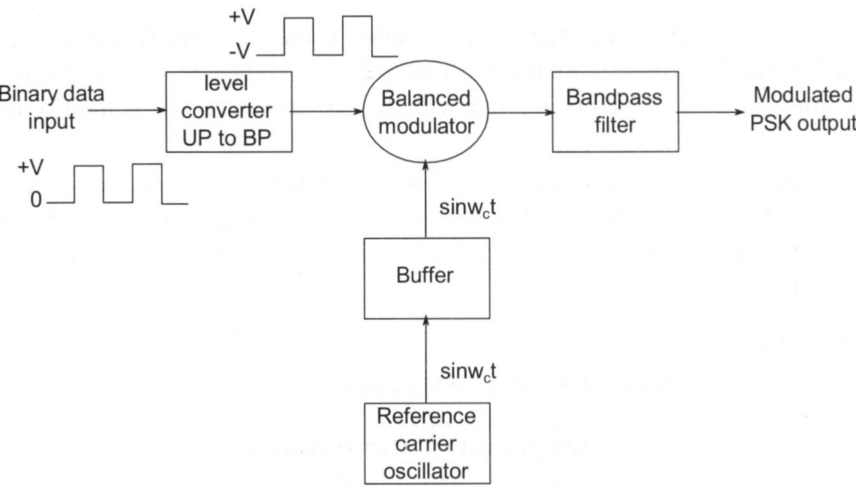

Fig. 8.10.1 Block diagram of PSK generation

Figure 8.10.1 shows the block diagram of PSK generation. Binary input is first converted from unipolar state to bipolar state. In unipolar state, logic 1 is represented by +V volts and logic 0 is represented by 0 volts whereas in bipolar state, logic 1 is represented by +V volts and logic 0 is represented by −V volts. The input is then applied to balanced modulator which acts as a phase reversing switch. When digital input is at logic 1, carrier is transmitted at the output in phase with the reference carrier oscillator and when digital input is at logic 0, carrier is transmitted at the output 180° out of phase with the reference carrier oscillator.

Role of balanced modulator

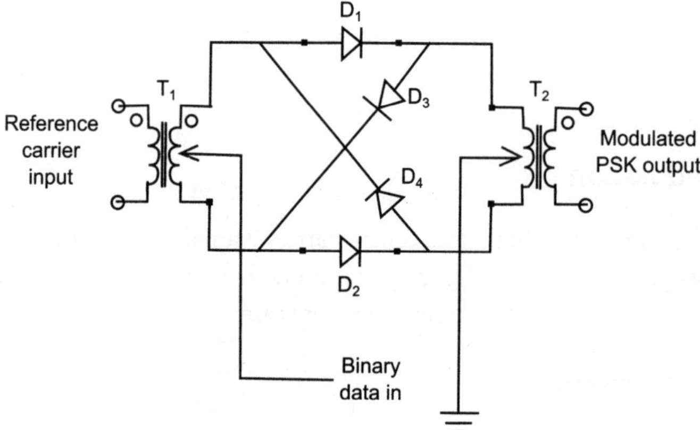

Fig. 8.10.2 Balanced ring modulator

Balanced ring modulator as shown in Fig. 8.10.2 has two inputs *i.e.* binary digital data and carrier signal which is in phase with the reference carrier oscillator.

1. When input data is at logic 1 (+V volts): Diodes D_1 and D_2 are forward biased and D_3 and D_4 are reverse biased. Carrier voltage is developed across transformer T_2 which is in phase with the carrier voltage across transformer T_1.

2. When input data is at logic 0 (-V volts): Diodes D_3 and D_4 are forward biased and D_1 and D_2 are reverse biased. Carrier voltage is developed across transformer T_2 which is 180° out of phase with the carrier voltage across transformer T_1.

Truth table

Table 8.10.1 Truth table for balanced ring modulator

Binary input	Output phase
Logic 0	180°
Logic 1	0°

Phasor and constellation diagram

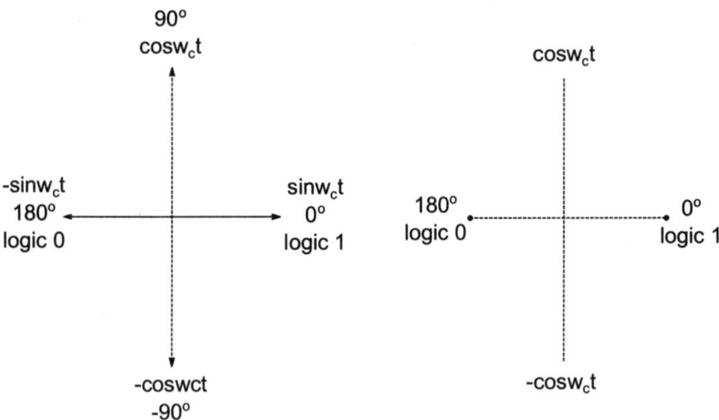

Fig. 8.10.3 Phasor and constellation diagram

Constellation diagram is same as phasor diagram except that instead of drawing the entire phasor, only the relative positions of the peak of phasors are shown.

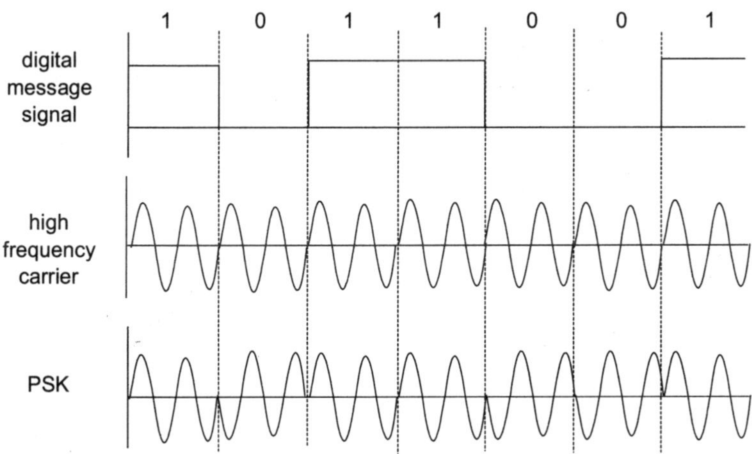

Fig. 8.10.4 Phase shift keying waveforms

PROCEDURE

1. Make the connections on PSK kit according to the block diagram shown in Fig. 8.10.1.

2. Connect DSO using probes across the message signal and carrier signal and take the traces.

3. Connect DSO across the PSK output and take the trace.

4. Change the input message signal and observe the output waveform.

OBSERVATIONS

Attach the traces for PSK waveform.

RESULT

Phase shift keying has been studied successfully.

DISCUSSION

PCM system is used to convert analog message signal into digital form *i.e.* string of 1's and 0's. Such a message signal can be easily transmitted as appropriate voltage levels from one point to another over copper wires. But due to the low frequency of the message signal, a large height antenna is required to transmit it over free space. Therefore, some kind of modulation is required in which digital message signal is modulated by a high frequency analog carrier signal. An example of such modulation scheme is PSK.

PSK has the best performance as compared to ASK and FSK with very good noise immunity. They are widely used in radio communications applications.

Other forms of PSK are QPSK (Quaternary Phase Shift Keying), O-QPSK (Offset Quadrature Phase Shift Keying), 8 PSK (8 point Phase Shift Keying), 16 PSK (16 point Phase Shift Keying), QAM (Quadrature Amplitude Modulation), 16 QAM (16 point Quadrature Amplitude Modulation), 64 QAM (64 point Quadrature Amplitude Modulation), MSK (Minimum Shift Keying) and GMSK (Gaussian filtered Minimum Shift Keying).

Experiment 11

FREQUENCY SHIFT KEYING

AIM

Study of Frequency Shift Keying.

APPARATUS REQUIRED

FSK kit, connecting wires, probes, digital storage oscilloscope.

THEORY

Digital Modulation

It is a process in which high frequency carrier signal is used to modulate the digital message signal. The digitally modulated analog carrier signal is then transmitted from one point to another in a communication system.

Frequency Shift Keying

It is the process in which frequency of the high frequency carrier signal is varied according to the binary information signal *i.e.* frequency of carrier signal represents the digital data. It is similar to frequency modulation.

FSK transmitter

A very simple method of generating an FSK signal is by the use of a voltage controlled oscillator as shown in Fig. 8.11.1.

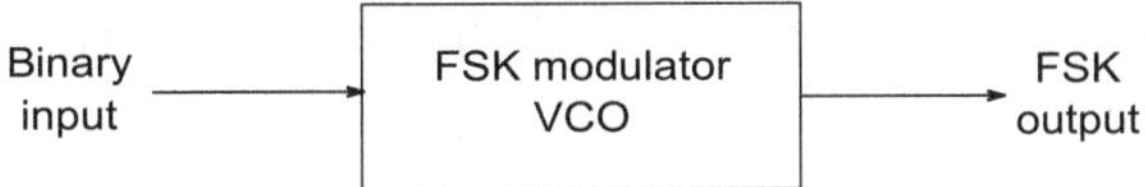

Fig. 8.11.1 FSK transmitter

Working

A VCO is a circuit that generates oscillations with frequency proportional to the applied input voltage. Initially, in the absence of any input signal, VCO operates at its center frequency which is set at the carrier frequency. The expression for the center frequency is given as

$$F_C = \frac{1}{2\Pi\sqrt{LC}}$$

$- \text{eqn 1}$

Where,

Fc: center frequency, L: inductance, C: capacitance of varactor diode.

The digital message signal is applied at the control voltage terminal of the VCO. As the message signal changes its state from 0 to 1 or 1 to 0, voltage changes and hence the reverse bias of varactor diode also changes. This changes the capacitance of varactor diode and hence VCO output frequency also changes (refer eqn1). Therefore, when input is at logic 1, it shifts the VCO frequency to a higher frequency known as mark frequency *i.e.* f_M and when input is at logic 0, it shifts the VCO frequency to a lower frequency known as space frequency *i.e.* f_S.

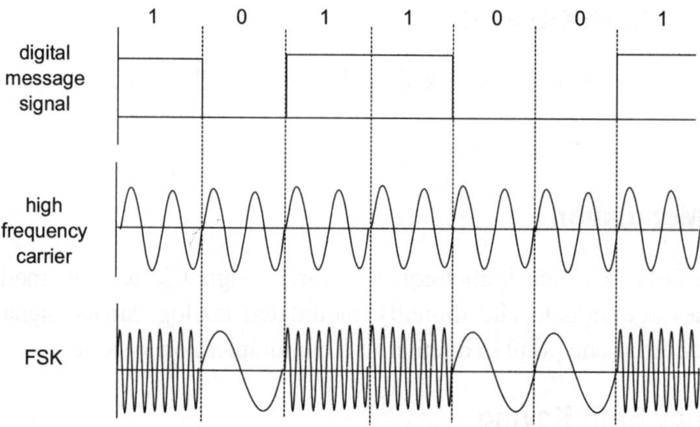

Fig. 8.11.2 Frequency shift keying waveforms

Let $v_{fsk}(t)$ be the frequency shift keying voltage which is given as

$$v_{fsk}(t) = V_c \cos[2\Pi (f_c + v_m(t)\, \Delta f)\, t]$$

Where,

$v_m(t)$: binary message signal, V_c: peak amplitude of analog carrier signal, f_c: frequency of analog carrier signal, Δf: peak shift in analog carrier frequency.

Above expression shows that peak shift in carrier frequency is directly proportional to the amplitude of the binary message signal.

When input is at logic 1 *i.e.* $v_m(t) = +1V$, expression for FSK voltage becomes,

$$v_{fsk}(t) = V_c \cos[2\Pi (f_c + \Delta f)\, t]$$

i.e. when input is high, carrier frequency is shifted up *i.e.* frequency increases. Mark frequency f_M is given as

$$f_M = f_c + \Delta f$$

When input is at logic 0 *i.e.* $v_m(t) = -1V$, expression for FSK voltage becomes,

$$v_{fsk}(t) = V_c \cos[2\Pi \, (f_c - \Delta f) \, t]$$

i.e. when input is low, carrier frequency is shifted down *i.e.* frequency decreases. Space frequency f_S is given as

$$f_S = f_c - \Delta f$$

Therefore, as the message signal changes from logic 1 to logic 0, the carrier frequency changes from mark frequency to space frequency.

Frequency deviation Δf

It is defined as half the difference of mark and space frequency and is given as

$$\Delta f = \frac{|f_m - f_s|}{2}$$

PROCEDURE

1. Make the connections on FSK kit according to the block diagram shown in Fig. 8.11.1.
2. Without any input, trace the output of VCO *i.e.* carrier signal using DSO.
3. Give message signal at the input of VCO.
4. Connect DSO at the input of VCO and trace the message signal.
5. Trace the FSK waveform at the output of VCO.
6. Change the input message signal and observe the output waveform.

OBSERVATIONS

Attach the traces for FSK waveform.

RESULT

Frequency shift keying has been studied successfully.

DISCUSSION

PCM system is used to convert analog message signal into digital form *i.e.* string of 1's and 0's. Such a message signal can be easily transmitted as appropriate voltage levels from one point to another over copper wires. But due to the low frequency of the message signal, a large height antenna is required to transmit it over free space. Therefore, some kind of modulation is required in which digital message signal is modulated by a high frequency analog carrier signal. An example of such modulation scheme is FSK.

Performance of FSK is better than ASK but poor as compared to PSK or QAM (quadrature amplitude modulation). Hence, it is hardly used for high performance digital radio systems. Another disadvantage of using FSK is its high bandwidth requirement. Use of FSK is restricted to low performance, low cost systems only.

Experiment 12

SINGLE SIDE BAND MODULATION and DEMODULATION

AIM

Study of Single Side Band Modulation and Demodulation.

APPARATUS REQUIRED

SSB Modulation kit, connecting wires, probes, digital storage oscilloscope.

THEORY

Amplitude Modulation

It is one of the modulation techniques in which amplitude of the high frequency carrier signal is varied according to the instantaneous value of the message signal. Therefore, amplitude of the carrier signal contains the information about the message signal. Expression for AM wave is given as

$$v_{AM} = V_c \sin \omega_c t + \frac{mV_c}{2} \cos(\omega_c - \omega_m)t - \frac{mV_c}{2} \cos(\omega_c + \omega_m)t \quad - \text{eqn 1}$$

Where, $v_{AM}(t)$: amplitude modulated wave, ω_c: angular frequency of carrier signal, ω_m: angular frequency of message signal, V_c: un-modulated carrier amplitude, m: modulation index

Total power in amplitude modulated wave

Total power is AM wave is sum of power in carrier signal, lower side band and upper side band.

$$P_{AM} = P_{carrier} + P_{LSB} + P_{USB}$$

Bandwidth

It is defined as the difference between upper and lower frequencies or it can be defined as the range of frequency over which an information signal is transmitted.

$$BW = 2f_m$$

Double Side Band Suppressed Carrier (DSBSC)

Equation 1 gives the expression for amplitude modulated wave and it shows the presence of three components - carrier signal with frequency ω_c, lower side band with frequency ω_c - ω_m and upper side band with frequency $\omega_c + \omega_m$. The first term which represents the carrier signal contains no information about the message signal and thus power present in first term is total waste. Therefore, carrier can be suppressed from an amplitude modulated wave to save on the power and this is known as double side band suppressed carrier. Expression for DSBSC wave is given as

$$v_{DSBSC}(t) = \frac{mV_c}{2}\cos(\omega_c - \omega_m)t - \frac{mV_c}{2}\cos(\omega_c + \omega_m)t \qquad - \text{eqn 2}$$

Total power in DSBSC wave

Total power is DSBSC wave is sum of lower side band and upper side band.

$$P_{DSBSC} = P_{LSB} + P_{USB}$$

Bandwidth

$$BW = 2f_m$$

Single Side Band (SSB)

Equation 2 gives the expression for double side band suppressed carrier wave and it shows the presence of two components - lower side band with frequency ω_c - ω_m and upper side band with frequency $\omega_c + \omega_m$. Both the terms carry the same information about the message signal. Therefore, one of the side bands can be suppressed to save the bandwidth as well as power requirement. This is called as single side band modulation. Expression for SSB wave is given as

$$v_{SSB}(t) = \frac{mV_c}{2}\cos(\omega_c - \omega_m)t$$

Or,

$$v_{SSB}(t) = \frac{mV_c}{2}\cos(\omega_c + \omega_m)$$

Total power in SSB wave

Total power in SSB wave is either equal to power of lower side band or upper side band.

$$P_{SSB} = P_{LSB} = P_{USB}$$

Bandwidth

$$BW = f_m$$

SSB generation

Filter method or Frequency discrimination method

Figure 8.12.1 shows the generation of SSB using filter method. Message signal and carrier signal are applied at the inputs of balanced modulator. Balanced modulator is used to produce double side band suppressed carrier waveform. DSBSC waveform is then passed through a crystal filter which is used to eliminate one of the side bands and hence producing single side band at the output.

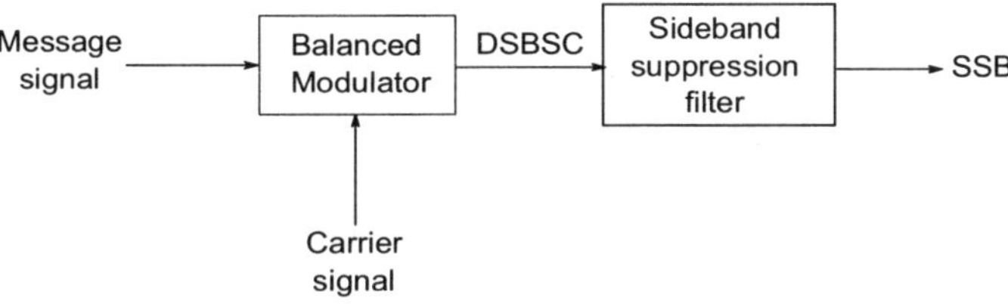

Fig. 8.12.1 SSB generation using filter method

SSB detection

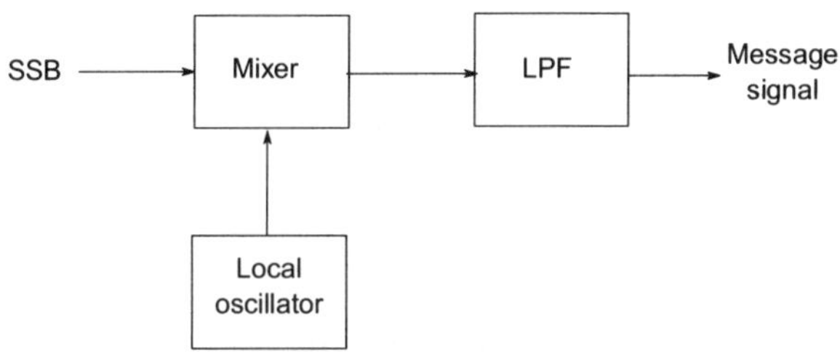

Fig. 8.12.2 SSB detection

Figure 8.12.2 shows the block diagram of SSB detection. SSB signal is multiplied with the locally generated carrier frequency. The local oscillator should produce the same frequency as that of the carrier signal used at the transmitter side otherwise the difference between the two can lead to error in the demodulated message signal. Let SSB contains the lower side band, then the output of mixer is given as

$$\cos(\omega_c - \omega_m)\,t \,.\, \cos \omega_c t$$

$$= \frac{1}{2} \left[\cos(2\omega_c t - \omega_m t) + \cos \omega_m t \right]$$

When the above signal is passed through the low pass filters, it rejects the first high frequency component and passes the second low frequency message signal. Therefore, at the output, message signal is produced.

PROCEDURE

1. Make the connections on SSB kit according to the block diagram shown in Fig. 8.12.1 and Fig. 8.12.2.

2. Connect DSO using probes across the message signal and carrier signal and take the trace. Measure the frequency of message signal f_m.

3. Connect DSO at the output of balanced modulator and filter and take the traces of DSBSC and SSB waveforms.

4. Connect SSB waveform to the input of demodulator circuit.

5. Connect DSO at the output of LPF in demodulator circuit and take the trace of demodulated signal and measure its frequency.

OBSERVATIONS TABLE

Table 8.12.1 Observation table for SSB

S.No.	f_m (Hz)	Demodulated signal frequency (Hz)
1.		
2.		
3.		
4.		

OBSERVATIONS

Attach the traces for single side band modulation.

RESULT

SSB Modulation and Demodulation have been studied successfully. The demodulated and modulating signal frequency comes out to be almost same.

DISCUSSION

SSB is widely used in those systems where power is the foremost requirement such as mobile communication systems, radio navigation, point to point communication, *etc.*

SSB receivers are complex in designing. It is difficult to tune the local oscillator in demodulator circuit to the exact frequency of the carrier signal used during transmission. The local oscillator frequency can be higher or lower than the carrier signal frequency which produces error in the demodulated message signal. To overcome this problem, pilot carrier SSB system is used. In this, the carrier signal is not completely suppressed. A pilot carrier *i.e.* low level carrier signal (16-26 dB below the normal carrier) is transmitted along with the SSB signal. This pilot carrier is used in the receiver circuit for detection of the message signal.